# Changing the Face of Engineering

# CHANGING THE FACE OF ENGINEERING

## The African American Experience

EDITED BY John Brooks Slaughter, Yu Tao, and Willie Pearson, Jr.

Johns Hopkins University Press | *Baltimore*

© 2015 Johns Hopkins University Press
All rights reserved. Published 2015
Printed in the United States of America on acid-free paper
9  8  7  6  5  4  3  2

Johns Hopkins University Press
2715 North Charles Street
Baltimore, Maryland 21218-4363
www.press.jhu.edu

Library of Congress Cataloging-in-Publication Data

Changing the face of engineering : the African American experience /
edited by John Brooks Slaughter, Yu Tao, Willie Pearson, Jr.
    pages cm
  Includes bibliographical references and index.
  ISBN 978-1-4214-1814-8 (hardcover : acid-free paper) —
  ISBN 1-4214-1814-2 (hardcover : acid-free paper) —
  ISBN 978-1-4214-1815-5 (electronic) — ISBN 1-4214-1815-0 (electronic)
   1. African American engineers.   2. Engineering—Study and teaching (Higher)—
United States.   3. African Americans—Education (Higher)   I. Slaughter, John Brooks,
1934–   II. Tao, Yu, 1978–   III. Pearson, Willie, 1945–
   TA157.C474 2015
   620.0089'96073—dc23        2015004323

A catalog record for this book is available from the British Library.

*Special discounts are available for bulk purchases of this book. For more information,
please contact Special Sales at 410-516-6936 or specialsales@press.jhu.edu.*

Johns Hopkins University Press uses environmentally friendly book materials, including
recycled text paper that is composed of at least 30 percent post-consumer waste, whenever
possible.

*With profound thanks to Ida Bernice Slaughter and our children, John Brooks II and Jacqueline Michelle, for their encouragement and support*

*To Saiying Tao, Adrian Zhang, and Bo Zhang*

*In memory of Vassie V. King, Bernice M. Leggon, James Turk, Dolly A. McPherson, and Maya Angelou*

# CONTENTS

# FOREWORD

*Shirley Ann Jackson*

Once in a great while a book emerges that embraces so much overlooked, discounted, or suppressed history, containing such important information, that it soars above peers. Such is *Changing the Face of Engineering: The African American Experience.*

This work constitutes the first comprehensive history of African Americans in the field of engineering. With contributions from many distinguished authors, the book fuses once-separate histories and offers a clearer view of the contributions African American engineers have made to the building of the United States—as well as to its civilization, culture, and economic vibrancy. It also records the efforts of many individuals and organizations to open the field of engineering to African American students and to encourage them to realize its benefits—for their careers, for their nation, for the world, for humankind.

Clearly, *Changing the Face of Engineering* owes a profound debt to Dr. Carter G. Woodson (1875–1950), the Harvard-educated historian, scholar, author, and activist who fathered the study of black history. Dr. Woodson devoted his life to raising societal awareness of the roles of African Americans in the development of the United States, believing deeply in the value of that history in inspiring the young.

This volume examining the history of African Americans in engineering comes at an especially auspicious time. The criticality of engineering in advancing progress, animating the economy, and improving the general quality of life is well understood. However, while US demand for engineers is at an all-time high, the total number of engineering graduates is historically low.

Our current cohort of engineers was born largely out of the then–Soviet Union's launch of the *Sputnik* satellite in 1957, which shocked the United States to attention and to action. Many young people were aroused and inspired by this event and took advantage of specially created scholarships and education programs in science, technology, engineering, and mathematics— the STEM fields. The nation needed them, and they responded. However, this cohort is fast retiring, and, as this volume documents, there are not enough engineering graduates to replace them.

In the 1990s I began to speak publicly of the nation's need to encourage and accommodate underrepresented individuals in the STEM fields. I coined the phrase "quiet crisis" to describe it. It is "quiet" because it unfolds gradually over time: it takes decades to foster, educate, and prepare human capital in STEM arenas. It is a "crisis" because a deficit of talent will cripple our national—even the international—capacity for discovery, for innovation, for meeting the world's human and economic needs. In the ensuing years since, numerous reports have documented this talent deficit.

Yet, as this volume correctly points out, many previous studies have obscured our understanding of the unique challenges that African Americans face in studying and working in engineering by considering all underrepresented groups as a whole and by combining engineering with the amalgamated STEM fields. For the first time, we can now focus on African Americans who benefit our nation and the world by applying scientific knowledge to practical purposes.

To understand the full import of this book, one must envision its place in the larger picture—the 30,000-foot view. The world in the early twenty-first century faces burgeoning challenges and complex, intersecting vulnerabilities. To address these challenges and vulnerabilities and their corresponding human needs, now and in coming decades, the current social order must invest in and secure adequate food production, sufficient accessible clean water, improvements to human health, national and global security, and abundant affordable energy. None of these goals will be reachable without a great deal of very imaginative engineering.

The United States has numerous advantages that foster technological innovation on a grand scale. It has a highly sophisticated educational system, a well-developed science infrastructure, government structures designed to support the scientific enterprise, a business community focused on innovation, and a financial system that provides ready access to capital for promising new ventures. We have a tradition of public and private sector collaboration, a thriving culture of entrepreneurship, and a long history—as a society—of taking great risks for great rewards.

Yet if, with these advantages, the United States is to revitalize fully an economy that sagged during the recent "great recession" and to continue its global leadership, it must recommit to and strengthen its investment in a vibrant innovation ecosystem. The key elements of such an ecosystem include federal investment in basic and applied research, corporate support of research and development, and, perhaps most important of all, academic research and academically prepared human capital. We require bright motivated individuals who ask the questions that may take decades to answer. We require *talent*.

When considering this talent, it is important to note that discovery and innovation are only rarely lone endeavors. More and more, they require cross-sector and multidisciplinary collaborations and input from diverse sources and resources. They demand the participation of groups of individuals who habitually have been overlooked, ignored, and underrepresented in all STEM arenas, and especially in engineering. These overlooked groups include African Americans, Latinos, women, persons with disabilities, armed forces veterans, and adult men and women reentering the workplace. These underrepresented individuals bring to the innovation equation a vibrant multiplicity of backgrounds, cultures, and experiences; a diversity of outlooks; and a variety of thoughts and practices that can spark new ideas, fire a better understanding of complexities and vulnerabilities, and foster new ways to address them. Diversity leavens and catalyzes innovation and discovery.

As the United States faces mounting global competition for economic and technological leadership in the twenty-first century, it urgently needs to exploit the full complement of engineering talent extant in its diverse population. The United States must redouble its special focus on groups traditionally overlooked, specifically discouraged, or simply not encouraged to pursue STEM fields, and who remain underrepresented. This is especially pressing now as, together, these groups are an expanding percentage of the US population, quickly becoming the "new majority." If we do not use the ideas raised by this volume as inspiration for future policies and programs, the United States will lose ground.

For the past few years, I have served on the President's Council of Advisors on Science and Technology (PCAST). A number of PCAST reports on STEM education and on transforming the US research enterprise recommend that the United States strengthen STEM education through empirically validated best practices. Some state and federal programs have attempted to enhance diversity in engineering. Businesses concerned with their own competitiveness often are willing to do more to increase the diversity of the engineering workforce. However, frequently, they do not know how.

*Changing the Face of Engineering: The African American Experience* provides all of us with a basis for our efforts moving forward to resolve the "quiet crisis." It offers a foundation on which we can build a much more reasonable and prosperous future.

# ACKNOWLEDGMENTS

This book project can be traced back to early 2012 when the three coeditors decided to collaborate to fill the void in the literature on an important yet understudied topic: the experiences of African Americans in engineering. Prominent researchers and professionals from academia, industry, and government were invited to contribute to this book. We express our deep appreciation to all the contributors who accepted our invitation and stayed committed throughout the process. Special thanks go to Shirley Ann Jackson, who graciously agreed to write the foreword.

The chapters included in this volume have been significantly improved by comments from an eminent group of external reviewers. For their critical feedback, we are grateful to the following individuals: Stephanie G. Adams (Virginia Polytechnic Institute and State University), Joan Burrelli (National Science Foundation), Karen Butler-Purry (Texas A&M University), Cinda-Sue Davis (University of Michigan), Henry T. Frierson (University of Florida), Norman Fortenberry (American Society for Engineering Education), Samuel Graham (Georgia Institute of Technology), Christine Grant (North Carolina State University), Wesley Harris (Massachusetts Institute of Technology), Peter Henderson (University of Maryland, Baltimore County), Kenneth Hill (Chicago Pre-College Science and Engineering Program, Inc.), Caesar Jackson (North Carolina Central University), Janice Jackson (National Equity Project), Judy Jackson (University of Kentucky), Nirmala Kannankutty (National Science Foundation), William A. Lester, Jr. (University of California, Berkeley), Kenneth Manning (Massachusetts Institute of Technology), Marsha Matyas (American Physiological Society), Timothy Pinkston (University of Southern California), Manu O. Platt (Georgia Institute of Technology), Francisco C. Rodriguez (Los Angeles Community College District), Eric Sheppard (Hampton University), Terry Russell (Terrence Russell, LLC), Janet Rutledge (University of Maryland, Baltimore County), Alvin Schexnider (Schexnider & Associates, LLC. ), Adrienne Stiff-Roberts (Duke University), James Stith (American Institute of Physics), and Bevlee A. Watford (Virginia Polytechnic Institute and State University).

Special thanks are due to Gary May (Georgia Institute of Technology), who was especially helpful in assisting the editors by recommending contributors and reviewers.

We thank Barbara Kline Pope (The National Academies) for bringing the book project to the attention of a few potential publishers. We sincerely thank Gregory M. Britton of Johns Hopkins University Press for immediately recognizing the importance and value of this book and Catherine Goldstead for her guidance and assistance.

The views and opinions expressed in this book are solely those of the authors and not their institutional affiliations. We owe a great deal of gratitude to the contributors for sharing their insights and knowledge and especially for their commitment to see the project through its numerous drafts. We have had enormous support from our respective institutions. The following are acknowledged at:

*University of Southern California*
- William Tierney
- Diane Flores
- Sonya Black Williams

*Stevens Institute of Technology*
- Lisa Dolling
- Debra Pagan
- Juanita Castillo

*Georgia Institute of Technology*
- LaDonna Bowen
- Grace Marriott
- Ivan Allen College Small Grants Program

However, we alone take full responsibility for any errors that remain.

# Introduction

JOHN BROOKS SLAUGHTER

From the time of their arrival in America, African American men and women have contributed significantly to the creation and development of many of the tools, machines, and devices that have propelled America's industrial progress and technological achievements. However, for much of this nation's history African Americans were viewed as being unlikely to contribute to our national capabilities in science and technology. Too often their inventions and scientific discoveries were neither recognized nor valued. Consequently, little emphasis was placed on efforts to increase their participation.

In spite of the lack of recognition and credit for their achievements, African Americans have made significant contributions to our knowledge of the universe and to the technologies that have made our lives safer, more comfortable, healthier, and more productive. It is a sad commentary that these accomplishments have rarely been brought to the attention of the public (Manning 1983, 1995; Pearson 1985, 2005; Thompson 2009). Perhaps most problematic and disappointing is the fact that most African Americans are unaware of the activities and achievements of African American scientists and engineers.

Early African American scientists, inventors, and engineers had to overcome professional and personal barriers to success in order to make the contributions that constitute the legacies they have left for the generations that have followed (Manning 1983, 1995; Pearson 1985, 2005; Thompson 2009). Their accomplishments refute the thinking that denies their intellectual capacities for scientific and technological achievements of the highest order. Even today, the contributions of African Americans (as well as other underrepresented racial/ethnic minorities) in science, technology, engineering, and mathematics (STEM) are largely unrecognized and underappreciated, despite the contributions of Black engineers and technologists such as

- Mark Dean, holder of three of the original nine patents for the IBM PC;
- Lt. Gen. (Ret.) Joseph N. Ballard, former commander of the US Army Corps of Engineers;
- Linda Gooden, former president of Lockheed Martin Information Technology; and
- Maj. Gen. (Ret.) Charles Bolden, military officer, astronaut, and administrator of the National Aeronautics and Space Administration (NASA).

Even taking into account these notable examples and those of other accomplished African American engineers who are less well known, participation of African American men and women in engineering is far from parity as compared to their proportion of America's population. Countless other barriers that have prevented, or at least impeded, their entry as full participants in our nation's STEM enterprises include the sordid history of racism, discrimination and exclusion encountered, and inadequate—and in some cases nonexistent—educational and employment opportunities.

Because many Black students attend schools with the least resources and the most poorly qualified teachers, Black-White achievement gaps in algebra and in advanced mathematics course taking (e.g., algebra II, precalculus/ analysis, trigonometry, statistics and probability, and calculus) continue to persist (NSB 2012). In a longitudinal study, Pearson and Miller (2012) found that students who begin algebra in grade 7 or 8 have a substantial advantage throughout secondary school. They found that students who did not begin algebra until grade 9 were significantly less likely to complete a calculus course in high school and, in turn, less likely to complete a baccalaureate in engineering. Chen's research (2009) confirms the strong correlation between the number of calculus courses completed and the completion of an engineering baccalaureate.

In recent years, federal and state policies have focused on reforming mathematics and science education. In particular, these efforts include promoting advanced course taking, enhancing teacher quality, raising graduation requirements, increasing student achievement, and reducing disparities in performance among subgroups of students (Peske and Haycock 2006; NSB 2012; Clark 2014).

Major changes have occurred in our economy largely as a result of globalization and technological innovation. Manufacturing has declined, while the information age requires more professional and high-tech skills from employees. The United States finds itself importing talent and exporting jobs, not just because it is less expensive to have the work performed by lower-wage skilled workers in developing countries but also because our nation's colleges and universities fail to produce enough native-born, well-qualified scientists and engineers. While outsourcing and offshoring may be here to stay, depending on foreign countries to fill our requirements for scientific and engineering skills is not a long-term and tenable practice. Exporting jobs and importing talent is not sound national policy. Certainly, we need to develop a more rational set of immigration policies for those wishing to study and work in science and engineering in this country. It is unquestionable that foreign-born scientists and engineers have added immensely to America's capacity for innovation and creativity. They have enhanced our capabilities in science and technology and have the potential to contribute much more. Simultaneously, it is imperative to increase opportunities for native-born students to prepare for and study these disciplines, especially engineering. Until it does so, America will be unable to retain its leadership position in scientific and technological innovation and keep its competitive edge in the global marketplace of ideas and artifacts (Fechter 1994; Friedman 2005; National Academies 2005, 2010, 2011; Jackson 2007).

Currently, the United States is vying with other nations for global leadership in science and technology. No longer do American corporations compete only with other domestic organizations for contracts or business opportunities; now they must be able to compete on the basis of cost and quality with companies throughout the world. Corporations, university and government research laboratories, and other organizations involved in research and development in STEM are also confronting the reality that new and creative approaches are needed to ensure an adequate supply of talented and skilled professionals in the future. Shirley Ann Jackson (2007), president of Rensselaer Polytechnic Institute, refers to this situation as the "quiet crisis," a condition in which the nation's need for engineers (and scientists) is mounting amid a potentially inadequate pool of skilled individuals to meet the demand. It is what I have called a "new" American dilemma, one that the nation needs to confront rapidly and vigorously.

Like the moral dilemma postulated by Gunnar Myrdal (1944), this new dilemma comes from our nation's failure to educate and develop a growing proportion of its potential STEM talent base at the same time that its need for innovation and capacity in science and engineering is escalating. It will continue to escalate until America more fully develops the vast potential offered by the rich diversity of its people.

Given the confluence of the rapid demographic changes that are occurring in America, the tremendous progress in science and technology that is taking place in developing countries, the serious shortcomings of our public education systems, shifting immigration policies, and the historic underrepresentation of sizable portions of our population, our nation must act quickly on a number of fronts to maintain a strong position of leadership in the STEM disciplines and to ensure a future of prosperity and security. Preeminence in innovation and entrepreneurship will accrue to those nations that are the most adept at quickly building and retaining talent. This is the challenge facing our nation today (BEST 2004; National Academies 2005, 2010, 2011; NSB 2012; NSF 2012).

The significance of the competitive situation facing our country is slowly being recognized and responded to by the highest branches of government and by corporations, foundations, and academic institutions. Widely read publications such as Thomas Friedman's *The World Is Flat* (2005) and the National Academies' *Rising above the Gathering Storm* (2005, 2010) have been the catalysts for government, corporations, and academic institutions at all levels to consider offshoring, outsourcing, and increasing H-1B visa allotments. However, our leaders seem to have lost sight of the fact that in America there are many persons (whose numbers are growing dramatically) for whom participation in science and engineering has been and, in too many instances, continues to be less likely for a variety of reasons (Jackson 2007; National Academies 2011; NSF 2012).

Recent trends in African Americans' and other underrepresented groups' participation in STEM education and careers are somewhat encouraging. Nevertheless, the net result of US citizen participation in STEM (especially in doctoral engineering education) may be that fewer qualified Americans will be available at a time when more job openings are expected. This highlights the urgency in developing policies that drive more innovative and evidence-based programs that facilitate a process of tapping more deeply into underutilized talent pools in order to meet future needs (Pearson and Fechter 1994; BEST 2004; National Academies 2011; Garrison 2013). This urgency is compounded by the fact that approximately 47% of all STEM majors switch to non-STEM majors or leave college without earning any degree (Chen 2009). Low rates of student retention within the discipline have heightened concerns in the engineering community about the structure, content, and delivery of engineering education (Fortenberry et al. 2007).

The demographic shifts in the US population are having a tremendous impact on the nation's social fabric and its political, economic, cultural, and educational institutions. Already, they have added immensely to America's capacity for ingenuity and creativity in general. They have enhanced the science and technology enterprise and have the potential to continue to do so.

As in the past, adequacy and equity continue to be hotly debated STEM human resources policy issues (e.g., possible shortages vs. possible excess production). Without seeking to resolve the issue of exactly how many engineers are needed, there is a broad consensus that sustaining a healthy flow of capable young adults into engineering careers is important for the health of the economy—regardless of the state of the labor market (Wulf 1998; National Academies 2005, 2010, 2011; NSB 2012). Significant demographic changes affect our ability to produce the next generation of scientists and engineers. Demographic trends indicate that the racial and ethnic groups that continue to be underrepresented in engineering are the very ones that are increasingly becoming a larger portion of the US population owing to increases in annual number of births and immigration (Leggon and Malcom 1994; Pearson and Fechter 1994; Vetter 1994; National Academies 2011).

Wulf (1998) argues that at a fundamental level, racial and ethnic minorities likely experience the world differently than most Whites. It is these experiential differences then that serve as the catalysts for creativity in engineering. He believes that a lack of such diversity diminishes and impoverishes the engineering profession. In a globalized market, to be competitive (i.e., successful), engineering designs must reflect the culture of a diverse customer base.

While the national rate of unemployment has averaged around 9% since 2009, the unemployment rate for engineers is a mere 2% (Gearon 2012). When the national unemployment rate hovered around 10%, the unemployment rate for engineers peaked at 6.4%. In 2013, the highest-paying salary for a new college graduate was in engineering. The average starting salary was $62,535 (up 4% from the previous year). Some starting salaries were reportedly as high as $93,500 (NACE 2013). Some analysts (Newman 2012; Yoder 2012; Stephens 2013) assert that because only 5% of all undergraduates complete an engineering degree, there is substantial market demand for their skills. In terms of life chances, quality of life, and intergenerational mobility, these findings reinforce the criticality of broadening the participation of African Americans in engineering education and careers.

Today, the United States can no longer ignore the reality that too few native-born youths are choosing to study engineering and that other nations are increasing their capabilities in science and technology at higher rates. The continued underrepresentation of African Americans in both education and employment impairs our future ability to compete successfully in the global marketplace of ideas and products. To adequately address the challenges of the technological issues related to inequality, energy independence, health care reform, global climate change, infrastructure upgrading, and national security, it is imperative to develop and utilize all of our human resources in the process.

Data indicate an inverse correlation between level of engineering education and the participation of African Americans in engineering education: the higher the level of education, the lower the level of participation. This has a significant effect on resulting employment patterns. The small number of doctoral recipients is one reason for the dearth of African American tenured and tenure-track faculty members in engineering, especially in predominantly White schools of engineering.

The editors of this volume share the belief that creativity is best developed in an environment of diversity—in a broad sense that encompasses race, ethnicity, gender, class, country of origin, physical ability, sexual orientation, political persuasion, religion, and cultural background. The editors strongly believe that the different life experiences and frames of reference exhibited by an amalgam of diverse individuals reflecting a broad cross section lead to the creativity needed to make innovative technological breakthroughs.

The editors agree with Fechter (1994) and Leggon and Malcom (1994) that some science and technology human resource policy issues may be independent of any current or projected state of the labor market. One exception is underrepresented groups, such as African Americans. Fechter argues that the mere presence of underrepresentation partially reflects structural barriers that prevent qualified individuals from underrepresented groups from pursuing scientific or engineering careers. Thus, underrepresentation is an indicator of underutilized talent. In brief, this underutilization can exist simultaneously with situations of abundance of science and engineering talent. This represents costs not only to individuals but equally at the societal level as well. Policies aimed at reducing this underrepresentation should not be driven solely by market conditions.

The origins of this volume date back two decades, when I first discussed with Willie Pearson, Jr., the need to address the lack of a scholarly book on the contemporary status of African Americans in engineering. Unfortunately, both of our careers would be consumed by administrative and professional obligations and job relocations. In 2012, we decided to commit to producing the volume. Fortunately, Yu Tao, whose research focuses on women and minorities in engineering, agreed to join us as a coeditor.

Since it would not be possible to cover all the experiences, roles, and contributions of each of the underrepresented groups and to do credit to any one of them, our goal in this book is to examine these features for African Americans, a population that has a long history of exclusion in engineering. The editors have further decided to narrow the coverage to the field of engineering in order to provide focus and sufficient depth to the subject. By exploring the story of African Americans in engineering, it is our hope that more young African American men and women will be inspired to pursue engineering as a career and, at the same time, encourage the engineering profession to provide greater and more inviting opportunities for them.

Although the editors acknowledge that some progress has been made in terms of the participation of African Americans in engineering education and careers, it has been slow and uneven. Fortunately, however, the significance of the topic has motivated a group of distinguished scholars from a range of disciplines and employment sectors to examine issues related to the status of African Americans in engineering. The results of their investigations are presented in the chapters that follow.

## References

BEST (Building Engineering and Science Talent). 2004. *A Bridge for All: Higher Education Design Principles to Broaden Participation in Science, Technology, Engineering, and Mathematics*. San Diego: Building Engineering and Science Talent.

Chen, X. 2009. *Students Who Study Science, Technology, Engineering, and Mathematics (STEM) in Postsecondary Education*. NCES 2009-161. Washington, DC: National Center for Education Statistics.

Clark, J. V., ed. 2014. *Closing the Achievement Gap from an International Perspective: Transforming STEM for Effective Education*. New York: Springer.

Fechter, A. 1994. "Future Supply and Demand: Cloudy Crystal Balls." In Pearson and Fechter 1994, 125–140.

Fortenberry, N. L., J. F. Sullivan, P. N. Jordan, and D. W. Knight. 2007. "Engineering Education Research Aids Instruction." *Science* 317:1175–1176.

Friedman, T. 2005. *The World Is Flat: A Brief History of the Twenty-First Century*. New York: Farrar, Straus and Giroux.

Garrison, H. 2013. "Underrepresentation by Race-Ethnicity across Stages of U.S. Science and Engineering Education." *CBE Life Sciences Education* 12 (3): 357–363.

Gearon, C. J. 2012. "You're an Engineer? You're Hired." *US News and World Report*, March 22.

Jackson, S. A. 2007. "Waking Up to the 'Quiet Crisis' in the United States." *College Board Review*, no. 210 (Winter/Spring).

Leggon, C. B., and S. M. Malcom. 1994. "Human Resources in Science and Engineering: Policy Implications." In Pearson and Fechter 1994, 141–151.

Manning, K. R. 1983. *Black Apollo of Science: The Life of Ernest Everett Just*. New York: Oxford University Press.

Manning, K. R. 1995. "The Society: Race, Gender and Science." *History of Science Society*. Notre Dame, IN: History of Science Society.

Myrdal, G. 1944. *An American Dilemma: The Negro Problem and Modern Democracy*. New York: Harper.

NACE (National Association of Colleges and Employers). 2013. www.naceweb .org/s05012013/salary-survey-top-paying-industries.aspx?terms=April%20 2013.

National Academies. 2005. *Rising above the Gathering Storm: Energizing and Employing America for a Brighter Economic Future*. Washington, DC: National Academies Press.

National Academies. 2010. *Rising above the Gathering Storm, Revisited: Rapidly Approaching Category 5*. Washington, DC: National Academies Press.

National Academies. 2011. *Expanding Underrepresented Minority Participation: America's Science and Technology Talent at the Crossroads*. Washington, DC: National Academies Press.

Newman, R. 2012. "Where the Jobs Are, and the College Grads Aren't." *US News and World Report*, May 14.

NSB (National Science Board). 2012. *Science and Engineering Indicators 2012*. Arlington, VA: National Science Foundation.

NSF (National Science Foundation). 2012. *Women, Minorities and Persons with Disabilities in Science and Engineering*. Arlington, VA: National Science Foundation.

Pearson, W., Jr. 1985. *Black Scientists, White Society and Colorless Science: A Study of Universalism in American Science*. New York: Associated Faculty Press.

Pearson, W., Jr. 2005. *Beyond Small Numbers: The Voices of African American Ph.D. Chemists*. New York: Emerald.

Pearson, W., Jr., and A. Fechter, eds. 1994. *Who Will Do Science? Educating the Next Generation*. Baltimore: Johns Hopkins University Press.

Pearson, W., Jr., and J. D. Miller. 2012. "Pathways to an Engineering Career." *Peabody Journal of Education* 87:46–61.

Peske, H. G., and K. Haycock. 2006. *Teaching Inequality: How Poor and Minority Students Are Shortchanged on Teacher Quality*. Washington, DC: Education Trust.

Stephens, R. 2013. "Aligning Engineering Education and Experience to Meet the Needs of Industry and Society." *Bridge* 43 (2), www.nae.edu/Publications -Bridge/81221/81233.aspx.

Thompson, G. 2009. *Unheralded but Unbowed: Black Scientists and Engineers Who Changed the World*. Baltimore: Career Communications Group.

Vetter, B. M. 1994. "The Next Generation of Scientists and Engineers: Who's in the Pipeline?" In Pearson and Fechter 1994, 1–19.

Wulf, W. A. 1998. "Diversity in Engineering." *Bridge* 28 (4), www.nae.edu/Publi
  cations/Bridge/CompetitiveMaterialsandSolutions/DiversityinEngineering
  .aspx.
Yoder, B. L. 2012. "Engineering by the Numbers." www.asee.org/papers-and
  -publications/publications/14_11-47.pdf.

# Historical Background

In part I, Percy A. Pierre lays the foundation for the book by providing a first-person account of the history and development of efforts to increase the participation of African Americans and other racial/ethnic minorities in engineering education and the engineering workforce. Specifically, he traces the origins of the more systematic efforts to diversify the engineering workforce to the federal policies related to affirmative action in the 1960s. Pierre chronicles the leadership roles that the Sloan Foundation, industry, the National Academy of Engineering, and senior administrators from academe played in galvanizing support for what was then characterized as the "minority engineering effort." By the 1970s, this coordinated effort resulted in funding minority-focused organizations to administer programs that supported students from precollege through graduate engineering education, scholarships, co-ops, and career opportunities. The author credits these early efforts with establishing the foundation for current programs and organizations that focus on broadening the participation of underrepresented minorities in engineering.

# CHAPTER ONE

# A Brief History of the Collaborative Minority Engineering Effort

## A Personal Account

PERCY A. PIERRE

IN THIS CHAPTER, I chronicle the history and development of the coalitions and resulting initiatives designed to address the underrepresentation of African Americans and other minorities in engineering. I focus mainly on the time period from the late 1960s to the early 1980s. The chapter is written as a first-person account because I was one of the principal architects of what became known as the "national minority engineering effort." Specifically, I provided leadership for the 1973 National Academy of Engineering (NAE) Symposium, which officially launched the effort (Pierre 1974).

The minority engineering effort started as a grassroots movement and evolved over time, as is the case for many successful movements. It began with parallel and somewhat disjointed efforts to increase the number of African American and other minority engineering students graduated by colleges and universities. To a large extent, the movement was in response to the civil rights movement and related Johnson and Nixon administrations' policies regarding affirmative action (see Farley 2000; Franklin 2005; Feagin and Feagin 2012).

In my opinion, the movement is arguably one of the most successful college-level science, technology, engineering, and mathematics (STEM) education programs targeting African Americans and other minorities. Since the early 1970s, African Americans and other minorities have shown considerable progress in earning engineering degrees. For instance, from 1974 to 2011, the percentage increases of African American engineering graduates at the bachelor's level, the master's level, and the doctoral level were 365% (743 to 3,457), 714% (153 to 1,246), and 1308% (12 to 169), respectively, while the percentage increases for all races/ethnicities at these degree levels were 104% (41,407 to 84,599), 187% (15,885 to 45,589), and 200% (3,362 to 10,086), respectively (Engineering Manpower Commission of Engineers Joint Council 1975; AAES EWC 2012). Yet, despite this progress, African Americans remain considerably underrepresented in engineering compared to their representation in both the general population and the college-age population (see chaps. 3 and 6 in this volume).

This chapter is organized around the following general themes: (1) origins of engineering efforts at historically Black colleges and universities (HBCUs), (2) minority engineering efforts at predominantly White colleges and universities, (3) emergent national coordinated effort, (4) the origin of the National Action Council for Minorities in Engineering (NACME), and (5) intervention programs created by the Sloan Foundation and supported by industry. Given the dynamic nature of the minority engineering effort, these themes are not discrete categories. They include intersectional and overlapping activities. It is important to point out that much of the original focus of the movement's activities was on African Americans, the largest racial minority at the time. As the movement evolved, other racial/ethnic minorities were included in the outreach efforts.

## Origins of Engineering at Historically Black Colleges and Universities

In the 1970s, most African Americans lived in the South (US Bureau of Census 1978; Franklin 2005). Most of those who attended college did so at HBCUs (Thomas 1981; Fleming 1984). Many of the public HBCUs in the South had programs to train technicians and technologists rather than engineers, often in associate's degree programs rather than baccalaureate programs. It is important to understand the difference between technical programs and engineering programs. The major difference in the curricula is proficiency in math and science. Advanced calculus is needed to fully describe and design dynamic systems, that is, systems that move. A student who graduates from high school with only algebra and geometry is not well prepared to start an engineering program (Pearson and Miller 2012). The student should have taken trigonometry and, ideally, precalculus. Improving math and science preparation at the high school and college levels has been the primary challenge for the minority engineering effort.

In 1910, Howard University (Washington, DC) established the first engineering program at an HBCU; it was the only engineering program at an HBCU until Tuskegee Institute (Alabama) developed one in 1941. In 1971,

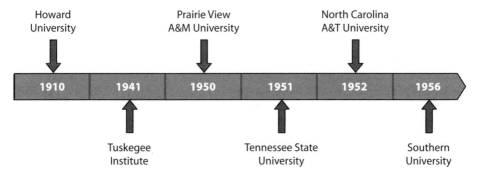

FIGURE I.I  HBCU engineering colleges
*Source*: Bradley (1973).

when I became dean of engineering at Howard University, there were five other engineering programs at HBCUs, as shown in figure I.I. (For a discussion of the contemporary status of engineering at HBCUs, see chap. 4 in this volume.)

One impetus for establishing engineering programs at HBCUs was a reaction to lawsuits by African Americans seeking to integrate Southern White public universities. For example, in Louisiana in 1950 and 1951, A. P. Tureaud represented plaintiffs in *Daryle Foster v. Board of Supervisors of Louisiana State University* (LSU), *Roy Wilson v. Board of Supervisors of LSU*, and *Payne v. LSU* in federal court. He won all three cases, which forced LSU to admit African Americans to some of its professional schools. Tureaud challenged the state's assertion that Louisiana operated a "separate but equal" system, with Southern University at Baton Rouge as the flagship Black university. Since there were few professional schools at any of the state-supported HBCUs, Tureaud argued that African Americans should be admitted to LSU. Although ultimately it lost (Hickman 1990; Emanuel and Tureaud 2011), the State of Louisiana took two actions designed to meet Tureaud's argument. First, it set up a scholarship fund to support sending African American students to out-of-state institutions for programs, such as engineering, that did not exist at HBCUs in Louisiana (see Pearson 2005). Second, it began creating such programs at HBCUs in an attempt to preserve the racially segregated system. Other Southern states established Black engineering schools in response to similar suits in their states.

Starting an engineering school and getting it accredited are separate and distinct issues. In 1960, Howard University had the only accredited engineering program at an HBCU, having been initially accredited in 1936 when engineering societies began accrediting programs. Graduating from an accredited program was more advantageous to students competing for top jobs in industry; otherwise, students would have to settle for jobs at lower levels. More importantly, accreditation was essential to colleges in assuring students a quality education. With assistance from the American Society for Engineering Education (ASEE), all of the existing engineering schools at HBCUs became accredited by the early 1970s. To accomplish this, ASEE established a committee to raise funds to assist HBCU engineering programs in attaining accreditation through the Engineer's Council for Professional Development (ECPD) (see www.abet.org).

Faculty quality is considered the most important factor in accreditation. One aspect of quality is having a faculty composed predominantly of doctorate holders. This was an area that needed to be improved at all HBCU engineering programs. HBCU administrators worked tirelessly to hire doctoral faculty, especially African Americans. Although all Howard University engineering faculty had at least a master's degree, more doctorate holders were needed. In 1971, six new African American faculty with doctorates were hired. After 1980, four of these became deans of engineering and played significant leadership roles in the minority engineering effort: Charles Watkins (City College of New York), Vascar Harris (Tuskegee Institute, now Tuskegee University), Eugene DeLoatch (Morgan State University), and James Johnson (Howard University).

Other important factors taken into account for accreditation were quality of the laboratories, curriculum, and student body. As a result of underfunding and limited resources, HBCUs faced considerable challenges in these areas. HBCUs developed plans to pursue public and private funding. These plans came to fruition with the help of ASEE, which is not an accrediting agency per se, but rather an organization that works to improve and maintain standards at their member institutions.

By the early 1970s, there were only six HBCUs offering engineering degrees (Howard University, North Carolina A&T State University, Prairie View A&M University, Southern University at Baton Rouge, Tennessee State University, and Tuskegee Institute). The ASEE HBCU Committee consisted of the deans of the six HBCU engineering schools, as well as faculty members from other universities. In addition to me at Howard University, the other deans included Austin Greaux (Prairie View A&M University), Dickey

Thurman (Southern University at Baton Rouge), Z. W. Dybczak (Tuskegee Institute), Willie Carter (Tennessee State University), and Reginald Amory (North Carolina A&T State University) (Bradley 1973). As the first coalition for increasing the representation of African Americans in engineering, the group worked closely together.

## HBCU Engineering and the Sloan Foundation

In 1972, the deans of engineering at HBCUs decided that the amount of money awarded by ASEE from industry sources was insufficient. A number of technological companies supported colleges of engineering, programs, and departments with small grants that usually went to institutions from which they recruited employees. Each of the HBCUs received small grants from such companies. Additionally, the ASEE HBCU Committee received small grants from industry to distribute to the HBCUs. Grants ranged from about $5,000 to $20,000 per company. An exception was Union Carbide, which contributed over $100,000 to ASEE.

In late 1972, the deans at HBCUs decided to pursue larger grants similar to the one awarded by Union Carbide. Because I had recently received a substantial grant ($330,000 for three years) from the Alfred P. Sloan Foundation to begin a program of professional master's degrees in urban systems engineering at Howard University, I suggested that deans at HBCUs approach the Sloan Foundation for support. To this end, I arranged a meeting between Sloan Foundation vice president Arthur Singer, deans of HBCU engineering colleges, and Francis X. ("Tim") Bradley of ASEE. At the meeting, the HBCU deans explained their needs and inquired whether the foundation would entertain a proposal of $300,000 over two years to assist the six schools.

Singer explained that the Sloan Foundation had a strong interest in science and engineering and in minority education. The Sloan Foundation had established a program to aid minorities in medicine and in MBA programs. While the Sloan Foundation had explored the possibility of establishing a "minorities in engineering program," ultimately it decided that foundation resources alone would not have a significant impact. Thus, we were surprised when Singer agreed to accept a proposal from us and asked us to double our request from $300,000 to $600,000. This was the Sloan Foundation's first step in overcoming its previous reluctance to become involved with the minority engineering effort. Later, the Sloan Foundation would partially fund an NAE symposium on the topic, which Singer agreed to attend.

## Minority Engineering Efforts at Predominantly White Engineering Colleges

Parallel to the work of HBCUs to increase African American enrollment, similar efforts were taking place at predominantly White institutions. During the 1960s and the early 1970s, companies experienced pressure from the federal government to hire African Americans and other minorities. Companies in the market for hiring engineers urged predominantly White colleges of engineering to recruit African American students (Willie and McCord 1973; Pearson 1985, 1988). A number of universities responded by hiring African American recruiters. While the HBCUs collaborated to work on this problem, there was no corresponding joint or coordinated effort on the part of predominately White universities. Each had its own program and sought support from industry individually. Together with the HBCUs, these universities helped lay a foundation for a national minority engineering movement.

In the late 1960s, Purdue University became an early leader in recruiting African American and other minority engineering students. Arthur J. Bond served as coordinator of programs for disadvantaged students at Purdue's College of Engineering. With the strong support of Purdue's president, Arthur Hansen, Bond managed several programs to recruit, mentor, and support African American and other minority students. African American students urged Bond to assist them in accomplishing their goal of establishing a national organization for African American engineering students. In 1974, this goal was realized with the founding of the National Society of Black Engineers (NSBE) in Atlanta, Georgia.

Robert Marshall, dean of engineering at the University of Wisconsin, hired Willie Nunnery to recruit African American students. Marshall would later become a key player with the NAE as chairman of its Commission on Education, which sponsored the NAE symposium. Thomas Martin, dean of engineering at Southern Methodist University (later president of Illinois Institute of Technology), strongly supported increases in the number of African American and other minority engineers and was one of the earliest participants in the minority engineering effort. Associate dean of engineering Joe Eisley, along with Keith Cooley, devised and directed a minority engineering program at the University of Michigan. Donald G. Dickason, director of admissions in engineering, worked to increase minority admissions at Cornell University. At Carnegie Mellon University, Norman Johnson directed the successful Carnegie Mellon Action Project (C-MAP)—now called the Carnegie Mellon Advising Resource Center (CMARC). Raymond Landis, professor of engineering, implemented a program at California State University, Northridge. W. Edward Lear, dean of engineering at the University of Alabama and later executive director of ASEE, established a program to recruit African American students.

Fred N. Peeples, dean of engineering at the University of Tennessee, managed a program based on the Cooperative Education Program as a focus for his minority efforts. One of the purposes of a co-op program is to help meet students' financial needs by providing work opportunities during the five-year degree program. Melvin Thompson directed the Engineering Opportunity Program at Newark College of Engineering, now the New Jersey Institute of Technology. Beginning in 1973, he served as staff officer of the NAE's Committee on Minorities in Engineering. James Bruce, associate dean of engineering at Massachusetts Institute of Technology (MIT), oversaw a program to recruit minority students while Paul Gray was dean. Gray later became president of MIT and was a strong supporter of increasing minority participation in engineering. These universities and individuals represented a significant groundswell of local activities that were ripe for coordination on a national level. All became key players in the collaboration that coalesced around the minority engineering effort and attended the NAE symposium. However, they needed a catalyzing force to bring them together, and that stimulus came from the General Electric Corporation (GE).

## The Dual-Degree Model

The dual-degree program provides opportunities for students at liberal arts colleges without engineering programs to pursue an engineering degree elsewhere (see also chap. 13 in this volume). Dual-degree programs require students to meet the requirements for a bachelor's degree at both institutions. Generally, this takes five years with a transfer to the second institution after three years, thus the 3/2 designation. The concept of 3/2 programs has a long tradition in engineering education.

In the 1960s, most African American students who were enrolled at HBCUs were at institutions without engineering programs. Although the dual-degree or 3/2 approach was attractive, most of the programs started at HBCUs in the 1960s and 1970s were not sustained. One notable exception is the Georgia Institute of Technology/Atlanta University Center (GT/AUC) dual degree, which is still in existence. After completing his PhD in engineering (1966) at Georgia Tech, Louis Padulo joined the mathematics department at Morehouse College. Shortly thereafter, he developed a dual-degree program between Georgia Tech and the AUC (consisting of Clark College,

Morris Brown College, Morehouse College, and Spelman College). Padulo was able to raise scholarship money for students transferring to Georgia Tech. In 1969, Padulo took a faculty position at Stanford University, where he initiated a program for minority engineering graduate students (see www .padulo.org/bio.html). Following Padulo's departure, Charles Meredith assumed responsibility for the GT/AUC program.

Other HBCUs also inaugurated dual-degree programs. In his 1974 study, Padulo surveyed such programs at Fisk University (Tennessee), Hampton Institute (now University, Virginia), LeMoyne–Owen College (Tennessee), Lincoln University (Pennsylvania), Morgan State College (now University, Pennsylvania), Norfolk State College (now University, Virginia), Shaw University (North Carolina), Tougaloo College (Mississippi), and Xavier University (Louisiana). In 1974, Padulo noted that the following HBCUs had recently started dual-degree programs: Grambling College (now Grambling State University, Louisiana), Wilberforce University (Ohio), Bishop College (Texas), North Carolina Central University, and Rust College (Mississippi).

In my opinion, the dual-degree concept proved not to be a major transferable model. However, an examination of the GT/AUC dual-degree program provides some critically important insight into understanding factors related to its success. First, and perhaps most important, was the commitment of the administrations at both Georgia Tech and the AUC. The second factor was the geographic proximity between the collaborating institutions (all located in Atlanta), which allowed students to begin taking courses at Georgia Tech before formally transferring. Third, the AUC colleges have strong math and science departments.

## Emergent National Coordinated Minority Engineering Effort

Thus far, I have noted the collaboration between the engineering colleges of the HBCUs and the role of African American engineering students in founding NSBE. In the early 1970s, another effort at collaboration was initiated by the ECPD. Under the leadership of its executive director, David Reyes-Guerra, ECPD also sponsored educational programs for high school students about engineering careers. One of those programs, Junior Engineering Technical Society (JETS), was designed to interest high school students in engineering.

ECPD staff noted that full-time, freshman engineering enrollments had fallen precipitously from 77,551 in 1967 to 51,925 in 1973 (Padulo 1974, 80). In the late 1960s, sharp shifts in federal funding priorities resulted in unemployment problems for the STEM workers (NAE 1973). The number of engineers laid off made headlines almost daily. Among the factors contributing to this decline were cuts in NASA's budget and the cancellation of the Super Sonic Transport that Boeing was building in Seattle. Minority students represented a resource that would help stem the decline in engineering enrollments. At the same time, companies came under tremendous pressure from the federal government to implement affirmative action policies and programs. Consequently, companies were prepared to fund ECPD to do something on a national scale.

In 1970, Calvin Conliffe, a Howard University alumnus and distinguished engineer in the jet engines division of GE in Cincinnati, Ohio, became a program officer of the GE Foundation in Connecticut. He instituted national programs to increase the number of African American and other minority engineers. With support from GE, ECPD initiated the Minority Engineering Education Effort (MEEE), under Conliffe's leadership. MEEE obtained $225,000 from industry (mainly GE) to support its successful Minority Introduction to Engineering (MITE) program. This program continues at some universities, including MIT. The MITE program supports high school students for two weeks of engineering study on a college campus. The program was not designed to remediate deficiencies in the education of minority students, but rather to convince reasonably prepared high school students to attend college better prepared to pursue an engineering major, particularly by taking more math courses before graduating from high school. To my knowledge, this was the first national precollege minority engineering program (see chap. 11 in this volume). The number of national efforts was growing, but there was still no nationally coordinated effort to tackle all of the issues that would have to be addressed to make significant progress.

## HOWARD UNIVERSITY'S STRATEGIC ROLE

In 1971, Howard University participated in most of the collaborative efforts—including the ASEE HBCU Committee and ECPD MITE. What Howard University did on its campus was similar to what was going on at other HBCUs and predominantly White universities. Because Howard University was strategically located in Washington, DC, it became a focal point for strategies to increase the number of African American and other minority engineers.

The principal challenge Howard University faced was that too few African Americans were graduating from high school with knowledge of and an interest in engineering. Consequently, the pool of African American and other minority students with prerequisites for an engineering curriculum was limited. In the spring of 1971, undergraduate engineering enrollment at Howard University amounted to fewer than 500 students. Of these, almost half were international students (mostly from Iran). For an engineering faculty dedicated to increasing the number of African American students, this posed an institutional challenge. Howard University's recruitment effort focused on identifying the best African American high school seniors in the country through the National Achievement Scholarship program for African American students, operated by the National Merit Scholarship Corporation. The limited scholarship funds available, mainly from industry, were targeted toward these students. The engineering recruitment committee developed explanatory literature, sent out mailings, visited local high schools, and operated summer programs.

As Howard University developed its recruitment program, other universities tried to do the same thing. Not surprisingly, with more universities competing for the small pool of students who were both prepared for and interested in engineering, it became difficult for each institution to achieve its enrollment goals. To expand the size and quality of the applicant pool, a major effort was needed to increase the number of high school students better prepared in math and science; in turn, this meant increasing the pool of math and science teachers. More African American and other minority high school students had to be motivated to take science and math high school courses, and the quality of those courses had to be ensured. Some argued that the only way to accomplish this was to reform all of public education at the K–12 levels. Clearly, the engineering community had neither the resources nor the influence to do that on its own. Operating outside of the educational system, strategies strongly influencing K–12 education for minorities were needed.

## CALL FOR A COORDINATED CORPORATE RESPONSE

In 1972, the Equal Employment Opportunity Commission (EEOC) established task forces to investigate possible discriminatory employment practices of GE, General Motors (GM), Ford, and Sears Roebuck. Under Title VII, charges were filed against these companies, resulting in settlements for entire classes of victims of discrimination (www.allgov.com). That summer, GE invited the six deans of engineering from the HBCUs and selected deans of engineering from predominantly White universities to a retreat at its education and training center at Crotonville, New York. GE's top management was at the conference, either presenting or presiding. Included were Fred J. Borch, chairman of the board and chief executive officer; J. Stanford Smith, senior vice president, corporate administrative staff; John F. Welch Jr., vice president and general manager, chemical and metallurgical division, and future CEO; and Lindon E. Saline, manager, corporate educational services. Also invited was Edward E. David Jr., science advisor to the US president (General Electric Company 1972). David's presence showed that GE was responding to the Nixon administration's emphasis on affirmative action and was putting its prestige behind GE's effort. In 1972, Reginald H. Jones, who had recently succeeded Borch as CEO of GE, did not attend the meeting, but he became the primary corporate leader of the minority engineering effort.

The conference was titled "Strategic Considerations in Engineering Education" (Smith 1972). Its principal focus was GE's new strategic plan on engineering education, including a discussion of minorities in engineering. Lindon Saline, director of the Crotonville Center, told me later that the impetus for this discussion of minority engineering was the Nixon administration's affirmative action policy encouraging companies receiving federal funds to hire more minorities. As a major federal contractor, GE knew that it had to be responsive to these new federal policies. Saline approached his boss, Smith, senior vice president of GE, with a plan.

Smith presented the plan to the deans in his speech, "Needed: A Ten-Fold Increase in Minority Engineering Graduates," at the Crotonville gathering (Smith 1972). The printed version, issued after the conference, differs from the plan offered at Crotonville in that the printed version did not include a focus on cooperative education (co-ops) as a means to fund the effort. The printed version recognizes that GE, as represented by Smith's remarks, had underestimated the difficulty of the problem.

The plan presented at the meeting identified the major barrier to minorities in engineering as economic, given the lower socioeconomic status of most minority students. In developing appropriate objectives in support of its goal, the company decided to focus on the financial gap. GE concluded that it was unrealistic for a scholarship fund to meet the needs of minority students. Instead, Saline's original plan was based on the experiences of GE and other companies in hiring large numbers of co-op students, who would spend several semesters working at companies where they could earn a significant part of the cost of their education. GE would lead a national effort to get other companies to create many more co-op opportunities.

While I was pleased that GE was taking the lead, I believed that their analysis was flawed. Although I agreed that the financial "gap" was a significant barrier, I was not convinced that the co-op approach would solve the problem. In my opinion, the chief barrier was the inadequate math and science preparation of African American and other high school students. I voiced this opinion at the meeting and was backed by only one or two other deans. I emphasized that if GE offered me multiple co-op positions, I would be unable to fill them. My comments were not well received by GE attendees. Accepting my criticism would negate the heart of their co-op strategy. I feared that GE would conclude that the problem was so intractable that it would revert to its company-centric effort, rather than leading a national effort to draw in other companies. Fortunately, GE decided on a different approach and asked for my assistance.

Saline informed me that GE recognized the problem as more difficult than originally imagined, that the approach needed to be educational and not limited to a financial one. Saline was charged with identifying an organization with educational interests which was willing to take the lead. This meant that GE would step back, but that they would lend their support to any organization that would lead the effort. Saline met with several national organizations, urging them to take the lead. He asked me and I agreed to join him in a meeting at the Conference Board in New York City. The Conference Board, a corporate-funded organization, undertakes studies for industry. The answer we got from them was the same that Saline got from all of the other organizations' studies he talked to—"no." The last name on his list of possible lead organizations was NAE.

In January 1973, Saline and I met with NAE representatives. The meeting also included Bob Marshall, chairman of the Commission on Education of NAE and dean of engineering at the University of Wisconsin–Madison; his assistant for minority recruiting, Willie Nunnery; James Mulligan, NAE executive director; Jean Moore, staff to the Commission on Education; Bob Mills of GE; and Willie Carter, dean of engineering at Tennessee State University. Saline and I asked NAE to take the lead in a national minority engineering effort. Marshall and Mulligan, speaking for NAE, did not believe that such an activity comported with the academy's charter. NAE, they explained, is an honorific organization chartered primarily to conduct studies for and provide advice to the federal government. Because it was not an agency, it would not be appropriate for the NAE to take the lead. Instead, Saline and I asked whether they would undertake a study. After an extended conversation, I asked whether they would sponsor a symposium on minorities in engineering. To this, they responded in the affirmative. I cannot emphasize enough the importance of this agreement.

Neither Saline nor I was interested in a symposium that would produce a report for dissemination with unpredictable consequences. Nevertheless, we saw this as a foot in the door which could lead to other opportunities. However, we did not give up our original aim to have the NAE lead the overall effort. I was asked to chair the planning committee and to assist Moore and Saline with fund-raising. One of the targets for funds was the Alfred P. Sloan Foundation. Most of the funding came from GE and other companies. What follows is a lesson on how to encourage an organization to do what it might ordinarily resist doing.

The NAE symposium on minorities in engineering was not designed as one of its typical symposia producing recommendations to disseminate for others to implement. Instead, we wanted to bring all of the stakeholders together to discuss what needed to be done and who would do it. The symposium was held in the Great Hall of the National Academies building. Official attendees included some 231 representatives of universities, industry, the federal government, and other organizations, as well as individuals with an interest in the subject. The affiliations of the attendees included the following: universities (106), industry (45), federal government (20), and other organizations (60). There were four presidents or chancellors of universities, 25 deans of engineering, and 20 managers or directors of major companies, in addition to those who worked in the field of the problem. This group came with great expectations that there would be significant action emerging from the discourse.

NAE's incoming president was Robert C. Seamans Jr., a former secretary of the US Air Force, who was very familiar with the Pentagon's efforts in affirmative action. He may have been predisposed to be positive about what we wanted to do because his son-in-law was Louis Padulo, founder and first director of the GT/AUC dual-degree program discussed earlier. Prior to the symposium, representatives of GE informed Seamans that if the NAE accepted the leadership of this effort, GE would put its resources behind it and would solicit other companies to do likewise. Seamans said he would attend the symposium and make a decision later. Saline and I decided that we would ask symposium participants to endorse NAE taking the lead. Saline wrote the following recommendations, which I presented at the symposium (NAE 1973):

1. That NAE establish a standing committee on minorities in engineering composed of a cross section of symposium delegates to include

engineering educators, members of minority groups, and representatives of industry and professional-technical organizations.

2. That NAE take the initiative in exploring with industry, government,
   and educators the feasibility of creating a national council on
   minorities in engineering.

The symposium produced many additional recommendations and af-
forded participants an opportunity to share common experiences of promising practices. Most importantly, it identified people who would work on
this problem for the next several decades with the support of industry and
the Sloan Foundation. Finally, it set the precedent of an annual gathering of
people interested in minorities in engineering. In early 1974, Saline edited an
issue of the *IEEE Transactions on Education* which captured the essence of
the symposium (Saline 1974).

## The Origin of NACME

Seamans agreed to the symposium's recommendations and set about creating the National *Advisory* Council on Minorities in Engineering (NACME).
In the November 1976 report on NAE's first 10 years, President Courtland
Perkins (1976) wrote, "In May of 1973, the Academy held a symposium on
minorities in engineering, in conjunction with its 13th annual meeting, and
under the aegis of the Commission on Education. As a direct result of the
symposium, a program for increasing participation of minorities in engineering was established which included three operating levels of volunteer efforts: a National *Advisory* Council on Minorities in Engineering
(NACME), a Committee on Minorities in Engineering, and four specific
subcommittees."

The original NACME was a committee of NAE, not the freestanding organization that it is today. In the early 1980s, it became an independent
organization with the same acronym (see discussion later in the chapter). The
first task in implementing the recommendations of the symposium was to
appoint a chair. Jones, chairman of GE, agreed to take on this role and to ask
several industrial colleagues to join the effort. Traditionally, companies prefer
to do their own thing rather than joining groups. Jones, however, was a very
influential member of the industrial establishment. He had helped organize
the Business Roundtable, a collaboration of companies working on problems
of common interest, and served as its chairman. He brought minority engineering to the roundtable as an issue of common interest. In fact, he sched-
uled roundtable meetings on the same day as the NACME meeting and es-

corted his colleagues from one session to the other at the National Academy of Sciences building. Prominent people from the educational community joined the council, including Theodore Hesburgh (president, University of Notre Dame), Paul E. Gray (president of MIT), and R. William Bromery (chancellor, University of Massachusetts). Additionally, various government agencies, such as the Department of Defense, NASA, and the Atomic Energy Commission, were represented on the NACME board.

NAE agreed that NACME's main order of business would be to mobilize resources to accomplish the goals set out by the symposium, and that most of those resources would have to come from the private sector. While government agencies were interested, they were unable to participate in the initial funding. In fact, government agencies contributed little financial support to the initial effort. NACME's first meeting, in January 1974, was held with the express purpose of funding the work of the newly established NACME. Jones secured commitments totaling $400,000 from industrial attendees. Subsequently, Melvin Thompson of the New Jersey Institute of Technology was hired as director of the NAE Committee on Minorities in Engineering, and Robert Finnell as his deputy.

The commitment of the Sloan Foundation to the minority engineering effort was a direct result of the NAE symposium on minority engineering and the creation of NACME. In the earlier meeting with the deans of HBCU engineering colleges, Singer, vice president of the Sloan Foundation, had said that the Sloan Foundation had not entered the minority engineering arena because the challenge was too big for it to have a singular significant impact. However, the Sloan Foundation's funding of HBCU colleges of engineering showed a clear interest in the movement. The Sloan Foundation also helped fund the NAE symposium, which Singer attended. After some internal discussions, the Sloan Foundation decided to become more involved. Sloan decided to commit 20% of its grant budget to this effort, about $12 million to $15 million over five to seven years. I was asked to come to New York City to discuss the decision.

I met with Robert Kreidler, executive vice president of the Sloan Foundation, and Singer. We agreed that this program should not follow the usual protocols of soliciting proposals and responding to them. A strategy was needed that would attack the fundamental problem of inadequate high school preparation in math and science for minority students and would leverage corporate interests identified by NACME. The Sloan Foundation would create programs and organizations to manage them with the understanding that industry and government would provide ongoing collateral support.

Kreidler and Singer asked me to become a program officer at Sloan and help run the program. As a relatively new dean of engineering at Howard University, I did not want to leave what I had just started, but I also did not want to say no to the Sloan Foundation, which had recently awarded me a large grant. I decided to make them an offer "that they *could* refuse." I told them I would accept a half-time post and commute from Washington, DC, on a weekly basis. They agreed. Through NACME, we realized that we had the corporate world ready and anxious to move. Moreover, we had significant money from the Sloan Foundation to seed any emergent opportunities. Universities were ready to respond. The challenge for me and my colleagues at Sloan was to help forge a direction that all could follow.

After the announcement of the new NACME program in 1973, Sloan had to decide what other activities it would fund. To gain consensus and to persuade industry and government to join in, the Sloan Foundation commissioned a study, which Padulo agreed to conduct. As a starting point for the Padulo study, we discussed goals and costs. Nearly all of the minorities involved insisted that "parity" in graduation rates for minorities compared to nonminorities should be a primary goal. Others argued for more readily achievable goals. Padulo resolved this conflict by plotting a range of goals and costs based on different assumptions about the growth of engineering enrollments. If engineering enrollments increased slowly, the number of additional racial/ethnic minority students needed to reach parity would be fewer and would cost less. A higher rate of growth would mean more minorities at a higher cost to achieve parity. But Padulo demonstrated that the cost of parity at the lower rate of growth of engineering enrollments was doable, which effectively quashed financial feasibility as a problem for the time being. Padulo also had to deal with the problem of defining who was a minority. The definition of underrepresented minorities was expanded from African Americans to other minority groups. We realized that programs might have to be framed for specific minority groups.

While the study was under way, the Sloan Foundation made some initial grants. Sloan officials recognized that the foundation should provide additional support to the HBCUs. Each of the six HBCUs with a college engineering program was awarded a grant of $100,000. Additionally, the University of New Mexico was awarded a special grant of $330,700 to support a program for Native Americans. The MEEE was awarded $225,000 to support its successful MITE program. Although these grants did not elicit large sums of additional funding from industry, future programs would.

The bulk of Sloan funds went into precollege activities, mainly through a consortia of universities, including the Committee on Institutional Cooperation (CIC) and the Southeastern Consortium for Minorities in Engineering (SECME) (discussed later in this chapter). Sloan's prior experience at the precollege level was primarily in curriculum projects—major curriculum projects, for example, in "new math," "new physics," and "new biology." The problem for African Americans and other minorities was not only the quality of the curriculum but also that few were registered in many of the math and science interventions targeted to improve student performance in math and science. We could improve the quality of the curriculum without expanding the talent pool; the difficult challenge was to build both quality and quantity of the talent pool.

Padulo's (1974) study produced two major strategic thrusts. The first was to establish a national scholarship fund for undergraduate minority engineering students. The second was to form a consortium of universities to develop precollege programs. In addition, there would be limited programs focusing on specific issues, such as graduate education. Padulo presented the recommendations of his study to NACME, which endorsed the creation of an independent scholarship fund on the understanding that Sloan would pay all of the initial administrative costs and that industry dollars would initially be used only for scholarships.

## Programs Created by Sloan and Supported by Industry

### NATIONAL SCHOLARSHIP FUND FOR MINORITIES IN ENGINEERING

The National Scholarship Fund for Minority Engineering Students (NSFMES) was set up as an independent organization—independent because if a minority scholarship fund was part of a larger organization, fund-raising would be in competition with fund-raising for other activities. Only an independent organization could sustain commitment over time.

Robert Finnel penned the charter for NSFMES. The fund was incorporated in Washington, DC, on October 30, 1974. The initial incorporators were Robert C. Seamans Jr., NAE president; Reginald H. Jones, board chairman of GE; and Richard J. Grosh, chairman of Committee on Minorities in Engineering of NAE and president of Rensselaer Polytechnic Institute (Ginsburg, Feldman, and Bress 1974). On November 7, Seamans submitted a proposal to the Sloan Foundation requesting initial funding of $436,310 for the scholarship fund, with NAE acting as the fiscal agent. Smith, at that time CEO of International Paper, agreed to chair NSFMES's board of directors. Hansen, president of Purdue, and I were charter members of the board, along with E. R. Kane, vice chairman of DuPont. Smith insisted that each scholarship awarded to a freshman would be for four years. While this lessened the number of scholarships given initially, it meant that all commitments could be met. Garvey Clark was appointed NSFMES's president.

During the first year of NSFMES, the Supreme Court was considering the *Bakke* case, which presented a threat to everything we were trying to do in the minority engineering effort. Allan Bakke sued the University of California Regents and its medical school, arguing that he was denied admission while less qualified African American students were admitted. NSFMES, with the help of NACME, filed an amicus brief arguing the need for affirmative action. In 1978, Bakke gained admission to medical school, but the Supreme Court did not outlaw race-based admission criteria (Hickman 1990).

Although NSFMES was not threatened by the court's decision then, more recent decisions and constitutional amendments passed by some states have limited the use of racial preferences. Usually, these cases come about when someone is denied admission to a professional school with a limited number of seats. In such cases, many qualified students are turned down. This has not been the case, however, for engineering schools. With few exceptions, over the past 30 years public universities have generally had space for all qualified engineering applicants.

## MATH, ENGINEERING, SCIENCE ACHIEVEMENT

In 1974 Ernest S. Kuh, dean of engineering at the University of California, Berkeley, submitted a proposal to the Sloan Foundation to work with Oakland Technical High School (California) to expand Oakland Tech's successful program to other high schools in the Berkeley area. Mary Smith, a math and science teacher at Oakland Tech, had started a math/science club to encourage more African American students to enroll in higher-level math and science courses. Engineering students from Berkeley worked with Smith as tutors and assistants. Wilbur Somerton, a Berkeley faculty member, wrote the proposal, seeking funds to (1) provide more support for the program at Oakland Tech and (2) expand the program to three other high schools in the Bay Area.

At the Sloan Foundation, we initially turned down this proposal because it was not consistent with our strategy of developing consortia of universities, rather than funding individual universities. However, when I called Kuh to tell him our decision, he asked me to come to Berkeley to see the program in action. I made the trip and was extremely impressed with what Smith had done. I became a convert. My task, then, was to convince my colleagues at the Sloan Foundation. One feature of Smith's program was cash awards to students who did well in math classes. My colleagues at the Sloan Foundation did not think that this was a good idea. Nonetheless, the Sloan Foundation agreed to fund this expansion and subsequent expansions. With state and private support, the program grew throughout California and beyond—to Colorado, Texas, and Maryland.

The funding of Math, Engineering, Science Achievement (MESA) illustrates that while the Sloan Foundation had an overall strategy, it was flexible and open to opportunities. A major asset was the commitment of the people involved in this particular proposal. The dean intervened, as did dedicated faculty members at Berkeley who wanted to make the program work. Smith was a creative and committed teacher. Thus, our funding decision was more about the people than about the proposal. In the book *The MESA Way*, the editors describe my visit as a "turning point," saying that "Pierre's introduction to MESA led to additional Sloan Foundation funding and his assessment of the program would bring major increases of donations in the future" (Somerton et al. 1994, 36).

## PRECOLLEGE CONSORTIA OF UNIVERSITIES

Beginning in the late 1960s, companies had made small grants to colleges of engineering to assist them in working with precollege-level students. However, the total amount of money was small, and the programs were not widely distributed. In order to launch a national effort, the Sloan Foundation needed a more organized approach. Consequently, it decided to organize regional consortia of universities.  ·

Among the first of these consortia was the CIC+Consortium. The "+" indicated the addition of five midwestern universities: University of Detroit, Illinois Institute of Technology, University of Notre Dame, Oakland University, and Wayne State University. Most of these were located in urban areas and had a long history of outreach to African American and other minorities. In November 1974, Singer and I met in Chicago with representatives of these universities to encourage them to submit a proposal. In April 1975, the Sloan Foundation made a grant of $750,000 to the consortium. Several other grants followed, for a total of $2,497,450. In 1983, CIC+reported that the Detroit Area Pre-College Engineering Program (DAPCEP) was by far its most successful program, working with five universities in southeastern Michigan. DAPCEP was eventually awarded major funding by the state of Michigan. Many of the precollege programs sponsored by CIC+no longer exist. Those that have survived did so through state funding. SECME is centered at Georgia Tech and focuses on working with high school teachers. It has survived, in part, because of excellent leadership and the strong commitment of Georgia Tech.

## THE NATIONAL CONSORTIUM FOR GRADUATE DEGREES FOR MINORITIES IN ENGINEERING AND SCIENCE (GEM) PROGRAM

While the Sloan Foundation believed that K–12 was critical, it also believed that progress could be made along all points of the pipeline. In January 1974, I published an article in the *IEEE Transactions on Education* arguing that it was timely to address minority graduate education (Pierre 1974). Later that year, I saw an opportunity to do something at the graduate level because companies hired many students with advanced degrees in their research establishments. I suggested that the research and development organizations partner with universities to provide fellowships to minority graduate students. Theodore Habarth suggested that we ask Father Hesburgh, president of the University of Notre Dame, to invite representatives of universities, industry, and nonprofit research organizations to convene at Notre Dame to discuss this idea. We thought that Father Hesburgh would be responsive, in part, because he was one of the first members of NACME and of the Civil Rights Commission.

In July 1974, the University of Notre Dame hosted a meeting of 40 representatives from 13 research centers, 14 universities, and five advocacy organizations to develop methods to increase minority representation. Singer and I presented our idea to the group. Several members from federally funded research and development centers (FFRDCs) indicated that, by federal regulations, they could provide internships but could not contribute to support for the schooling of students. This problem was overcome with the assistance of NACME, which solicited and received a ruling from the US Office of Management and Budget (OMB) allowing federal funds to be used for fellowships as part of a recruitment effort.

Father Hesburgh provided some university resources to assist in the establishment of GEM and appointed Robert Gordon, vice president for academic studies, to coordinate the effort. Joe Hogan, dean of engineering at Notre Dame, became GEM's first president and board chairman. Habarth, who was still employed at the Applied Physics Laboratory, was elected vice president.

In 1976, the Sloan Foundation awarded a grant to the University of Notre Dame to fund GEM (see www.gemfellowship.org). Initially, the consortium consisted of eight industrial and research laboratories and 19 universities.

The GEM office was housed in the College of Engineering at Notre Dame. GEM recruited African American and minority students and encouraged them to apply to member universities. Each industrial member provided an internship for each student selected, as well as a grant to GEM to cover some of its administrative costs and some of the students' costs through a fellowship. The admitting university would provide funds for all stipends and fees not covered by the GEM fellowship. For the first two years, Hogan managed GEM with the assistance of Notre Dame staff. Perhaps the most important contribution that he made to GEM was to hire Howard Adams in 1978. Adams was a charismatic leader who built GEM into a strong organization. In 1995, Adams left GEM to pursue other opportunities.

THE "NEW" NACME

In 1980, there were three centers of gravity with significant corporate support in the minority engineering effort: (1) NAE, which housed NACME; (2) MEEE at ECPD, with its precollege programs; and (3) NSFMES, the scholarship fund. All were supported in some manner by industry. NAE was still producing reports, while MEEE focused on recruiting activities. Industry sponsors and other supporters decided that it was time to consolidate these activities. A "Transition Council" under the leadership of Smith, chairman of International Paper and the GE presenter at the original 1972 Crotonville meeting, was set up to consider consolidation. On June 17, 1980, Smith reported the Transition Council's recommendation to integrate the national-level work in order to find ways of making our efforts and resources more productive and our support broader based. The new organization was to be located in New York City, where MEEE and NSFMES also resided. The original incorporators were Howard C. Kaufman, J. Stanford Smith, William S. Sneath, and Raymond Wingard. Although the new organization was called NACME, this time the acronym stood for the National *Action* Council for Minorities in Engineering. In 1981, Richard Neblett of Exxon became its interim president. On March 3, 1981, NACME created the Management Advisory and Review Committee (MARC) under the chairmanship of Saline. MARC consisted of individuals from companies representing the principals who sat on NACME's board.

Today, there is an array of national and regional organizations working to increase the number of underrepresented racial/ethnic minority engineers. Most of them have their origins in the 1970s, when grassroots activities met with committed and powerful leadership. The chapters that follow expand and provide context for many of the issues raised here.

## Acknowledgments

I want to thank the Alfred P. Sloan Foundation for support of research on the minority engineering effort which is a major part of this chapter. I also want to thank Ted Greenwood for his support and for giving me access to the Sloan Foundation archives. Finally, I would like to thank the editors for providing the forum for telling this story and for their many substantive contributions while editing this chapter. I particularly want to thank Willie Pearson, Jr., for his help in researching some issues and editing the final product.

## References

AAES EWC (American Association of Engineering Societies, Engineering Workforce Commission). 2012. *Engineering and Technology Degrees.* Reston, VA: American Association of Engineering Societies.

Bradley, Francis X. 1973. *The Traditionally Black Engineering Colleges: Gateway to Engineering Professions.* Proposal to the Alfred P. Sloan Foundation, September 1. Washington, DC: American Society for Engineering Education.

Emanuel, Rachel L., and Alexander P. Tureaud Jr. 2011. *A More Noble Cause: A. P. Tureaud and the Struggle for Civil Rights in Louisiana.* Baton Rouge: Louisiana State University Press.

Engineering Manpower Commission of Engineers Joint Council. 1975. *Engineering and Technology Degrees, 1974.* New York: Engineering Manpower Commission of Engineers Joint Council.

Farley, John E. 2000. *Majority-Minority Relations.* 4th ed. Upper Saddle River, NJ: Prentice Hall.

Feagin, Joe, and Clairece Booher R. Feagin. 2012. *Race and Ethnic Relations.* 9th ed. Upper Saddle River, NJ: Prentice Hall.

Fleming, Jacqueline. 1984. *Blacks in College: A Comparative Study of Students' Success in Black and White Institutions.* San Francisco: Jossey-Bass.

Franklin, John Hope. 2005. *Mirror to America: The Autobiography of John Hope Franklin.* New York: Farrar, Straus and Giroux.

General Electric Company. 1972. "Strategic Considerations in Engineering Education." Agenda of the Engineering Education Management Conference, Crotonville, New York, July 23–27.

Ginsburg, Feldman, and Brees (law firm). 1974. "Articles of Incorporation of the National Fund for Minority Engineering Students." Filed with the Recorder of Deeds, Washington, DC.

Hickman, Darrell K. 1990. "Realizing the Dream: *United States v. State of Louisiana.*" *Louisiana Law Review* 50:583–607.

NAE (National Academy of Engineering). 1973. *Symposium on Increasing Minority Participation in Engineering*. Proceedings. Commission on Education. Washington, DC: National Academy of Engineering.

Padulo, Louis. 1974. *Minorities in Engineering: A Blueprint for Action*. New York: Alfred P. Sloan Foundation.

Pearson, Willie, Jr. 1985. *Black Scientists, White Society, and Colorless Science: A Study of Universalism in American Science*. Millwood, NY: Associated Faculty Press.

Pearson, Willie, Jr. 1988. "The Role of Colleges and Universities in Increasing Black Representation in the Scientific Professions." In *Toward Black Undergraduate Student Equality in American Higher Education*, edited by Michael T. Nettles, 105–124. Westport, CT: Greenwood Press.

Pearson, Willie, Jr. 2005. *Beyond Small Numbers: The Voices of African American Ph.D. Chemists*. New York: Elsevier.

Pearson, Willie, Jr., and J. Miller. 2012. "Pathways to an Engineering Career." *Peabody Journal of Education* 87:46–61.

Perkins, Courtland D. 1976. *The National Academy of Engineering: The First Ten Years*. Washington, DC: National Academy of Engineering.

Pierre, Percy. 1974. "Minority Graduate Education." *IEEE Transactions on Engineering Education* 17:59–61.

Saline, Lindon. 1974. "A National Effort to Increase Minority Engineering Graduates." *IEEE Transactions on Engineering Education* 17:1–2.

Smith, J. Stanford. 1972. "Needed: A Ten-Fold Increase in Minority Engineering Graduates." Address to Engineering Education Conference, Crotonville, New York, July 25.

Somerton, Wilbur H., Mary Perry Smith, Robert Fennell, and Ted W. Fuller, eds. 1994. *The MESA Way: A Success Story of Nurturing Minorities for Math/Science Based Careers*. San Francisco: Caddo Gap Press.

Thomas, Gail E., ed. 1981. *Black Students in Higher Education: Conditions and Experiences in the 1970s*. Westport, CT: Greenwood Press.

US Bureau of Census. 1978. Current Population Reports Series, P-20, no. 361.

Willie, Charles V., and Arline S. McCord. 1973. *Black Students at White Colleges*. New York: Praeger.

## Websites

www.ABET.org
www.allgov.com
www.padulo.org/bio.html

# PART II

# Educational Systems

The five chapters in this section address critical aspects of higher education pathways in the context of the African American experience in engineering education. In chapter 2, Cheryl B. Leggon breaks new ground with her interviews of African American deans of engineering at historically White universities. The interviewees are uniquely positioned to set the tone of their departments and to hold faculty members and department chairs accountable for an overall institutional commitment to diversity in general and to efforts to recruit and retain African Americans in particular. The chapter explores the deans' perceptions of strategies to increase the representation of African Americans in engineering.

In chapter 3, Yu Tao and Sandra L. Hanson use an intersectionality approach to analyze national samples of engineering doctorate degree holders. Specifically, their findings highlight the distinctiveness of the doctoral portion of the engineering education pathway and the complex interaction of race/ethnicity and gender. They find a male advantage in earning engineering doctorates across race/gender groups which is smallest among African Americans.

In chapter 4, Lindsey E. Malcom-Piqueux and Shirley M. Malcom present recommendations for practice, policies, and programs to increase access, retention, and success in engineering careers for African American men and women. The authors examine the institutional pathways to engineering degrees for African Americans and discuss the implications of changing patterns of participation in higher education on the engineering pathways. They identify multiple factors that may account for the failure of engineering to keep pace with the overall growth of African Americans in higher education, and they discuss the roles of major trends that likely contribute to the underparticipation of African Americans in engineering. To increase African Americans' participation in engineering, the authors suggest a comprehensive approach that includes changes in federal, state, and institutional policies; improvements in educational practice; and rigorous research and evaluation to augment understanding of the ways in which institutions, programs, and individual faculty members can better promote access to and success in engineering.

In chapter 5, Tafaya Ransom examines the contributions of historically Black colleges and universities (HBCUs) to degree production in engineering, with particular attention to the role of gender. The chapter underscores the importance of HBCUs in the education and degree production of African American engineers at the bachelor's, master's, and doctoral levels.

In the final chapter of part II, Sybrina Y. Atwaters, John D. Leonard II, and Willie Pearson, Jr., examine trends in undergraduate engineering education for African American citizens and permanent residents. They draw on three major databases to explore gender differences in enrollment, field choice, persistence and retention, degrees awarded, and baccalaureate-origin institutions. The authors use a case study of the Georgia Institute of Technology to explore institutional-level data on trends in enrollment, degrees awarded, retention, and performance among select engineering disciplines.

# CHAPTER TWO

# African American Engineering Deans of Majority-Serving Institutions in the United States

## CHERYL B. LEGGON

ENGINEERING IS ONE OF THE drivers of economies worldwide. Not only does engineering enhance the lives of people in a variety of ways; it also is a major factor in the extent to which a nation is competitive in the global economy: "This nation's prosperity, security, and quality of life are direct results of leadership in the engineering achievements that drive society forward" (National Academy of Engineering 2012).

Engineers are "viewed as an important engine for . . . innovation and economic growth" (Mather and Lavery 2012). However, reports from government commissions express concerns about the United States' ability to continue to produce engineers (and scientists) needed to fuel innovation and maintain competitiveness (BEST 2004; National Research Council 2007, 2010). The pool from which the United States traditionally drew its engineering talent has consisted mostly of non-Hispanic White males. However, over the past 20 years, this talent pool has decreased in size because of two major factors: (1) decreases in the number of non-Hispanic White males in the US population overall, and (2) decreases in interest in engineering careers among this group. Historically, the United States met perceived shortages in the engineering talent pool by "importing" talent from abroad. However, this short-term "solution" is increasingly less viable because of competition for the same talent from nations of the Pacific Rim and the European Union.

In 2011, John Brooks Slaughter, president and CEO of the National Action Council for Minorities in Engineering (NACME), addressed concerns about the adequacy of the US engineering talent pool: "The solution to America's competitiveness problem is to activate the hidden workforce of young men and women who have traditionally been underrepresented in STEM careers—African Americans, American Indians, and Latinos"

(NACME 2011). African American engineers have a long record of contributions that have enhanced US competitiveness in the global economy (Johnson and Watson 2005). However, data on engineering degrees earned by African Americans are not very encouraging. For more details, see chapters 4, 5, and 6 in this volume.

Research indicates that historically Black colleges and universities (HBCUs) have been the baccalaureate origins for many African American engineers earning doctorate degrees in the United States (see chaps. 4, 5, and 6 in this volume). However, recent data reveal that increasingly African Americans are earning engineering baccalaureate degrees from non-HBCUs (see chaps. 4 and 6 in this volume). Trend data show little growth in the percentage of engineering faculty comprised by African Americans: from 2002 to 2007, African Americans' percentage increased from 2.0% to 2.5%, and it has remained at 2.5% since 2007. In addition, there is an inverse correlation between academic rank and the percentage of African Americans in that rank. As of fall 2011, among tenured/tenure track engineering faculty, African Americans were 3.5% of assistant professors, 3.2% of associate professors, and only 1.9% of full professors.

Although there has been some research on the impacts and role(s) of faculty on increasing African American participation in engineering, there is a paucity of research on African American deans in majority-serving (predominantly White) colleges of engineering in the United States. This chapter presents results from an exploratory study of African American engineering deans in majority-serving institutions.

## Overview

The purpose of this exploratory study is as follows:

1. To ascertain the deans' perceptions of ways to increase and enhance the participation of African Americans in engineering: as students, faculty, and administrators, and in leadership roles in the profession.
2. To explore the deans' perception of the role(s) of professional associations in augmenting African American participation in engineering.
3. To explore career pathways to leadership positions in academic engineering. This complements and extends research on race and diversifying science and engineering faculties (Leggon 2010).

## Data and Methods

This exploratory study utilizes mixed methods. Data are derived primarily from semi-structured in-depth interviews, which are useful methodologically in two major ways: (1) to explore issues based on people's interpretation of their experiences, and (2) to generate new theory. Additional data came from systematic analyses of documents—both electronic and hard copy—including resumes, position papers, and websites. Data from the interviews illuminate and augment our understanding of promising and effective practices to increase African Americans' participation in engineering careers.

Using the technique of purposive sampling, the author created an initial list of African American deans at majority-serving institutions in the United States. Purposive sampling is well suited to selecting particular individual cases for in-depth investigation (Zikmund 2003). In-depth telephone interviews were conducted in two waves: from February to April 2013 and during August 2013. Although the second wave was not planned initially, it was added to enable the author to interview a newly appointed African American engineering dean of a majority-serving institution—who is also the first woman to hold such a position.*

Eight individuals were identified. The author reviewed the profile of each person on the list, and, as a result, two persons were eliminated from the original list: one was not African American, and the other was no longer a dean. The author invited each of the remaining six deans to participate in a telephone interview; one was unable to participate because of scheduling conflicts. In sum, qualitative data are derived from in-depth interviews with five of six African American engineering deans at majority-serving institutions. The total number of interviewees comprises 80% of the deans listed and verified; this is not methodologically problematic because the purpose of this exploratory study is to identify issues for further study rather than to confirm hypotheses. Four of the five interviewees are deans at public institutions. By doctoral training, two of the interviewees are mechanical engineers, one is an electrical engineer, and one is a chemical engineer.

Consistent with the requirements for conducting research with human subjects, participants in the interviews were briefed on the purpose of the exploratory study and on their right to end the interview at any time without penalty or pressure to continue. In addition, participants were informed that data in all forms would be secure and confidential to the extent possible.

---

* It is noteworthy that she consented to be interviewed, knowing that it would not be difficult to identify her.

All of the interviewees gave their permission to tape the interviews—which varied in length from 20 to 90 minutes.

Results from this exploratory study have the potential to inform the knowledge base, initiatives, policy, and programs designed to increase the participation of African Americans in engineering—not only as students but also as faculty, administrators, and leaders in the profession. Increasing African Americans' participation in engineering enhances the quality and vitality of the engineering enterprise, which, in turn, can fuel innovation and maintain US competitiveness.

## Themes

Data from the interviews were systematically analyzed to identify common themes across interviews and between interviews and relevant research findings. Results and themes are organized around two of the primary interview questions:

1. What do you think are the most effective ways to increase the participation of African Americans in engineering as students, faculty, and administrators, and in leadership roles in the profession?
2. What is your perception of the role(s) of professional associations in augmenting African Americans' participation in engineering?

Responding to how to effectively increase the participation of African Americans in engineering as students, interviewees pointed out the criticality of "identifying the gatekeepers"—that is, those who control access, entry, and progress in various stages of the science, technology, engineering, and mathematics (STEM) education and career pathways—and getting rid of inappropriate gatekeepers. In fact, one interviewee noted that this gatekeeper issue begins as early as secondary school, and offered an example: "High school counselors who never encourage or direct African American students towards fields like science and engineering—that says a lot because at that impressionable age, students really do put a lot of weight on what teachers and counselors say . . . and then they internalize what they can and can't do and keep that with them."

Gatekeepers can also be barriers to the participation of African Americans on engineering faculties. The evaluation processes for tenure and promotion and the messages they send are problematic, as expressed by one dean: "I think a big obstacle is the tenure and promotion process. . . . We've made this whole tenure and promotion evaluation process so mysterious, nebulous, and open-ended that it can be a turn off to people who want to have certain defined things happen in their lives at certain times and who have really specific goals." Other responses to this question cluster into the following themes: institutionalization, accountability, strategic organization of targeted initiatives, and the role of engineering in society.


## INSTITUTIONALIZATION

It should be pointed out that even though the interview protocol asked about increasing the participation of African Americans in engineering, all of the interviewees tended to respond in terms of various groups that are underrepresented—not only African Americans but also, for example, Hispanics and women. This response suggests that the deans' strategy is based on the canary-in-the-mine phenomenon—that is, what works to broaden the participation of one underrepresented group can work for other underrepresented groups. Interviewees noted the crucial role of institutional environment in recruitment and retention: "If the environment . . . is so unwelcoming and inhospitable, they never get the sense of belonging . . . so the idea of staying in the field or even finishing the degree sometimes just goes out the window because of the environment they were trained in."

If an individual's experiences in engineering are limited to one institution—either attending the same institution for both graduate and undergraduate studies or teaching at only one institution—they may overgeneralize from those experiences and assume that the environment, and hence engineering, is the same in all institutions. The same applies to limiting one's experience to a single sector; this is important because as an "applied discipline" engineering provides career pathway flexibility insofar as one can work in the private for-profit sector and/or the public sector and still have the option to work in academe at various points in one's career.

All of the interviewees emphasized the criticality of institutional commitment. This is consistent with findings from reports from the National Research Council (2005) and the National Science Foundation (2005) that successful interventions to improve the participation of underrepresented minorities (URMs) must be institutionalized. Institutionalization means that interventions to diversify an institution must not be tangential or marginal, but must be incorporated as an integral component of the standard operating procedures of that institution, as described by one interviewee: "It's a matter of having the proper support network interventions in place, as well as financial assistance." It is important to note that this support network can be extended to include alumni. One interviewee mentioned that the Black alumni provide support in a variety of ways—especially in terms of career assistance and mentoring.

ACCOUNTABILITY

Commitment to diversity is necessary—but not sufficient. When diversity is an integral component of the standard operating procedure of an institution, faculty members and administrators must be held accountable in terms of what they have—and have not—done to address that commitment. Therefore, diversity becomes one criterion on which the performance of faculty and administrators is assessed. In the words of one interviewee, "The leadership is empowered in a way that could affect change by holding their own people accountable." In the opinion of another interviewee, accountability is critical at every level of an institution: "I think that the leadership plays a huge role in setting the tone for the level of commitment and providing environments that are inclusive and supportive so that people can see that 'yes there is a place for me . . .' and a lot of it does start with the leadership." Those at the highest level of an institution set the tone—either positive or negative—for those below them: "At the highest level of an institution . . . provosts must be committed and force accountability on the ones who report to them. . . . So that when we get to the level where the hiring really takes place, which is the department level, the department search committees have to be committed and be held accountable. . . . But if they know that none of the leadership above them is actually serious about the recruitment and retention of African Americans, then they don't have to be serious either."

Projections of faculty retirements indicate that over the next 15 to 20 years, there will be many faculty opportunities available. However, it is up to the dean to be vigilant and very aggressive about recruiting women and minorities into these faculty positions. How an institution operationalizes its diversity efforts depends on its size and structure. For example, although hierarchical structures may enable the dean to exert more control over faculty, they also have the potential to foster behavior that could be perceived as being somewhat autocratic. Nevertheless, one interviewee asserts that "at the dean's level you can set the tone for the department and extract that level of accountability that puts some onus on the institutional environment and not on the individual."

It is not unusual for a college or university to have programs designed to recruit URMs and/or women to the institution as undergraduate students, graduate students, and faculty members. Interview data suggested that some existing programs could broaden their goals. For example, programs that focus on strengthening students' skills in science and mathematics can also help to develop their interest in engineering as a career. Clearly, some of these programs could serve as feeders to others. Unfortunately, it is also not unusual for these programs to be scattered and not coordinated with one another.

When asked about effective ways to enhance African American students' retention in engineering, one dean asserted that "there has to be a clear focus on academics as the number one priority. I think we often confuse some of the things we do that are sort of around academics with helping retention. . . . You have recruitment activities and other kinds of things that are not specifically on tutoring, peer mentoring, bridge programs, and things that really affect academic performance."

One such effective practice utilizes peer-led team learning (PLTL). PLTL is a model of teaching undergraduate engineering (as well as mathematics and science) in which peer-led workshops are an integral part of a course— and of an engineering curriculum (Loui and Robbins 2008). In the PLTL model, upperclassmen lead workshops that combine tutorials and mentoring to help underclassmen in a particular course. An interviewee pointed out that one of the most important outcomes of these programs—especially those that include a mentoring component—is that they can help to establish and nurture a sense of community and a culture of support among students, as depicted in figure 2.1.

When asked specifically about the most effective ways that deans can increase the participation of African Americans in engineering, one interviewee said, "Deans, for the most part, with anything we do, are just facilitators and the providers and allocators of resources to achieve certain outcomes. You should be able first to recognize and quantify what the issues and problems are and then direct the resources at your disposal for solutions to those problems. For us, that means provide programmatic support for the various enrichment programs."

Another interviewee pointed out that in terms of retention, "activities such as recruitment are confounded with things that really affect academic performance." Interviewees emphasized the connection between retention on the one hand and training and support on the other. In the words of one

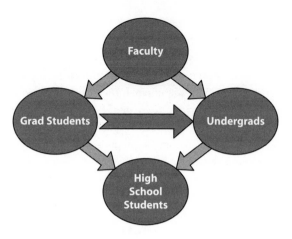

FIGURE 2.1  Schematic of PLTL model

interviewee, retention is "environmentally induced"—for example, if some-one who may lack particular aspects of training is put in a supportive envi-ronment, they are likely to thrive; conversely, if someone who is well trained is put in an unsupportive environment, they are likely to deteriorate. This is consistent with research findings that HBCUs tend to have the kind of sup-portive environments that enable students to thrive and persist in other en-vironments (Fleming 1984; Allen 1992; Leggon and Pearson 1997; Leggon, Tao, and Pearson 2009). Creating and sustaining supportive environments necessitate close and continuous monitoring of the institutional environment. Similar to a continuous process improvement model, periodic institutional climate surveys can provide data so that administrators and faculty can make whatever adjustments are warranted.

During discussions of retention, an important issue arose: that many Af-rican Americans—especially African American women—who earn a PhD do not practice engineering per se, but work in outreach roles. One interviewee attributed this to their experiences in graduate school: "The environment they were trained in was so unwelcoming and inhospitable that they never got a sense of belonging in that community . . . they go practice some other field because engineering became so distasteful because of their experience." Interviewees disagreed about the meaning of this: some said that this could send the wrong message to African American women pursuing the engi-neering doctorate; others said that even though these women were not practicing engineers, they were not necessarily "lost to engineering." One interviewee said that this is a woman's issue in general, and not limited to African American women in particular, adding, "I think it's okay that people who get engineering degrees don't all stay in the practice of engineering. I think engineering is—some call it the liberal arts degree of the new millennium—a good foundation for many careers, whether it be in engineering or law or medicine or business or what have you. So I think it is a good preparation for a career in many different areas."

## CONCEPTUALIZATION

Interview data indicated that it might be useful to think of increasing the participation of African Americans in engineering as an ongoing process instead of as a static end-state. Just as having undergraduates mentor high school students can be a useful recruitment tool for undergraduate institutions, so too can having graduate students mentor undergraduates be a recruiting mechanism for graduate school. Interviewees reported that they make concerted efforts to utilize existing programs and activities on their campus to help increase the participation of African Americans in engineering; consequently, it is critical to have external evaluators conduct periodic evaluations of diversity interventions individually and in terms of the extent to which they are coordinated with one another. Data from these evaluations could provide feedback to facilitate making, in real time, whatever adjustments are warranted.

## ENGINEERING AND SOCIETY

The deans agreed that one major barrier to the recruitment and retention of African Americans in engineering careers is the lack of awareness about the benefits of careers in engineering. African Americans are less likely than their nonminority peers to have family members who are or have been engineers. Consequently, their exposure to engineers and engineering careers is limited. At various times during the course of the interviews, each dean mentioned that most African American students want to use their knowledge to give back to their community. This is consistent with research findings on career choice that indicate that one important criterion in career choice for African Americans is the extent to which they perceive that a particular career enables them to make contributions to the community. As the interviewees pointed out, African American students are more likely to see what their contributions to the community would be if they were a medical doctor, lawyer, or teacher than if they were an engineer. Consequently, some African American students may feel that they have to leave their community to pursue a career in engineering. As one interviewee said, "When kids are asked what they want to do when they grow up, engineering is not something that just rolls off everybody's tongue." Therefore, professors have an obligation not only to prospective students but also to their parents to provide accurate information on what an engineering career is and how it can benefit society in general and the African American community in particular. In the words of one interviewee, "I think professors need to be aware of that and when they're talking to prospective students or their parents, they need to be able to tell them. . . . You're going to be able to develop some things that can be helpful for the whole country, world, and in the long run you're going to be a good citizen wherever you live, and you may not be living in the same community in which you lived before—but you can still give back." Similar sentiments were echoed by another interviewee: "I think pushing the idea that engineers are helping to create a more sustainable world, a more comfortable world, a safer world for everyone, would be more appealing."

Data from programs targeted to increase the participation of URMs in science and engineering indicate that having programs and/or internships in place that clearly demonstrate the contributions that engineering can make to society tend to make engineering more attractive and appealing to African American students. One such program is Engineers without Borders; another is Agua Clara, which is a project designed to build water purification systems in Honduras and teach Hondurans how to run them.

### PARTICIPATION ON ENGINEERING FACULTIES

Interview data indicate that respondents tend to view faculty diversity as being almost inextricably intertwined with student diversity. As one respondent put it, "African American faculty and African American students— chicken and egg." However, respondents were quick to point out that "all faculty—not just African American faculty—must be committed to increasing the pool and hire enough people in the pool to be influential in growing the next generation." Increasing the pool is more than just increasing student numbers and decreasing (if not eliminating) leaks along the pathway. In the words of one interviewee, "To move students forward requires intervention—which starts with undergraduates doing research."

The presence of African Americans among engineering faculty is important for all students, but it is especially important for African American engineering students. At the undergraduate level it can be an important factor in both selecting one engineering program over another and deciding whether or not to choose a career in academic engineering. One interviewee pointed out that "graduate students may be thinking only about their trials and tribulations getting the PhD and say 'I'm not going through something like that again.' . . . They are turned off by the capricious nature of academe. . . . It's important that there be a counter message about the opportunities. . . . You can be your own boss. . . . You can do things that most people do not get to do. . . . Such as, for example, travel the world." Interviewees agreed on the criticality of mentoring for engineering faculty. In the words of one interviewee, "At the core success of every faculty member is great mentoring. African Americans coming from a disadvantaged background in systems with built-in biases are especially vulnerable. . . . Trying to overcome systematic bias that's telling them (African Americans) 'you do not belong in science or engineering.' "

Not all deans agreed about the significance of mentor and protégé being of the same race. One dean pointed out that "academia can be very intimidating to graduate students . . . while being told 'you can do it,' here are some people who look like you doing it." However, another interviewee said, "We have to be able to find mentors, and those mentors need not look like us all the time either. I always tell people that your mentor does not have to be another African American. Your mentor can be someone of any background who can be helpful to you. So we have to be receptive to it and then those of us who are in the leadership roles have to be willing to be the mentors."

One respondent discussed how essential it is for engineering colleges to develop and implement a well-defined mentoring plan for faculty: "A faculty job is like having a startup company. . . . In six years from now, I want to see how you did. That's a poorly defined problem." One effective mentoring practice has been to require that a faculty member have an internal mentor and an external mentor. The internal mentor "gives a sense of specific expectations of the department. . . . Here's the bar, this is what you need to do to be successful." The external mentor serves as a sounding board and confidant precisely because they are not in the protégé's department. Having more than one mentor at a time has been an effective practice for women in engineering (Leggon, Eller, and Frehill 2011). Developing African American junior engineering faculty members requires a multilevel approach: at the macro level, accountability from administrative leadership, and at the micro level, empowering individuals.

## ADMINISTRATION AND LEADERSHIP

It is noteworthy that all interviewees emphasized the importance of leadership—which they defined as commitment to recruiting and retaining

African American faculty. Concomitant with leadership is accountability, as reflected in the comments of one interviewee: "Leadership needs to be accountable. For example . . . several (engineering) departments have openings for faculty positions and none of them have identified any members of underrepresented groups—much less African Americans, and they come up with this slate. . . . There is no reason why the dean can't push back and reject the slate, which does not happen often enough."

One interviewee cautions those who plan to become an engineering dean not to look only at administrative positions, because the institutions where the vacancies occur might not be the caliber of institution they would choose otherwise. When asked, "Did you decide early on that you wanted to be in administration?" one interviewee said that they did not know of anyone who did that—especially people who are really successful—and shared their own experiences:

> I think it sort of evolves and you take advantage of opportunities as they present themselves and you experiment and you listen to people's advice. I certainly thought I would be in some sort of leadership role. I didn't know if that would be in academic leadership or more in the research side or industry or government. But I didn't set out 25 years ago and say, I want to be a Dean some day. I don't know that you can really do that. But at each step along the way, it's important to have yourself prepared for the opportunity as it comes along. And to be getting the right advice and take some risks occasionally, career-wise. And it worked out for me.

Although an individual may not have planned to be a dean, sometimes they are identified by top administrators in their institution as a promising candidate should a position become available, and then they are groomed for the position by top-level administrators. This anticipatory socialization can include an extensive shadowing experience with an institution president or engineering dean, or participating in an executive leadership program. Another interviewee became dean when the previous dean became provost. Although there was an external search conducted, the search committee concluded that the internal candidates were better than the external candidates and that this interviewee was the best. The interviewee decided to accept the offer because "I saw the potential to have an impact. . . . A lasting legacy at my institution." The desire to have an impact—both short- and long-term—was another recurring theme in the interviews.

Interviewees agreed that the prerequisite to getting into the administrative ranks is to be successful as a faculty member. One interviewee noted that in most engineering programs, there were two big departments—

mechanical and electrical. Being the chair of either of these departments puts an individual "on the radar for all engineering Dean searches in the United States." Another path to becoming dean is by serving as the director of graduate studies in an engineering program—through which one comes to be viewed by others as an administrator. Opportunities and recognition play critical roles, as discussed by one interviewee: "For African Americans in particular, I think that there are insufficient opportunities to shine and be in roles where people can see what you can do."

Interview data suggest a phenomenon that I termed "the downward opportunity spiral." African Americans rarely get the opportunities to hold the positions and provide leadership at higher levels. African Americans tend to be more likely to serve on committees than to chair the committee—unless it deals with diversity issues. However, because African Americans are in short supply in academe in general—and on engineering faculties in particular—they are frequently called on to represent their unit on several committees. This kind of service is time-consuming and at many institutions does not count heavily toward tenure and promotion. One interviewee explains, "There is so much work that African American faculty tend to do that gets buried . . . unrecognized . . . unrewarded . . . there are so many ways in which (African Americans) are called upon to work, but their contributions remain invisible."

In addition to the volume of committee work, African American faculty members frequently take on additional work with their community and their students insofar as they help to develop, recruit, and retain talent, and serve as role models. However, there is no reason for deans not to give credit for that type of work—that is, to change the current reward structure.

Interview data indicate that the role of dean can be conceptualized in a variety of ways. In the words of one interviewee,

> Administration is a sort of paid service. Once you understand what's expected of you. . . . At the end of the day, you're expected to do two things: You are expected to lead and serve. So some people who just want to lead . . . everyone else should serve them, and it almost never works that way. You really want faculty to come to you to do things to support them. Then once you get enough of that happening, then it's easier for you to go to them with things that they need to do to support you. Understand, it's give-and-take. When I go into a conversation with you and you are one of my best faculty members, let me figure out what I need to do in order to help you. It's a much better situation than the other way around when I'm saying "look I need you to come and do this for me."

When discussing their role as administrators, interviewees described their relationships with their faculty members as peer relationships rather than in terms of hierarchy. In other words, these deans viewed their faculty as part of a team. One interviewee shared their philosophy of being a dean: "Once you are a successful faculty member, you must look at certain traits in people to be able to do that job (as administrator). It is a very difficult job that requires you to be more about promoting others than yourself. This job requires you to have ability to be an agent to look for win-win solutions, because at the end of the day you don't have the ability to fire. You have a lot more carrots than sticks." The deans agreed that diverse leadership teams make better decisions; this is supported by research findings indicating that management teams consisting of people from diverse backgrounds bring a variety of perspectives to bear on an issue or problem, which results in better solutions. Therefore, once a good, effective management team has been assembled, it is important to keep the best people on that team by "telling people what is expected of them. Not only do people need to understand what is expected of them, they also need to know how they're doing."

## PROFESSIONAL ASSOCIATIONS

Interviewees were asked, "What are the roles of professional associations in increasing the participation of African Americans in engineering?" Although this question was posed in terms of all professional associations, one interviewee's initial response was in terms of a predominantly African American professional association, the National Society of Black Engineers, which, since its inception, has been at the forefront of efforts to increase African Americans' participation in engineering through "raising awareness, providing role models and career opportunities, stressing academic excellence, and providing gathering places for students and professionals to learn about each other and hear about common issues and discuss best practices." Similarly, responses indicate that the minority professional associations raise awareness and maintain involvement by providing role models and advice. Moreover, interviewees acknowledged that many, if not most, of the majority professional associations have some diversity activity and/or diversity committee. However, they also noted that these organizations vary in the extent to which these activities are diffused or focused. One interviewee described their perception of the role of professional organizations: "I see the role of professional organizations as providing best practices and information to the universities. The best thing about those organizations is that they kind of have a bird's-eye view. . . . They are able to be a repository of best practices and then translate those practices to institutions. The biggest benefit is that they can tell you what is working nationally and keep you from reinventing the wheel."

Interviewees acknowledged the contributions of some professional engineering societies—especially in terms of mentoring and professional socialization. However, interviewees also noted that some of these groups tend to be primarily focused on relationships between students and industry. Moreover, because these organizations do not share the same constraints as universities, they could have a much larger role insofar as they can provide opportunities for mentoring: "There are not enough minority faculty members to begin with, and a subset of these belongs to those organizations. Yet, there is an incredible need for mentoring."

In addition, professional associations can provide opportunities for presenting research, through which African Americans can become known in the field of engineering and get the visibility and recognition—on national and international levels—that they may not be getting in their department. Through workshops, seminars, and networking opportunities, professional societies also provide both the means and mechanisms for professional development.

There was a consensus among the interviewees that in order for professional associations to address issues of broadening the participation of African Americans in engineering, it is first necessary to have African Americans active in those organizations—and the earlier the better, because professional associations can provide socialization, which, in turn, impacts throughout an individual's career. Indeed, these associations can provide leadership training and opportunities: "By doing service work in an organization, you will get leadership positions relatively early. It starts from a mentality to serve, and not everybody has that mentality." Once again, the deans discussed how leadership and service are—or at least should be—inextricably intertwined.

## Conclusion

The words of one interviewee succinctly presented a major conclusion of this exploratory study: "Increasing African Americans' participation in engineering takes work—nothing happens just because an African American is the Dean." Data from the interviews suggest that interviewees viewed diversifying the students and diversifying the faculty as being inextricably intertwined—although many institutions treat them as separate issues.

In order to recruit and retain African Americans in engineering, engineering faculty and administrators have an obligation to inform students (and their parents) about the valuable contributions that engineering makes to communities, to nations, and globally. Not only would this make an engineering career more attractive and meaningful, but it would also facilitate African Americans making career choices that are better informed.

Interviewees agreed that increasing the participation of African Americans on engineering faculties must be a core value of an institution and, consequently, be institutionalized as an integral component of—instead of tangential to—the institution's standard operation. Diversifying the faculty and the student body in general, and increasing the participation of African Americans in particular, should be one criterion used to evaluate the performance of both faculty members and administrators; accountability is crucial. Periodic assessments conducted by external evaluators are necessary to determine what is working and why, and what is not working and why. Programs, interventions, and initiatives that focus on enhancing participation should be assessed individually and as components of a systematic and systemic effort to broaden participation. Evaluations identify factors, conditions, and issues that provide feedback to facilitate making whatever adjustments are deemed warranted. Engineering faculty and administrators should not be solely responsible for imparting information about engineering and engineering careers. All members of the engineering community can disseminate information about the variety of engineering projects and career pathways.

Through mentoring, professional socialization, and providing opportunities for leadership, professional engineering associations can and do play major roles in increasing African American participation in engineering in academe as well as industry (although the extent and focus of these activities vary across associations). Professional associations can enhance African American engineers' recognition in the profession by providing opportunities to present their research. Interviewees agreed that success as an engineering faculty member and researcher is essential to become an engineering dean. Interviewees were unanimous in asserting that efforts to increase African American participation in engineering should be approached on both the organizational and individual levels. Individuals must take it upon themselves not only to join professional associations but also to work on committees and projects that provide leadership experience and enhance recognition. Individuals get out of these professional associations what they contribute to them. African American engineers' work in professional associations better positions them to take advantage of opportunities—including pursuing an administrative path.

Further research will provide rich data that can be used to develop and implement a robust research agenda on African American deans of engineering in majority-serving institutions. Two of the interviewees suggested a study of African American engineering deans at both majority-serving institutions and minority-serving institutions. The importance of further research was summed up by one interviewee, who said, "It would be nice if people focus on Deans more. I mean the students are important, but it's also important to manage the other professionals involved in increasing African American participation in engineering careers."

## Acknowledgments

The author wishes to express gratitude and appreciation to Dr. Gary S. May, dean of the College of Engineering and professor in the School of Electrical and Computer Engineering at the Georgia Institute of Technology, for his insights. Also, the author is grateful to the interviewees, who generously shared their time, experiences, insights, perceptions, and wisdom.

## References

Allen, Walter R. 1992. "The Color of Success: African-American College Student Outcomes at Predominantly White and Historically Black Public Colleges and Universities." *Harvard Educational Review* 62 (1): 26–45.

BEST (Building Engineering and Science Talent). 2004. *A Bridge for All: Higher Education Design Principles to Broaden Participation in Science, Technology, Engineering and Mathematics.* www.bestworkforce.org/PDFdocs/BEST_High _Ed_Rep_48pg_02_25.pdf.

Fleming, Jacqueline. 1984. *Blacks in College: A Comparative Study of Students' Success in Black and in White Institutions.* San Francisco: Jossey-Bass.

Johnson, Keith V., and Elwood Watson. 2005. "A Historical Chronology of the Plight of African Americans Gaining Recognition in Engineering and Technology." *Journal of Technology Studies* 31 (2): 81–93.

Leggon, Cheryl B. 2006. "Women in Science: Racial and Ethnic Differences and the Differences They Make." *Journal of Technology Transfer* 31:325–333.

Leggon, Cheryl B. 2010. "Diversifying Science and Engineering Faculties: Intersections of Race, Ethnicity, and Gender." *American Behavioral Scientist* 53 (7): 1013–1028.

Leggon, Cheryl B., Troy Eller, and Lisa M. Frehill. 2011. "Women in Engineering: The Illusion of Inclusion." *Journal of the Society of Women Engineers* 5:84–95.

Leggon, Cheryl B., and Willie Pearson, Jr. 1997. "The Baccalaureate Origins of African American Female Ph.D. Scientists." *Journal of Women and Minorities in Science and Engineering* 3 (4): 213–224.

Leggon, Cheryl B., Yu Tao, and Willie Pearson, Jr. 2009. "Baccalaureate Origins of African American Female Doctoral Scientists and Engineers: Revisited." Unpublished manuscript.

Loui, Michael C., and Brett A. Robbins. 2008. "Assessment of Peer-Led Team Learning in an Engineering Course for Freshmen." In *Proceedings of the Thirty-Eighth ASEE/IEEE Frontiers in Education Conference*, Saratoga Springs, NY, October 22–25, F1F-7 to F1F-8.

Mather, Mark, and Diana Lavery. 2012. "U.S. Science and Engineering Labor Force Stalls, but Trends Vary across States." Population Reference Bureau.

NACME (National Action Council for Minorities in Engineering). 2011. "African Americans in Engineering." *NACME Research and Policy Brief* 1 (4).

National Academy of Engineering. 2012. Speech by Charles M. Vest, President of NAE, November 13.

National Research Council. 2005. *Assessment of NIH Minority Research and Training Programs, Phase 3*. Washington, DC: National Academies Press.

National Research Council. 2007. *Rising above the Gathering Storm: Energizing and Employing America for a Brighter Economic Future*. Washington, DC: National Academies Press.

National Research Council. 2010. *Rising above the Gathering Storm, Revisited: Rapidly Approaching Category 5*. Washington, DC: National Academies Press.

National Science Foundation. 2005. *Broadening Participation through a Comprehensive Integrative System*. Final workshop report. Washington, DC: National Science Foundation.

Zikmund, William G. 2003. *Business Research Methods*. 7th ed. Mason, OH: Thomson South-Western.

# CHAPTER THREE

# Engineering the Future

## African Americans in Doctoral Engineering Programs

YU TAO *and* SANDRA L. HANSON

A MAJOR CHALLENGE for the United States in the twenty-first century is maintaining a competitive science and engineering labor force. This challenge resonated with Americans when presented in *Rising above the Gathering Storm* and the follow-up reports (National Academy of Sciences 2007, 2010). The authors of the reports suggest that America is at a critical crossroads where competitiveness will rest on the ability to develop and invest in a strong and talented science, technology, engineering, and mathematics (STEM) workforce. The juxtaposition of this need to remain competitive and a science culture that is not always welcoming to African Americans and other underrepresented groups in STEM has been noted (Pearson and Bechtel 1989; Hanson 2009; Committee on Science, Engineering, and Public Policy 2011). Thus, the Committee on Science, Engineering, and Public Policy (2011) refers to a different crossroads. Economic growth and security depend on a talented, *diverse*, well-educated STEM workforce (Noeth, Cruce, and Armstrong 2003; COSSA 2008; Espenshade and Radford 2009; American Association for the Advancement of Science 2010). The absence of racial/ethnic minorities and women in high-level STEM positions robs employers of diverse strategies, skills, and competence that translate into economic gain in an age of global markets (Congressional Commission on the Advancement of Women and Minorities in Science, Engineering, and Technology Development 2000). Research on the development of all STEM talent is important given the increased demand for well-trained technical workers, scientists, and engineers in the United States and the implications of this development for critical areas of the US economy.

*Note*: The authors are joint authors listed in reverse alphabetical order.

Hanson (2012) argues that providing opportunities for all youths in STEM works to the good of all—to the common good. We cannot have good science without the development of all talent in STEM, regardless of race/ethnicity or sex. Nevertheless, race/ethnicity and sex continue to be factors in who will participate in these fields. Instead of using diversity as a competitive edge, STEM in the United States (and elsewhere) often sets up barriers for racial/ethnic minorities and women.

Remaining competitive in a global economy requires a supply of workers across all areas of STEM training. Engineering has played and continues to play a critical role in technological progress. Advances here have been key in revolutionizing life as we know it in a postmodern global society (Committee on Science, Engineering, and Public Policy 2011). In the past century, engineering has been critical in advances in almost every area of life (e.g., medical, transportation, environment, communications, security, and space exploration). These advances have contributed to massive improvements in our standard of living and way of life (National Academy of Engineering 2008; Committee on Science, Engineering, and Public Policy 2011). It is this historically important area of STEM that is the focus of this volume. The focus on African Americans is important given their underrepresentation in engineering. Although there has been considerable progress in achievement of STEM degrees and occupations by African Americans in some areas of STEM, this progress has been minimal in engineering.

The United States employs more engineers than almost any industrialized country, and the growth of the US engineering workforce has been considerable over the past decades (Tang 2000; Brookings Institution 2014). Engineering jobs come with considerable power, influence, status, and pay. The fact that African Americans remain underrepresented in engineering even as engineering has become one of the largest and fastest-growing professions in the United States is problematic. We are in an age of international competition for science and engineering talent. Over half of engineering doctorates awarded in US higher education institutions are earned by individuals with temporary visa status (NSF 2012). The trend of less overall interest in engineering and demographic change involving shifting race/ethnic population sizes may, however, create an opportunity for underrepresented minorities in engineering (Tang 2000). Although engineering has historically been culturally and demographically dominated by Whites, African Americans and other minorities have always been present and have made important contributions. It was especially the case during World War II that increasing opportunities became available for African American scientists in engineering, but the historical racism in engineering did not disappear (Pearson and Bechtel 1989; Pearson and Fechter 1994; Slaton 2010).

It is in engineering doctoral programs that the underrepresentation of African Americans is most extreme. In a 2008 National Action Council for Minorities in Engineering (NACME) report it was suggested that the underrepresentation of African American and other minorities in engineering education and occupations is a "new American dilemma" and a "quiet crisis" that needs to be acknowledged and addressed (see also the introduction and chap. 12 in this volume). The report defines a moment in history when the demand for engineers is at an all-time high and the number of African American engineering graduates is at a historically low point. Progress for African Americans in engineering education has been minimal (NACME 2008). Issues involving racism, lack of role models, lack of peer support, and shortage of financial assistance have worked in concert to keep African Americans underrepresented in PhD engineering programs (Pearson and Bechtel 1989; Pearson and Fechter 1994; Slaton 2010; National Research Council 2011). These programs are critical in the pipeline to engineering occupations. In his introduction to the 2008 NACME report, John Slaughter refers to the shortage of African Americans and other minorities in engineering as an indicator of "what is at best benign neglect, and at worst active discrimination" (NACME 2008, 5).

In this chapter, we argue that diversity in engineering is critical to scientific innovation and advance. We consider the diverse factors leading to the completion of the PhD in engineering and success (once achieving the PhD) in gaining engineering occupations for African Americans and compare these patterns across race/ethnic groups. The unique experiences of African American men and women are also examined. More specifically, we analyze three sets of research questions among US citizens and permanent residents. The first set of research questions examines patterns in the achievement of doctorates in engineering for African Americans (including comparisons across race/ethnic groups, sex, and year). The second set of questions again focuses on African Americans and examines how the race/ethnic and sex patterns of influence on achievement of the doctoral degree in engineering and the transition from engineering doctorates to engineering occupations are affected when other factors are taken into account. More specifically, we ask the following questions (with a special focus on African Americans): Do racial/ethnic differences in the achievement of engineering doctorates (as opposed to doctorates in other fields) exist when background characteristics (e.g., class and sociodemographic) are taken into account? What background characteristics have an impact on earning engineering doctorates for African Americans, and are the characteristics associated with getting a doctoral degree in engineering for African Americans the same as for other race/ethnic groups? Similar multivariate questions are posed for the achievement of engineering occupations (among those who have received engineering doctorates) with a focus on African Americans. Finally, we investigate a third set of research questions that look specifically at how African American men and women differ in achieving engineering doctorates and occupations and whether this sex effect varies across race/ethnic groups.

This research provides a unique application of the intersectionality approach to descriptive and multivariate analyses of national samples of African American engineering doctorates over the past decade. The research questions and analyses focus on African Americans but acknowledge complexity in the way that race/ethnicity and sex affect graduate engineering outcomes in a multivariate context that considers social class and demographic variables, as well as time period.

## Background

### RACE AND EDUCATION IN THE UNITED STATES

Research showing the impact of race-biased educational systems on young African American and other minority students' achievement and attitudes is extensive (Cooper 1996; Freeman 1997). Thus, it is not a cultural orientation that works to the disadvantage of these youths in US schools. Rather, it is structured disadvantage that impacts the young person's hopes and achievements. For many (especially in higher education), racism is a central component of their educational experience (Feagin, Vera, and Imani 1996). The negative school experiences of many African American youths contribute to lower participation in higher education (Freeman 1997). This lower participation is problematic for developing science talent among African Americans.

### MINORITIES IN STEM

Considerable research shows high interest in math and science among minority students in general. Wenner (2003) found that poor, minority, inner-city elementary students had lower test scores in science than did their White counterparts but had more positive attitudes about science and more time in science instruction. A report by the American Council on Education provides evidence that African American (and Hispanic) students are as likely as Asian American and White students to enter college interested in STEM fields (Anderson and Kim 2006). Given the salience of evidence on race as a continuing factor in the US education (and STEM education) system, this research is guided by a framework that acknowledges the role of race discrimination and the background and experiences of African American students in affecting their success in graduate engineering programs.

## AFRICAN AMERICANS IN ENGINEERING

Although African Americans represented approximately 13% of the US population in 2010, they represented only 2% of those getting doctorates in engineering (NACME 2012). Figures from the National Science Foundation (NSF) show that African Americans who earn PhDs in engineering are less represented at universities with the highest research productivity than are other race/ethnic groups (NSF 2013). They also are underrepresented within areas of engineering which are often associated with the highest salaries (e.g., aerospace engineering). The underrepresentation of African Americans in engineering has been recognized as a dilemma, one that will have an impact on our ability to maintain a globally competitive science workforce (see also chap. 12 in this volume). There has been little progress for African Americans in gaining engineering degrees and employment in the past decades. African Americans constitute 12% of the overall workforce but just 5% of the US engineering workforce and just 2.5% of tenured and tenure-track engineering faculty (NACME 2012). Race (and racism) has been recognized as a continuing factor for African Americans in STEM. The relatively larger shortage of African Americans in engineering (relative to some other STEM areas) has led some to question the distinctive race culture of engineering (Stanford Humanities Lab 2006; Hanson 2009, 2012; Slaton 2010; Moreira 2012).

## INTERSECTION OF RACE AND SEX

In this research, we acknowledge the importance of distinct experiences of African American men and women in graduate engineering education. Insights on the intersection of race and sex come from the intersectionality framework. Although the framework was first used to examine the intersection of race/ethnicity and gender for African American women, it has been expanded to include all race/ethnic and sex groups (Mann and Huffman 2005) at both individual and institutional levels (Browne and Misra 2003). The term *intersectionality theory* was first coined by Kimberle Crenshaw in

1989 when describing race and gender effects that are not additive or summative (DeFrancisco and Palczewski 2007). As with other race/ethnic groups, African American men and women do not have the same experiences in STEM. In engineering, African American women are more underrepresented than African American men at each stage in the engineering pipeline from education through occupations (NACME 2008; NSF 2013). Data from NSF show that African American women constituted 31% of African American graduate students in engineering programs in 2010. In combined samples of men and women (across race/ethnic groups), 23% were female, suggesting higher proportions of women (compared to men) among African Americans in these graduate programs than in other race/ethnic groups (NSF 2013). These trends provide preliminary support for the following conclusions: African Americans have unique experiences in engineering training and occupations, African American men and women have distinct experiences in engineering training and occupations, and the intersection of race and sex creates unique experiences for African American men and women compared with other race/ethnic groups. Nevertheless, few researchers have applied the intersectionality approach to the achievement of engineering degrees and occupations.

Our empirical examination of race/ethnicity and sex influences on engineering doctoral degrees incorporates the intersectionality approach. In addition, we take into account family characteristics involving class, sociodemographic status, and educational characteristics. Success in STEM education has been shown to be affected by family socioeconomic resources (see, e.g., Hanson 1996; Herzig 2004; Pearson 2005; Choobbasti 2007; Rochin and Mello 2007). Given this and the wider literature on the impact of family socioeconomic background on education and occupation outcomes in general (Blau and Duncan 1967; Charles, Rosicgno, and Torres 2007) and in the context of race/ethnicity (Roscigno 2000; Hill 2008; O'Connor, Diane, Robinson 2009), we include measures of mother's and father's level of education in our models as indicators of family socioeconomic status. Use of parent's education is standard in social science research since it is a good indicator of family socioeconomic status. Although it would be ideal to have multiple indicators of status, data on income and occupation are often not available (as is the case with the data examined here). We also include measures of marital and child status since they are important considerations in educational and occupational outcomes and also vary by race/ethnicity and sex (Xie and Shauman 2003; Bryant et al. 2010). In addition to class and sociodemographic characteristics, we include educational variables, Carnegie Classification, and institutional control (public vs. private university) since they can affect doctoral degree production or the preparation of baccalaureate degree recipients who continue on to earn a doctoral degree in science and engineering (NSF 2006; Burrelli, Rapoport, and Lehming 2008).

## Methods

### DATA SET FOR DESCRIPTIVE ANALYSIS:
### IPEDS COMPLETIONS SURVEY BY RACE

To answer the first set of research questions regarding how the achievement of an engineering PhD varies by race/ethnicity, sex, and year (descriptive results), we use the Integrated Postsecondary Education Data System (IPEDS) Completions Survey by Race from the National Center for Education Statistics 2000 and 2010. The IPEDS Completions Survey has been used by researchers to examine a wide variety of issues involving postsecondary education and degrees, including research questions that focus on who completes science and engineering degrees (e.g., Shapiro 2001; Sonnert and Fox 2012). This data set provides statistics regarding doctoral degrees by race, sex, field, Carnegie Classifications, and so on, and is available on the NSF's WebCASPAR website. We chose 2010 and 2000 in order to show the decade of change or lack of change in participation of these groups over time. The 2000 data included 91 African Americans, 1,948 Whites, 380 Asian Americans, and 86 Hispanics receiving doctoral degrees in engineering in 2000. The 2010 data included 155 African Americans, 2,311 Whites, 519 Asian Americans, and 197 Hispanics who received engineering doctorates in 2010. See the appendix for more details on the sample.

### DATA SET FOR MULTIVARIATE ANALYSIS:
### SURVEY OF DOCTORATE RECIPIENTS

The second and third sets of research questions inquire about the impact of race/ethnicity and sex on the achievement of doctorate degrees in engineering in the context of multiple controls (e.g., class and sociodemographic characteristics). They also address the attainment of occupations in engineering for those who received engineering doctorates. To answer these questions, we use the 2008 Survey of Doctorate Recipients (SDR) data from NSF (the most recent year available when the chapter was started).

SDR is ideal for research that focuses on doctorate recipients because it tracks doctoral degree earners' progress throughout their careers from the time they received their first doctoral degree. In addition, SDR oversamples certain groups, including women, underrepresented minorities, and early-career individuals, so that the larger sample size allows for meaningful statistical analysis for these groups. The SDR sample used in this chapter includes US citizens and permanent residents who received a doctoral degree by 2008. The sample includes 28,165 doctorate recipients (1,661 African Americans, 20,282 Whites, 4,417 Asian Americans, and 1,805 Hispanics). Other racial/ethnic groups are not included owing to their small size. Among the 28,165 doctorate recipients, 4,509 (of which 201 were African Americans) received their first doctorate in engineering and 3,021 (of which 128 were African Americans) were employed in engineering occupations when the survey was conducted in 2008. For more details on this sample see table 3A.1 on page 85.

## VARIABLES AND MEASURES

This chapter is focused on the achievement of the engineering doctorate for those with a doctoral degree and engineering occupations for those with an engineering doctorate. In the IPEDS data, the engineering doctorate is measured as the doctoral degree received in 2000 or 2010. In the 2008 SDR data, the engineering doctorate is measured as the first doctoral degree, and the engineering occupation is the field of the principal job at the time of the survey (the week of October 1, 2008). Both are in the broad field of engineering.* Both are dichotomous variables (1 = engineering; 0 = nonengineering). The racial/ethnic groups discussed in this chapter include African Americans, Whites, Asian Americans, and Hispanics.

Our first group of research questions inquires about African Americans' share of engineering doctorates, racial/ethnic and sex differences in the area of engineering, the type of university that conferred the degree (Carnegie Classifications), and how the above variables and patterns changed over time (from 2000 to 2010). The areas of engineering examined in this chapter include aerospace, chemical, civil, electrical, mechanical, materials, in-

---

* In SDR, the first doctoral degree instead of the most recent degree is used in this chapter because cohorts in SDR are sampled based on the first science, engineering, or health doctoral degree they received. Also, this variable can indicate the trajectory or progress of the doctorate recipient's career better than other variables, such as the most recent doctoral degree, which could be obtained toward the end of one's career and does not necessarily reflect the progress of one's career since the receipt of the degree. Furthermore, the most recent doctoral degree, if different from the first doctoral degree, may not be in engineering. Nevertheless, most individuals have only one doctoral degree, and the number of individuals whose first doctoral degree is in engineering and the number of individuals whose most recent degree is in engineering are quite close.

dustrial, and other engineering. The Carnegie Classifications include research universities with very high research activities and other types of universities (e.g., research universities with high research activities, doctoral/research universities, master's colleges and universities, baccalaureate colleges, schools of engineering, and medical schools and medical centers). When discussing data on the types of universities, we only distinguish institutions with very high research activities from the rest because the majority of engineering doctorate recipients received their doctoral degrees from the former. Because SDR 2008 uses 2005 Carnegie Classifications, we use the 2005 edition throughout this chapter for consistency.

The second and third sets of research questions examine the effect of race/ethnicity and sex on obtaining an engineering doctorate and an engineering occupation for those with the engineering doctorate in a multivariate context where class and sociodemographic factors are considered. The questions also consider distinct processes for African American men and women. The key independent variables are race/ethnicity and sex. The interaction effect of race/ethnicity and sex is also included given the intersectionality framework, which argues that the effects of race/ethnicity vary across sex groups and the effects of sex also vary across racial/ethnic groups. Other independent variables include class variables (mother's and father's education) and demographic variables (age, age squared, marital status, and having at least one child). When addressing changes in access to engineering occupations (but not in earning an engineering degree), we also examine educational variables, including Carnegie Classifications, institutional control, and cohorts. Institutional control refers to whether the doctoral institution is a public or a private university. We group the sample into four cohorts, namely, those who received their doctorates before 1995, from 1995 to 1999, from 2000 to 2004, and from 2005 to 2007. We do not include the areas of engineering when examining the second and third sets of research questions owing to the small number of respondents belonging to some racial/ethnic groups in certain areas of engineering.

MODELING AND ANALYSIS

We use cross tables to answer the first set of research questions given the categorical level of measurement and the descriptive nature of the questions. We use logistic regression to answer the second set of research questions, on racial/ethnic and sex differences in earning an engineering doctorate (as opposed to other PhDs) and an engineering occupation in a multivariate context.* Some logistic equations examine the effects of race/ethnicity and sex when various independent variables are controlled. Others examine the effects of sex and other independent variables for each racial/ethnic group— African Americans, Whites, Asian Americans, and Hispanics. For the first set of logistic regressions, we add independent variables hierarchically. Model 1 includes the four racial/ethnic groups, with African Americans as the reference group. Model 2 introduces the sex variable, and model 3 further includes the race/ethnic–sex interaction term. Additional models add class variables (model 4) and then sociodemographic variables (model 5). This series of equations will help us understand the effect of race/ethnicity and sex on the achievement of an engineering doctorate by examining how this effect changes as groups of other independent variables are included. The key issue is the extent to which a race/ethnicity effect (involving the comparisons between African Americans and others) exists above and beyond the effects of sex and other factors that influence the achievement of engineering doctorates. The second set of logistic regressions investigates how these independent variables affect the achievement of engineering doctorates differently for each racial/ethnic group with the same dependent variable and all independent variables included in model 5 except for race/ethnicity and the race/ethnicity–sex interaction term. We use the same two sets of logistic regressions in our examination of the attainment of engineering occupations (for those with an engineering PhD), but we limit the SDR sample to those who received the first doctorate in engineering as opposed to all fields to focus on engineering doctorate recipients. Finally, these analyses include one more model (model 6) that further controls on educational variables (e.g., on institution where respondent received degree) since the respondents in this analysis have received their engineering doctoral degrees.

Modeling for the third set of research questions addressing the distinct experiences of African American men and women in achieving engineering degrees and occupations involves separate logistic regression models for African American men and women. We include class and sociodemographic characteristics in the engineering doctorate models and, additionally, educational characteristics in the engineering occupation models. The special focus here is on the differential effect of sex across race/ethnic groups in engineering doctorate and occupation models.

---

* When analyzing logistic regression results for answering the second and third groups of research questions, we convert logistic coefficients into odds ratios for easier interpretation. A value greater than 1 means a positive effect, and a value less than 1 indicates a negative effect. The farther the coefficient is from 1, the greater the effect, either positive or negative.

# Findings

RACE, SEX, AND THE ACHIEVEMENT OF ENGINEERING
DOCTORATES—A DESCRIPTION

The first two tables address our first set of research questions involving a descriptive analysis of race/ethnicity, sex, and engineering doctorates granted in 2000 and 2010. Table 3.1 shows the percent of doctorates awarded to US citizens and permanent residents by race/ethnicity (and sex within race/ethnic groups) for 2000 and 2010. African Americans received 155 engineering doctorates (or 5% of these degrees) in 2010.* This increased from 91 (or a 4% share) in 2000. African Americans (and Hispanics) are underrepresented in the percent of doctorates earned. Each of these groups represents over 10% of the US population, but their representation among those with engineering doctorates is less than half of this. Whites earned the largest percentage of engineering doctorates in both 2000 (1,948, or 78%) and 2010 (2,311, or 73%). Asian Americans are also overrepresented in the percent of engineering doctorates they receive. It should be noted that although the numbers of African Americans (and Hispanics) achieving the doctorates were small, the increases between 2000 and 2010 for these groups were larger than for the other groups.†

Figures in table 3.1 also show sex trends within race/ethnic groups for the engineering doctorate data. Figures show that in both years, African American males earned a majority of the engineering doctorates awarded to African Americans. This sex trend held for all race/ethnic groups. However, in 2000, it was African American females who earned the largest share of doctorates in engineering (within race groups). These women earned 32% of the doctorates awarded to African Americans in that year. Figures for other groups were considerably less.

Figures in table 3.2 provide information on race/ethnicity, sex, and subfield of engineering degree for 2000 and 2010. Summary figures for the race

---

*The percentages by race (but not by sex) are calculated with the total of the four races (Blacks, Whites, Asians, and Hispanics) but not the total of all races as the denominator since the data included in this paper do not include American Indians and Other/Unknown races. The percentages by race in table 3.2 are also calculated in this way.

†A more positive perspective of interpreting the data is that the increase for African Americans from 2000 to 2010 is 70% [(155−91)/91], 19% for Whites, 37% for Asian Americans, and 129% for Hispanics. Nevertheless, when we compare the growth of these groups against each other, we find that even though African Americans and Hispanics had much greater growth rates than the other two groups, their representation among engineering doctorate recipients of all racial/ethnic groups in 2010 (5% and 6%, respectively) was still low.

TABLE 3.1    Engineering doctorates awarded to US citizens and permanent residents by race/ethnicity and sex, 2000 and 2010

| Race/Ethnicity | 2000 | 2010 |
|---|---|---|
| African Americans | | |
| Total | 91 (4%) | 155 (5%) |
| Male | 68% | 66% |
| Female | 32% | 34% |
| Whites | | |
| Total | 1,948 (78%) | 2,311 (73%) |
| Male | 83% | 78% |
| Female | 17% | 22% |
| Asian Americans | | |
| Total | 380 (15%) | 519 (16%) |
| Male | 79% | 68% |
| Female | 21% | 34% |
| Hispanics | | |
| Total | 86 (3%) | 197 (6%) |
| Male | 81% | 67% |
| Female | 19% | 33% |

SOURCE: IPEDS Completions Survey on WebCASPAR (NSF Population of Institutions).

data from table 3.2 (not presented in the table) show that African Americans are particularly underrepresented in some areas of engineering. These are often the areas associated with the highest status and salaries (e.g., aerospace engineers and chemical engineers). In 2000, African Americans earned just 3% of the doctorates in aerospace engineering. This figure did not change in 2010. Whites were overrepresented in the earning of these doctorates, while African Americans and Hispanics were underrepresented. The degree of underrepresentation for African Americans (and Hispanics) is greater in some areas of engineering than others. African Americans have their highest representation in industrial engineering, especially in 2010 (15%).

In addition, figures in table 3.2 show percentages for men and women by race/ethnicity in each area of engineering. As with the general data on engineering doctorates shown in table 3.1, the data here show that men are more likely to get doctorates in each area of engineering. A number of exceptions do occur, however. In both 2000 and 2010, there were a number of degree areas where African American women earned as many (or more) degrees as African American men. For example, in 2000, while the numbers were small for both men and women, there were more African American women earning degrees in chemical engineering than African American men. In 2010, African American women earned half of the doctorates in industrial engineering.

TABLE 3.2 Engineering doctorates awarded to US citizens and permanent residents by race/ethnicity, subfield, and sex, 2000 and 2010

| Subfield | 2000 | | | | | 2010 | | | | |
|---|---|---|---|---|---|---|---|---|---|---|
| | Female | | Male | | Total | Female | | Male | | Total |
| **African Americans** | | | | | | | | | | |
| Aerospace engineering | 1 | 25% | 3 | 75% | 4 | 1 | 33% | 2 | 67% | 3 |
| Chemical engineering | 5 | 63% | 3 | 38% | 8 | 5 | 36% | 9 | 64% | 14 |
| Civil engineering | 2 | 25% | 6 | 75% | 8 | 2 | 17% | 10 | 83% | 12 |
| Electrical engineering | 2 | 9% | 21 | 91% | 23 | 5 | 18% | 23 | 82% | 28 |
| Mechanical engineering | 7 | 29% | 17 | 71% | 24 | 4 | 27% | 11 | 73% | 15 |
| Materials engineering | 1 | 20% | 4 | 80% | 5 | 8 | 44% | 10 | 56% | 18 |
| Industrial engineering | 2 | 33% | 4 | 67% | 6 | 8 | 50% | 8 | 50% | 16 |
| Other engineering | 9 | 69% | 4 | 31% | 13 | 19 | 39% | 30 | 61% | 49 |
| Total | 29 | 32% | 62 | 68% | 91 | 52 | 34% | 103 | 66% | 155 |
| **Whites** | | | | | | | | | | |
| Aerospace engineering | 14 | 13% | 94 | 87% | 108 | 10 | 10% | 86 | 90% | 96 |
| Chemical engineering | 65 | 24% | 210 | 76% | 275 | 88 | 27% | 243 | 73% | 331 |
| Civil engineering | 53 | 22% | 185 | 78% | 238 | 73 | 31% | 162 | 69% | 235 |
| Electrical engineering | 53 | 12% | 383 | 88% | 436 | 50 | 11% | 385 | 89% | 435 |
| Mechanical engineering | 31 | 12% | 230 | 88% | 261 | 41 | 13% | 272 | 87% | 313 |
| Materials engineering | 37 | 19% | 154 | 81% | 191 | 51 | 24% | 158 | 76% | 209 |
| Industrial engineering | 18 | 23% | 61 | 77% | 79 | 16 | 24% | 50 | 76% | 66 |
| Other engineering | 69 | 19% | 291 | 81% | 360 | 180 | 29% | 446 | 71% | 626 |
| Total | 340 | 17% | 1,608 | 83% | 1,948 | 509 | 22% | 1,802 | 78% | 2,311 |

(continued)

TABLE 3.2 *continued*

| Subfield | 2000 | | | | | 2010 | | | | |
|---|---|---|---|---|---|---|---|---|---|---|
| | *Female* | | *Male* | | *Total* | *Female* | | *Male* | | *Total* |
| **Asian or Pacific Islander** | | | | | | | | | | |
| Aerospace engineering | 1 | 11% | 8 | 89% | 9 | 5 | 42% | 7 | 58% | 12 |
| Chemical engineering | 14 | 28% | 36 | 72% | 50 | 25 | 38% | 40 | 62% | 65 |
| Civil engineering | 8 | 22% | 29 | 78% | 37 | 13 | 35% | 24 | 65% | 37 |
| Electrical engineering | 22 | 18% | 98 | 82% | 120 | 34 | 22% | 122 | 78% | 156 |
| Mechanical engineering | 12 | 18% | 55 | 82% | 67 | 7 | 16% | 38 | 84% | 45 |
| Materials engineering | 8 | 30% | 19 | 70% | 27 | 17 | 40% | 26 | 60% | 43 |
| Industrial engineering | 0 | 0% | 13 | 100% | 13 | 4 | 27% | 11 | 73% | 15 |
| Other engineering | 16 | 28% | 41 | 72% | 57 | 69 | 47% | 77 | 53% | 146 |
| Total | 81 | 21% | 299 | 79% | 380 | 174 | 34% | 345 | 66% | 519 |
| **Hispanics** | | | | | | | | | | |
| Aerospace engineering | 0 | 0% | 2 | 100% | 2 | 3 | 43% | 4 | 57% | 7 |
| Chemical engineering | 3 | 17% | 15 | 83% | 18 | 14 | 42% | 19 | 58% | 33 |
| Civil engineering | 1 | 9% | 10 | 91% | 11 | 8 | 31% | 18 | 69% | 26 |
| Electrical engineering | 1 | 6% | 16 | 94% | 17 | 9 | 21% | 33 | 79% | 42 |
| Mechanical engineering | 1 | 10% | 9 | 90% | 10 | 4 | 17% | 19 | 83% | 23 |
| Materials engineering | 4 | 57% | 3 | 43% | 7 | 7 | 37% | 12 | 63% | 19 |
| Industrial engineering | 1 | 33% | 2 | 67% | 3 | 4 | 33% | 8 | 67% | 12 |
| Other engineering | 5 | 28% | 13 | 72% | 18 | 16 | 46% | 19 | 54% | 35 |
| Total | 16 | 19% | 70 | 81% | 86 | 65 | 33% | 132 | 67% | 197 |

SOURCE: IPEDS Completions Survey by Race on WebCASPAR.

In tables not shown here, we examine types of institutions where African Americans are receiving engineering doctorates (for the years 2000 and 2010).* African Americans are underrepresented at the very high research universities at a similar rate to their general underrepresentation in engineering doctorate programs. Four percent of the engineering doctorates from very high research activity universities went to African Americans in both survey years. Hispanics were also underrepresented. Findings show a unique trend for African American women. In 2000, African American women earned 33% of the engineering doctorates awarded at very high research activity universities to African Americans. This number was considerably higher than for women in any other race/ethnic group. In both years it was White women who earned the smallest share (e.g., 23% in 2010).

RACE, SEX, AND THE ACHIEVEMENT OF ENGINEERING
DOCTORATES—MULTIVARIATE PROCESS

The last four tables, based on the 2008 SDR data, present findings regarding the second and third sets of research questions involving the effect of race/ethnicity and sex on the achievement of engineering doctorates and occupations in a multivariate context. With regard to our second set of research questions, results in tables 3.3 and 3.4 show findings from multivariate analyses examining the effect of race/ethnicity and sex on achieving an engineering doctorate. The results in model 1 in table 3.3 show that race/ethnicity has an effect. African Americans have lower odds of earning a doctorate in engineering than Whites and Asian Americans—Whites have 17% greater odds and Asian Americans have 190% greater odds than African Americans. Model 2 in table 3.3 introduces the sex variable, and the difference between African Americans and Whites disappears, but the difference between African Americans and Asian Americans remains significant, although reduced. Models 3–5 in table 3.3 further introduce the interaction effect of sex and race, parental education, and demographic variables, respectively, and the difference between African Americans and Asian Americans remains significant.

Results in table 3.4 show details on the specific effects of independent variables on earning engineering doctorates for African Americans and other race/ethnic groups (second set of research questions), as well as effects for African American men and women separately (third set of research

---

* Readers who are interested in the tables or other data discussed but not presented in this chapter can contact the authors for more details.

TABLE 3.3   Odds ratios and standard errors showing effects of race and sociodemographic characteristics on engineering doctoral degree outcomes among doctoral degree recipients (US citizens and permanent residents)

| Characteristics | Model 1 | Model 2 | Model 3 | Model 4 | Model 5 |
|---|---|---|---|---|---|
| Race and sex variables | | | | | |
| Race (reference: African American) | | | | | |
| White | 1.17* (0.09) | 0.96 (0.08) | 1.05 (0.10) | 1.02 (0.10) | 1.09 (0.10) |
| Asian American | 2.90*** (0.24) | 2.55*** (0.21) | 2.87*** (0.28) | 2.72*** (0.27) | 2.61*** (0.26) |
| Hispanic | 1.11 (0.11) | 1.07 (0.11) | 1.19 (0.14) | 1.15 (0.14) | 1.14 (0.14) |
| Female | | 0.26*** (0.01) | 0.36*** (0.06) | 0.37*** (0.06) | 0.36*** (0.06) |
| Interaction effect (reference: female*African American) | | | | | |
| Female*White | | | 0.76 (0.13) | 0.75 (0.13) | 0.72 (0.13) |
| Female*Asian American | | | 0.63* (0.12) | 0.62* (0.12) | 0.61** (0.11) |
| Female*Hispanic | | | 0.69 (0.16) | 0.68 (0.16) | 0.66 (0.16) |
| Class variables | | | | | |
| Mother's education | | | | 0.83*** (0.03) | 0.76*** (0.03) |
| Father's education | | | | 1.31*** (0.06) | 1.25*** (0.05) |
| Demographic variables | | | | | |
| Age | | | | | 0.94*** (0.01) |
| Age squared | | | | | 1.00*** (0.00) |
| Marital status (reference: single and never married) | | | | | |
| Married | | | | | 1.02 (0.06) |
| Other marital status | | | | | 0.76*** (0.06) |
| Having at least one child | | | | | 1.12** (0.05) |
| Constant | 0.14*** (0.01) | 0.23*** (0.02) | 0.21*** (0.02) | 0.20*** (0.02) | 1.42 (0.46) |
| N | 28,165 | 28,165 | 28,165 | 28,165 | 28,165 |

SOURCE: Survey of Doctorate Recipients, 2008.
NOTE: *** $p < 0.001$; ** $p < 0.01$; * $p < 0.05$.

TABLE 3.4   Odds ratios and standard errors showing effects of sociodemographic characteristics on engineering doctoral degree outcomes among doctoral degree recipients (US citizens and permanent residents) by race/ethnicity and by sex among African Americans

| Characteristics | African Americans | | | Whites | Asian Americans | Hispanics |
| --- | --- | --- | --- | --- | --- | --- |
| | Total | Women | Men | | | |
| Sex variable | | | | | | |
| Female | 0.32*** (0.06) | | | 0.26*** (0.02) | 0.23*** (0.02) | 0.24*** (0.04) |
| Class variables | | | | | | |
| Mother's education | 0.86 (0.17) | 0.99 (0.36) | 0.80 (0.19) | 0.74*** (0.04) | 0.74*** (0.07) | 1.12 (0.19) |
| Father's education | 1.70** (0.33) | 1.15 (0.40) | 2.03** (0.48) | 1.19*** (0.06) | 1.27** (0.11) | 1.47* (0.26) |
| Demographic variables | | | | | | |
| Age | 1.05 (0.07) | 1.53* (0.31) | 1.07 (0.09) | 0.92*** (0.01) | 0.98 (0.03) | 0.87* (0.05) |
| Age-squared | 1.00 (0.00) | 0.99* (0.00) | 1.00 (0.00) | 1.00*** (0.00) | 1.00 (0.00) | 1.00* (0.00) |
| Marital status (reference: single and never married) | | | | | | |
| Married | 0.97 (0.23) | 0.58 (0.23) | 1.36 (0.43) | 0.94 (0.07) | 1.53** (0.21) | 0.68 (0.16) |
| Other marital status | 0.80 (0.22) | 0.38 (0.19) | 1.30 (0.46) | 0.72*** (0.07) | 1.24 (0.23) | 0.34*** (0.1) |
| Having at least one child | 1.04 (0.20) | 1.72 (0.64) | 0.85 (0.20) | 1.12* (0.06) | 1.10 (0.10) | 1.22 (0.21) |
| Constant | 0.14 (0.21) | 0.00* (0.00) | 0.06 (0.11) | 2.94** (1.11) | 0.68 (0.46) | 5.91 (7.55) |
| N | 1,661 | 819 | 842 | 20,282 | 4,417 | 1,805 |

SOURCE: Survey of Doctorate Recipients, 2008.
NOTE: *** $p < 0.001$; ** $p < 0.01$; * $p < 0.05$.

questions). First, we examine effects across race/ethnic groups. Results show that women have lower odds of earning an engineering doctorate than their male counterparts across racial/ethnic groups, but this sex difference is the lowest among African Americans—African American women have 32% as great odds of achieving an engineering doctorate as their male counterparts, higher than the odds of women in other race/ethnicities, especially Asian American women (23%). In addition to sex, class variables affect the four racial/ethnic groups differently. Father's education (some college education or higher) increases the odds of achieving engineering doctorates for all four racial/ethnic groups. The impact is the greatest for African Americans—father's education increases the odds of earning an engineering doctorate by 70% for African Americans (but only by 19% for Whites). Mother's education has an effect on Whites and Asian Americans but not on African Americans. In addition, sociodemographic variables have some effects for the other three racial/ethnic groups but not for African Americans.

Results in table 3.4 on the separate processes for African American men and women reveal that very few characteristics affect chances of getting the doctorate for African American men or African American women. However, when there is an effect, it is significant for men but not women (or vice versa). Among African American women, neither parent's education has an impact on their chances of achieving the doctorate in engineering. Age increases their odds of earning this doctorate (at a decreasing rate). Among African American men, father's education doubles their odds of earning an engineering doctoral degree, but their sociodemographic characteristics do not have an impact. Thus, the only class effect for African Americans in achieving a doctorate in engineering involves the influence of father's education for African American men. No education variable (type of institution and cohort) has an effect on African American men or women.

## RACE/ETHNICITY, SEX, AND THE ACHIEVEMENT OF ENGINEERING OCCUPATIONS—MULTIVARIATE PROCESS

Results in table 3.5 address our multivariate questions concerning the effect of race/ethnicity on the odds of achieving an engineering occupation (for those with engineering doctorates). Interestingly, African Americans do not differ from other race/ethnic groups in achieving an engineering occupation, regardless of the additional variables included in the model. Results in table 3.6 examine the effects of the independent variables on achieving engineering occupations for each race/ethnicity. For African Americans, no variable, including sex, has an impact on the chance of achieving an engineering

TABLE 3.5  Odds ratios and standard errors showing effects of race and sociodemographic characteristics on engineering occupation outcomes among engineering doctoral degree recipients (US citizens and permanent residents)

| Characteristics | Model 1 | Model 2 | Model 3 | Model 4 | Model 5 | Model 6 |
|---|---|---|---|---|---|---|
| Race and sex variables | | | | | | |
| Race (reference: African American) | | | | | | |
| White | 0.94 (0.14) | 0.92 (0.14) | 0.89 (0.15) | 0.88 (0.15) | 1.13 (0.20) | 1.19 (0.21) |
| Asian American | 0.98 (0.15) | 0.96 (0.15) | 0.93 (0.17) | 0.93 (0.17) | 0.93 (0.17) | 0.96 (0.18) |
| Hispanic | 0.99 (0.19) | 0.98 (0.19) | 0.99 (0.22) | 0.99 (0.22) | 1.05 (0.24) | 1.08 (0.25) |
| Female | | 0.87 (0.07) | 0.78 (0.25) | 0.77 (0.24) | 0.64 (0.20) | 0.65 (0.21) |
| Interaction effect (reference: female* African American) | | | | | | |
| Female*White | | | 1.14 (0.38) | 1.14 (0.38) | 1.00 (0.34) | 0.99 (0.34) |
| Female*Asian American | | | 1.15 (0.41) | 1.14 (0.40) | 1.18 (0.42) | 1.19 (0.43) |
| Female*Hispanic | | | 0.89 (0.40) | 0.89 (0.40) | 0.89 (0.40) | 0.82 (0.38) |
| Class variables | | | | | | |
| Mother's education | | | | 1.05 (0.08) | 0.89 (0.07) | 0.91 (0.07) |
| Father's education | | | | 1.05 (0.08) | 0.89 (0.07) | 0.89 (0.07) |
| Demographic variables | | | | | | |
| Age | | | | | 1.07** (0.03) | 1.11*** (0.03) |
| Age squared | | | | | 1.00*** (0.00) | 1.00*** (0.00) |
| Marital status (reference: single and never married) | | | | | | |
| Married | | | | | 1.12 (0.13) | 1.12 (0.13) |
| Other marital status | | | | | 0.93 (0.14) | 0.93 (0.14) |

(continued)

TABLE 3.5 *continued*

| Characteristics | Model 1 | Model 2 | Model 3 | Model 4 | Model 5 | Model 6 |
|---|---|---|---|---|---|---|
| Having at least one child | | | | | 0.99 (0.08) | 1.00 (0.08) |
| Educational variables | | | | | | |
| Research Carnegie Classification: research universities with very high research activities | | | | | | 0.98 (0.08) |
| Public universities | | | | | | 1.38*** (0.09) |
| Cohort (reference: receiving an engineering doctorate prior to 1995) | | | | | | |
| Cohort 2005–2007 | | | | | | 1.32 (0.24) |
| Cohort 2000–2004 | | | | | | 1.29 (0.17) |
| Cohort 1995–1999 | | | | | | 0.90 (0.09) |
| Constant | 1.23 (0.17) | 1.28 (0.18) | 1.32 (0.22) | 1.26 (0.22) | 0.65 (0.39) | 0.17* (0.15) |
| N | 4,509 | 4,509 | 4,509 | 4,509 | 4,509 | 4,509 |

SOURCE: Survey of Doctorate Recipients, 2008.
NOTE: *** $p < 0.001$; ** $p < 0.01$; * $p < 0.05$.

occupation (among those with engineering doctorates). For the other three race/ethnic groups, the type of doctoral institution (public vs. private) has an impact. Additionally, for Whites, sex and age also have an impact.

Results in table 3.6 also address our questions focusing on occupational processes for African American men and women separately (third set of research questions). They show effects of background characteristics on the odds of achieving an engineering occupation for African American men and women. None of the class, sociodemographic, or educational variables affect the odds of employment in engineering occupations for either sex. As shown in our discussion of samples, this lack of significant effects could in part be due to the small sample size. However, readers are reminded that in the much larger sample of African men and women examined in table 3.4 (in engineering doctorate models) there was a similar pattern with very few effects of background characteristics.

## Conclusion and Discussion

### SUMMARY OF FINDINGS

In general, descriptive findings show little progress for earned doctorates in engineering for African Americans (first set of research questions). African Americans are particularly underrepresented in some areas of engineering and at high research universities. Findings also show that African American men earn more of these doctorates than do African American women, but women's share is increasing. Additionally, African American women gain an equal or larger share of the doctorates within race groups compared with other groups of women.

Multivariate results (second set of research questions) confirm the effects of race/ethnicity and sex on achieving a doctorate and occupation in engineering, with a disadvantage for African Americans relative to Whites and relative to Asian Americans. The gap between African Americans and Whites disappears after sex and other variables are added to the model, but the African American–Asian American difference remains when sex and other background characteristics are added into the model. For those with an engineering doctorate, there is no difference between African Americans and other race/ethnic groups in obtaining an engineering occupation. Analyses focusing on distinct processes of achievement for African American men and women (third set of research questions) show that the process of earning an engineering doctorate varied slightly for African American men and women, but the process of achieving an engineering occupation does not differ for African American men and women. In addition, women's disadvantages

TABLE 3.6 Odds ratios and standard errors showing effects of sociodemographic characteristics on engineering occupation outcomes among engineering doctoral degree recipients (US citizens and permanent residents) by race/ethnicity and by sex among African Americans

| Characteristics | African Americans | | | Whites | Asian Americans | Hispanics |
|---|---|---|---|---|---|---|
| | Total | Women | Men | | | |
| Sex variable | | | | | | |
| Female | 0.55 (0.21) | | | 0.63*** (0.07) | 0.85 (0.15) | 0.51 (0.19) |
| Class variables | | | | | | |
| Mother's education | 0.76 (0.32) | 0.15 (0.17) | 1.15 (0.57) | 0.93 (0.09) | 0.89 (0.13) | 0.95 (0.32) |
| Father's education | 0.72 (0.29) | 2.15 (2.27) | 0.42 (0.21) | 0.91 (0.09) | 0.80 (0.12) | 1.33 (0.47) |
| Demographic variables | | | | | | |
| Age | 1.00 (0.18) | 0.63 (0.61) | 1.05 (0.24) | 1.16*** (0.04) | 1.04 (0.06) | 0.99 (0.13) |
| Age-squared | 1.00 (0.00) | 1.00 (0.01) | 1.00 (0.00) | 1.00*** (0.00) | 1.00 (0.00) | 1.00 (0.00) |
| Marital status (reference: single and never married) | | | | | | |
| Married | 1.95 (0.97) | 2.32 (2.15) | 1.77 (1.19) | 0.96 (0.14) | 1.49 (0.37) | 1.05 (0.50) |
| Other marital status | 1.81 (1.06) | 0.39 (0.62) | 2.91 (2.23) | 0.83 (0.15) | 1.06 (0.34) | 0.82 (0.49) |
| Having at least one child | 0.61 (0.25) | 0.35 (0.35) | 0.83 (0.42) | 1.07 (0.11) | 0.91 (0.13) | 1.59 (0.56) |
| Educational variables | | | | | | |

| | | | | | | |
|---|---|---|---|---|---|---|
| Carnegie Classification: research universities with very high research activities | 0.83 (0.34) | 0.24 (0.25) | 1.03 (0.48) | 0.98 (0.11) | 0.94 (0.15) | 1.18 (0.47) |
| Public universities | 0.65 (0.22) | 0.36 (0.27) | 0.6 (0.26) | 1.34*** (0.11) | 1.51** (0.19) | 2.43** (0.76) |
| Cohort (reference: receiving an engineering doctorate prior to 1995) | | | | | | |
| Cohort 2005–2007 | 0.93 (0.67) | 0.04 (0.07) | 6.15 (6.91) | 1.52 (0.37) | 1.35 (0.47) | 0.39 (0.28) |
| Cohort 2000–2004 | 1.44 (0.78) | 0.76 (0.88) | 1.64 (1.12) | 1.23 (0.21) | 1.53 (0.37) | 0.57 (0.32) |
| Cohort 1995–1999 | 0.96 (0.47) | 0.20 (0.23) | 1.59 (0.98) | 0.95 (0.12) | 0.94 (0.17) | 0.50 (0.23) |
| Constant | 10.42 (46.17) | 21243187 (446322578.9) | 1.41 (8.32) | 0.08* (0.09) | 0.58 (0.97) | 2.94 (10.63) |
| N | 201 | 57 | 144 | 2,807 | 1,261 | 240 |

SOURCE: Survey of Doctorate Recipients, 2008.
NOTE: *** $p < 0.001$; ** $p < 0.01$; * $p < 0.05$.

relative to their male counterparts on earning engineering doctorates vary considerably across race/ethnic groups, with the smallest disadvantage in the African American sample. Nevertheless, once they received the degree, women are not disadvantaged in the acquiring of an engineering occupation, with the exception of White women. The findings also suggest the equalizing effect of an engineering doctorate for men and women.

## DISCUSSION

One of the key findings from this research is that African Americans are disadvantaged relative to some other groups in the attainment of engineering doctorates but not occupations. The gap between African Americans and Whites in earning engineering doctorates disappeared when sex was entered into the model. This finding suggests that the White–African American difference in earning engineering doctorates may be more of a gender issue than a race/ethnicity issue. This finding could be due to the sampling of SDR—the SDR sample includes individuals with doctoral degrees in science, engineering, and health (SEH) fields but not non-SEH fields, such as education. There are greater gender differences in degree achievement in SEH fields than in non-SEH fields (even eclipsing race/ethnic differences). Thus, our findings on gender and race effects could be affected by this sample characteristic. The absence of the African American–White race effect when sex is controlled may also have to do with oppositional processes where race/ethnicity in general favors Whites and sex in general favors males but African American females have an advantage that is larger than that for White women. The interaction effect should control on this trend, but the small cell sizes (and larger measurement errors) for the African American group could be a factor. On the other hand, the African American–Asian American difference continued when sex and other variables were entered into the model. This finding confirms literature on Asian Americans showing their greater representation in engineering than any other racial/ethnic minority group (e.g., NSF 2012). It is also consistent with our descriptive findings that revealed a continuing Asian American advantage over African Americans in achieving engineering doctorates over the past decade.

We also find a unique process of degree achievement for African Americans. The odds of earning an engineering doctorate for this racial/ethnic group are affected by fewer class and sociodemographic variables than is the case for other racial/ethnic groups (especially Whites and Hispanics). These results confirm our arguments that race/ethnicity continues to structure the process of educational achievement in the United States, and generalizations across race/ethnic groups (even underrepresented ones) cannot be made. To some extent the lack of class effects and demographic effects may be a result of an incomplete set of measures in our research. For example, the SDR data on class are limited to parents' education. Other researchers may find class effects when they use additional measures of class (e.g., including household income, earnings, poverty, and neighborhood measures). To the extent that these findings on class and demographic effects hold, however, solutions that work for other racial/ethnic groups may not work for African Americans.

One of the notes of optimism in this research is the finding showing a smaller gender gap in engineering doctorates for African Americans relative to other race/ethnic groups. The intersectionality approach argues that race/ethnic and sex effects are not additive. An important element of the unique view of gender used by the practitioners of intersectionality involves agency and historical context (Collins 1999). Researchers who have applied this approach to success in science tell us that we should expect the unexpected when it comes to interest and engagement in science among African American women. Although they have a double disadvantage involving race/ethnicity and gender in the White male STEM culture, African American women's unique history and gender culture provide them resources to "swim against the tide" (Hanson and Johnson 2000; Hanson 2009).

What are the implications of our findings for policy makers and researchers? The presence of race/ethnicity effects in achieving engineering doctorates (especially African American disadvantages) and the absence of race/ethnicity effects in achieving engineering occupations (once the doctorate is achieved) indicate that an engineering doctorate helps reduce or eliminate disadvantages in engineering associated with race/ethnicity. These findings suggest that research and policy efforts should focus on the engineering doctorate process as a critical gateway to engineering occupations since those with a doctorate experience a more level playing field in the engineering workplace. A recent report by the Committee on Science, Engineering, and Public Policy (2011) makes recommendations for how diversity can be increased in postsecondary and doctoral engineering programs. These recommendations include efforts by the federal government to increase the efficiency and effectiveness of outreach programs such as the TRIO program, which targets youths from disadvantaged backgrounds. They also include changes in higher education institutions, including more summer STEM programs, targeted outreach to minorities, support of need-based financial aid, and greater inclusion of minorities as doctoral assistants. We believe that early interventions that involve minority students in STEM education in a college context can better attract and prepare them for STEM education at the postsecondary level. Early involvement in research activities among minority students can also encourage them to pursue a career in engineering research. While this chapter does not focus on engineering baccalaureate degree recipients, the bachelor's degree is essential in encouraging and retaining African Americans' and women's participation in engineering careers. Doctorate recipients are an elite group and survivors of many selection processes, while many with baccalaureate degrees still have to overcome multiple hurdles, including funding for graduate school (National Research Council 2011), and this issue is also addressed in more detail in other chapters of this volume. In other words, policy makers need to interpret the findings of this chapter along with literature on the transition from earning a bachelor's degree to earning a doctoral degree in engineering.

Another major finding from this chapter involves the effect of sex on engineering outcomes. Sex is a factor affecting the doctorate process for all race/ethnic groups, and the nature of the effect varies across groups. Additionally, sex has a significant influence on the engineering occupation outcome for Whites. These results show the complex intersection of race/ethnicity and sex in the study of STEM processes. Researchers who fail to take into account racial/ethnic and sex interactions might conclude a general sex effect in STEM processes since White women are the largest group in most probability samples. Research that does not acknowledge the intersection of race/ethnicity and sex in STEM achievement might also miss important progress that African American and other minority women are making in STEM in catching up with their male counterparts. Some have suggested that gender cultures among Whites are not the same as gender cultures among other race/ethnic groups, and minority women (including African American women) have an interest and success in science that is not necessarily mirrored among White women (Hanson 2009).

Our findings regarding African American men and women suggest that the process leading to success for African Americans later in the engineering pipeline may involve factors other than their class and sociodemographic status. Class and sociodemographic characteristics may be more important in early engineering success than in later engineering success. It is hoped that future researchers consider additional influences on the achievement of engineering doctorates for African Americans such as teachers, peers, and cognitive factors (e.g., stereotype threat).

## Conclusion

This chapter examines racial/ethnic and sex differences in achieving engineering doctorates (as opposed to other doctorates) and in achieving engineering occupations among those with a doctoral degree. The sample used in this study is limited to US citizens and permanent residents who had already received a doctoral degree. Although there is some research available on earlier sections of the engineering pipeline, it is important that researchers look carefully at race, sex, and background factors in affecting each stage of the achievement process in this pipeline.

We have argued that diversity in engineering is a resource that has not been taken advantage of in the US education and occupation system. There has been growth in the underrepresented minority population in the United States and growth in the US engineering workforce, but our findings show little progress for African Americans in access to engineering doctorates. African Americans represent 14% of the US population but received just 4% of the engineering doctorates in 2000. This number remained virtually unchanged in 2010. Findings also suggest that programs and policies attempting to reduce inequality in graduate engineering cannot use simple race-based or sex-based strategies. We conclude that the "quiet crisis" that NACME (2008) warns of is an ongoing crisis.

## *Appendix*

### IPEDS DATA SET

IPEDS is a system of interrelated surveys conducted by the US Department of Education's National Center for Education Statistics (NCES). The IPEDS Completions Survey is conducted annually, and institutions participating in Title IV report degrees and other formal awards (e.g., certificates) conferred by the institutions by degree level (associate's, bachelor's, master's, doctor's, and first-professional) and length of program for some. The IPEDS Completions Survey data have been reported by race/ethnicity since 1990. The data in the IPEDS Completions Survey by Race come from the Higher Education General Information Survey (HEGIS) and IPEDS conducted by NCES.* HEGIS is a predecessor of IPEDS and covers data from 1966 to 1985. IPEDS started in 1986 and collects

---

* For more information regarding the IPEDS Completions Survey by Race on WebCASPAR, refer to https://ncsesdata.nsf.gov/webcaspar/Help/dataMapHelpDisplay .jsp;jsessionid = 6E4E6F0CED04761FB32C92B7884584E8.prodas2?subHeader=Data SourceBySubject&type=DS&abbr=DEGSRACE&noHeader=1&JS=No.

institutional data annually.* The IPEDS Completions Survey is available on NSF's WebCASPAR. WebCASPAR, the integrated science and engineering resources data system, provides statistical data sets from NSF and NCES for public use. WebCASPAR allows for a selection of an NCES or NSF population of institutions when one retrieves the IPEDS data. Before 1996, the two sets of populations of institutions were the same and included postsecondary institutions accredited at the college level. Current NCES institutions cover institutions based on the degree-granting status but also eligibility for Title IV federal financial aid. While these two populations of institutions still overlap significantly, this chapter uses the NSF population for descriptive data because this chapter also uses an NSF data set to answer other research questions.

IPEDS covers all postsecondary institutions with a Program Participation Agreement with the Office of Postsecondary Education (OPE) at the US Department of Education (or participating in "Title IV"). Since it is mandatory for these institutions to complete the IPEDS survey, the response rate is nearly 100%. In other words, the IPEDS Completions Survey includes almost all doctorate recipients receiving their doctorates in any field in a given year from an institution that participates in "Title IV." These doctorate recipients include US citizens, permanent residents, and individuals on temporary visas. Degree completion is one of the components that IPEDS currently covers. The other components of IPEDS include institutional characteristics, human resources, 12-month enrollment, fall enrollment, graduation rates, finance, and student financial aid.

SDR DATA SET

SDR is a longitudinal survey conducted for NSF and the National Institutes of Health (NIH) every two to three years. Variables included in this data set include demographic (e.g., sex, race, age, marital status, children), class (e.g., parental education), educational (e.g., field of the doctorate, year of receiving the doctorate, Carnegie Classifications of the doctoral institution), and occupational characteristics (e.g., labor force status, field of employment). While the public-use version is available, the restricted-use version, which requires a license from NSF, is more appropriate for this study. The restricted-use version includes important variables that are not available in the public-use version, such as some racial/ethnic categories, including African Americans, as well as some demographic characteristics, such as marital status. SDR data provide important information for researchers and policy makers regarding the career experiences and outcomes of doctorate recipients in SEH fields in multiple years (e.g., Stephan and Levin 1991; Turner, Myers, and Creswell 1999; Ginther and Hayes 2003; Mason and Goulden 2004; Mason, Goulden, and Frasch 2009; Kim, Wolf-Wendel, and Twombly 2011).

---

* For more information regarding HEGIS and IPEDS, refer to http://nces.ed.gov /ipeds/glossary/?charindex=H, http://nces.ed.gov/ipeds/glossary/?charindex=I, and http:// nces.ed.gov/ipeds/.

TABLE 3A.1  Engineering doctorates and occupations achieved by US citizens and permanent residents by race/ethnicity and sex in the sample, 2008

| Race/Ethnicity or Sex | Number, All Doctorates | Engineering Doctorate | | Engineering Occupation | |
|---|---|---|---|---|---|
| | | Number | Percentage (of All Doctorates) | Number | Percentage (of Engineering Doctorates) |
| All races (Whites, Asian Americans, Hispanics, and African Americans) | 28,165 | 4,509 | 16 | 3,021 | 67 |
| Whites | 20,282 | 2,807 | 14 | 1,827 | 65 |
| Asian Americans | 4,417 | 1,261 | 29 | 785 | 62 |
| Hispanics | 1,805 | 240 | 13 | 153 | 64 |
| African Americans | 1,661 | 201 | 12 | 128 | 64 |
| Women | 819 | 57 | 7 | 37 | 65 |
| Men | 842 | 144 | 17 | 91 | 63 |

SOURCE: Survey of Doctorate Recipients, 2008.

While this chapter uses the SDR data, the readers are reminded that the use of NSF data does not imply NSF endorsement of the research, research methods, or conclusions contained in this report.

## Acknowledgments

We would like to thank Nirmala Kannankutty and Darius Singpurwalla of the National Science Foundation for their assistance with the use of Survey of Doctorate Recipients data in this chapter.

## References

American Association for the Advancement of Science. 2010. "Support for Women Scientists Grows as Agencies Seek Pathways for Development, Diplomacy." September 21. www.aaas.org/news/support-women-scientists-grows-agencies -seek-pathways-development-diplomacy.

Anderson, Eugene, and Dongbin Kim. 2006. *Increasing the Success of Minority Students in Science and Technology*. Washington, DC: American Council on Education.

Blau, Peter M., and Otis Dudley Duncan. 1967. *The American Occupational Structure*. New York: John Wiley & Sons.

Brookings Institution. 2014. "The Importance of the Science and Engineering Workforce for Future Growth." April 23. www.brookings.edu/blogs/techtank /posts/2014/04/23-stem-workforce.

Browne, Irene, and Joya Misra. 2003. "The Intersection of Gender and Race in the Labor Market." *Annual Review of Sociology* 29:487–513.

Bryant, Chalandra M., K. A. S. Wickrama, John Bolland, Barlynda M. Bryant, Carolyn E. Cutrona, and Christine E. Stanik. 2010. "Race Matters, Even in Marriage: Identifying Factors Linked to Marital Outcomes for African Americans." *Journal of Family Theory and Review* 2:157–174.

Burrelli, Joan, Alan Rapoport, and Rolf Lehming. 2008. "Baccalaureate Origins of S&E Doctorate Recipients." *Info Brief.* www.nsf.gov/statistics/infbrief/nsfo 8311/nsfo8311.pdf.

Charles, Camille Z., Vincent J. Rosicgno, and Kimberly C. Torres. 2007. "Racial Inequality and College Attendance: The Mediation Role of Parental Investments." *Social Science Research* 36:329–352.

Choobbasti, Heydar Janalizadeh. 2007. "The Social Origins of Eminent Scientists: A Review, Comparison and Discussion." *Research in Social Stratification and Mobility* 25:233–243.

Collins, Patricia H. 1999. "Moving beyond Gender: Intersectionality and Scientific Knowledge." In *Revisioning Gender*, edited by Myra Marx Feree, Judith Lorber, and Beth B. Hess, 261–284. Thousand Oaks, CA: Sage.

Committee on Science, Engineering, and Public Policy. 2011. *Rising above the Gathering Storm: Developing Regional Innovation Environments*. Washington, DC: National Academies Press.

Congressional Commission on the Advancement of Women and Minorities in Science, Engineering, and Technology Development. 2000. *Land of Plenty? Diversity as America's Competitive Edge in Science, Engineering, and Technology*. Washington, DC: Author.

Cooper, Robert. 1996. "Detracking Reform in an Urban Californian High School: Improving the Schooling Experiences of African American Students." *Journal of Negro Education* 65:190–208.

COSSA (Consortium of Social Science Associations). 2008. *Enhancing Diversity in Science*. Washington, DC: COSSA.

DeFrancisco, Victoria P., and Catherine H. Palczewski. 2007. *Communicating Gender Diversity: A Critical Approach*. Los Angeles: Sage.

Espenshade, Thomas J., and Alexandria W. Radford. 2009. *No Longer Separate, Not Yet Equal: Race and Class in Elite College Admission and Campus Life*. Princeton, NJ: Princeton University Press.

Feagin, Joe R., Hernan Vera, and Nikitah Imani. 1996. *The Agony of Education: Black Students at White Colleges and Universities*. New York: Routledge.

Freeman, Kassie. 1997. "Increasing African American's Participation in Higher Education: African American High School Students' Perspectives." *Journal of Higher Education* 68:523–550.

Ginther, Donna K., and Kathy J. Hayes. 2003. "Gender Differences in Salary and Promotion for Faculty in the Humanities, 1977–1995." *Journal of Human Resources* 38:34–73.

Hanson, Sandra L. 1996. *Lost Talent: Women in the Sciences*. Philadelphia: Temple University Press.

Hanson, Sandra L. 2009. *Swimming against the Tide: African American Girls in Science Education*. Philadelphia: Temple University Press.

Hanson, Sandra L. 2012. "Science for All? The Intersection of Gender, Race, and Science." *International Journal of Science in Society* 3:113–136.

Hanson, Sandra L., and Elizabeth P. Johnson. 2000. "Expecting the Unexpected: A Comparative Study of African American Women's Experiences in Science during the High School Years." *Journal of Women and Minorities in Science and Engineering* 6:265–294.

Herzig, Abbe H. 2004. "Becoming Mathematicians: Women and Students of Color Choosing and Leaving Doctoral Mathematics." *Review of Educational Research* 74:171–214.

Hill, Catherine. 2008. "Race and Income Matter More Than Gender: New Research from AAUW Sheds Light on Key Predictors for Student Success." *Student Affairs Leader* 36:4. http://connection.ebscohost.com/c/articles/335 25932.

Kim, Dongbin, Lisa Wolf-Wendel, and Susan Twombly. 2011. "International Faculty: Experiences of Academic Life and Productivity in U.S. Universities." *Journal of Higher Education* 82:720–747.

Mann, Susan A., and Douglas J. Huffman. 2005. "The Decentering of Second Wave Feminism and the Rise of the Third Wave." *Science and Society* 69:56–91.

Mason, Mary A., and Marc Goulden. 2004. "Marriage and Baby Blues: Redefining Gender Equity in the Academy." *Annals of the American Academy of Political and Social Science* 596:86–103.

Mason, Mary A., Marc Goulden, and Karie Frasch. 2009. "Why Graduate Students Reject the Fast Track." *Academe* 95:11–16.

Moreira, Belinda. 2012. "STEM Industries Need to Hack Sexism and Racism Now." November 28. http://dev.policymic.com/articles/19700/stem-industries -need-to-hack-sexism-and-racism-now.

NACME (National Action Council for Minorities in Engineering). 2008. *Confronting the "New American Dilemma—Underrepresented Minorities in Engineering: A Data-Based Look at Diversity*. White Plains, NY: Author.

NACME (National Action Council for Minorities in Engineering). 2012. "African Americans in Engineering." *NACME Research and Policy* 2:1–2. www.nacme .org/publications/research_briefs/NACMEAfricanAmericansinEngineering .pdf.

National Academy of Engineering. 2007. *Grand Challenges for Engineering*. Washington, DC: Author.

National Academy of Sciences. 2007. *Rising above the Gathering Storm: Energizing and Employing America for a Brighter Economic Future*. Washington, DC: National Academies Press, 2007.

National Academy of Sciences. 2010. *Rising above the Gathering Storm, Revisited: Rapidly Approaching Category 5*. Washington, DC: National Academies Press.

National Research Council. 2011. *Expanding Underrepresented Minority Participation: America's Science and Technology Talent at the Crossroads*. Washington, DC: National Academies Press.

Noeth, Richard J., Ty Cruce, and Matt T. Armstrong. 2003. *Maintaining a Strong Engineering Workforce: ACT Policy Report*. Iowa City: ACT.

NSF (National Science Foundation). 2013. *Women, Minorities, and Persons with Disabilities in Science and Engineering*. Arlington, VA: NSF.

NSF (National Science Foundation), Division of Science Resources Statistics. 2006. *U.S. Doctorates in the 20th Century*, by Lori Thurgood, Mary J. Golladay, and Susan T. Hill. Arlington, VA. www.nsf.gov/statistics/nsf06319/pdf/nsf06319.pdf.

NSF (National Science Foundation), National Center for Science and Engineering Statistics. 2012. *Doctorate Recipients from U.S. Universities: 2011*. Special Report NSF 13-301. Arlington, VA. www.nsf.gov/statistics/sed/.

O'Connor, Carla, Lori Diane, and Shanta R. Robinson. 2009. "Who's at Risk in School and What's Race Got to Do with It?" *Review of Research in Education* 33:1–34.

Pearson, Willie, Jr. 2005. *Beyond Small Numbers: Voices of African American Chemists.* Bingley, UK: Emerald Group.

Pearson, Willie, Jr., and H. Kenneth Bechtel. 1989. *Blacks, Science and American Education.* New Brunswick, NJ: Rutgers University Press.

Pearson, Willie, Jr., and Alan Fechter. 1994. *Who Will Do Science? Educating the Next Generation.* Baltimore: Johns Hopkins University Press.

Rochin, Refugio I., and Stephen F. Mello. 2007. "Latinos in Science: Trends and Opportunities." *Journal of Hispanic Higher Education* 6:305–355.

Roscigno, Vincent. 2000. "Family/School Inequality and African American/Hispanic Achievement." *Social Problems* 47:266–290.

Shapiro, Douglas T. 2001. "Modeling Supply and Demand for Arts and Sciences Faculty: What Ten Years of Data Tell Us about the Labor Market Projections of Bowen and Sosa." *Journal of Higher Education* 72:532–564.

Slaton, Amy. 2010. *Race, Rigor and Selectivity in U.S. Engineering: The History of an Occupational Color Line.* Cambridge, MA: Harvard University Press.

Sonnert, Gerhard, and Mary Frank Fox. 2012. "Women, Men, and Academic Performance in Science and Engineering: The Gender Difference in Undergraduate Grade Point Averages." *Journal of Higher Education* 83:73–101.

Stanford Humanities Lab. 2006. "Is There Racial Discrimination in Engineering?" http://humanitieslab.stanford.edu/76/288.

Stephan, Paula E., and Sharon G. Levin. 1991. "Inequality in Scientific Performance: Adjustment for Attribution and Journal Impact." *Social Studies of Science* 21:351–368.

Tang, Joyce. 2000. *Doing Engineering: The Career Attainment and Mobility of Caucasian, Black, and Asian-American Engineers.* New York: Rowman & Littlefield.

Turner, Caroline S. V., Samuel L. Myers Jr., and John W. Creswell. 1999. "Exploring Underrepresentation: The Case of Faculty of Color in the Midwest." *Journal of Higher Education* 70:27–59.

Wenner, George. 2003. "Comparing Poor, Minority Students' Interest and Background in Science with That of Their White, Affluent Peers." *Urban Education* 38:153–172.

Xie, Yu, and Kimberlee A. Shauman. 2003. *Women in Science: Career Processes and Outcomes.* Cambridge, MA: Harvard University Press.

# CHAPTER FOUR

# African American Women and Men into Engineering

## Are Some Pathways Smoother Than Others?

LINDSEY E. MALCOM-PIQUEUX *and* SHIRLEY M. MALCOM

SINCE THE MID-TWENTIETH CENTURY, the participation of African Americans in higher education has markedly increased. In 1950, slightly more than 75,000 African Americans were enrolled in higher education (US Bureau of the Census 1968); this number grew to just above one million by 1976 and swelled to nearly three million by 2012 (National Center for Education Statistics 2013). Multiple social, political, and economic factors fueled this rapid growth in college participation among African Americans. The civil rights movement and the end of de jure segregation expanded educational opportunity for African Americans, while a range of enabling legislation that provided federal financial aid, strengthened historically Black colleges and universities (HBCUs), prohibited racial and gender discrimination, and permitted targeted recruitment and admissions facilitated college access for African Americans (Anderson 2002). Although recent changes in the political and legal landscape have severely curtailed affirmative action, such policies positively affected college participation among African Americans over decades of implementation (Bowen and Bok 1998).

Increased access to higher education for African Americans has led to a corresponding rise in educational attainment. As the number of African Americans enrolled in college tripled in the 35 years between 1976 and 2011, the number of bachelor's degrees conferred to African Americans also nearly tripled, from 58,636 in 1976–1977 to 185,518 in 2011–2012 (National Center for Education Statistics 2013). However, these gains in bachelor's degree attainment are not present at the same levels across all academic disciplines. Though the number of African Americans who earned bachelor's degrees in engineering rose from 1,385 to 3,327 between 1977 and 2012, this rise did not keep pace with growing college enrollment (National Action

Council for Minorities in Engineering 2008; National Center for Education Statistics 2013). Further, over the past decade, the proportion of African American bachelor's degree earners who earned degrees in engineering has slowly—but steadily—declined (see also chaps. 3 and 6 in this volume). As a result, African Americans earned just 3.8% of all engineering degrees in 2012, after reaching a peak of 5.4% in 2004. The underrepresentation of African Americans in engineering is clear, given that they earned 9.5% of all bachelor's degrees and currently represent 15.1% of the college-age population (US Census Bureau 2011; US Department of Education 2014a, 2014b).

What might account for the failure of engineering to keep pace with the overall growth of African Americans in higher education? In this chapter, we explore this question by examining the institutional pathways to engineering degrees for African Americans and discussing the implications of their changing patterns of participation in higher education on these pathways. In particular, we consider the role of the following trends on engineering degree attainment among African Americans:

1. Since the late 1960s, more African American women than men attend college; the "female advantage" continues to grow, and women currently outnumber men nearly 2 to 1 (US Bureau of the Census 1968; National Center for Education Statistics 2012). Yet women are less likely than men to major in engineering: African American men enrolled in engineering programs outnumbered their female counterparts by more than 3 to 1 (National Center for Education Statistics 2012).

2. In fall 2012, nearly 56% of African American undergraduates attended either a community college or a for-profit institution, compared to 46.8% of undergraduates from all racial/ethnic groups (National Center for Education Statistics 2013).

3. In 1960, the vast majority of African Americans enrolled in higher education attended HBCUs. By 2012, 14.5% of African American undergraduates enrolled in four-year institutions attended HBCUs (Allen 1992; US Department of Education 2014b).

4. In fall 2006, 7.2% of African American freshmen indicated an intention to pursue an engineering degree; however, just 3.1% of African American first-time, first-year undergraduates enrolled in engineering degree programs in the same year (National Science Foundation 2013; US Department of Education 2014b). Six years later, in 2012, just 1.8% of African Americans awarded bachelor's degrees earned those degrees in engineering (US Department of Education 2014a).

Based on our analysis of engineering degree attainment data and the extant literature, we argue that each of these trends likely contributes to the underparticipation of African Americans in engineering. Consequently, increasing the numbers of African American engineering degree earners requires a multifaceted approach at the undergraduate level. We conclude this chapter by offering recommendations for practice, policies, and programs to increase access, retention, and success in engineering for African American women and men and for smoothing the pathways to engineering through HBCUs and predominantly White institutions (PWIs).

## The African American Female Undergraduate Majority and Engineering

In recent years, researchers and policy makers have placed increasing attention on the "disappearing Black male" in higher education (e.g., Esters and Mosby 2007; Jackson and Moore 2008; Harper and Harris 2012). Perhaps surprising to many, however, is the fact that African American women have long outnumbered their male counterparts in terms of college enrollments. Few data regarding African American college enrollments prior to the 1960s are available; however, the US Bureau of the Census reported that in 1950, 53% of African Americans enrolled in higher education were men (US Bureau of the Census 1968). By 1960, women outnumbered men among African American college students (53% to 47%). From that point forward, the gender gap in enrollment among African American undergraduates continued to grow. In 1976, women were 56% of African Americans enrolled in colleges and universities; by 2012, African American women in college outnumbered African American men nearly 2 to 1 (African American women constitute 63% of all African Americans in college; US Department of Education 2014b).

This growing female majority among African American undergraduates has important implications for engineering bachelor's degree attainment. Women, including African American women, are less likely to indicate an intention to major in engineering. Data from the University of California, Los Angeles (UCLA), Higher Education Research Institute's 2011 National Survey of the American Freshman indicate that while 15.6% of African American men planned to major in engineering, just 3.2% of African American women planned to pursue an undergraduate degree in engineering (see table 4.1; National Science Foundation 2013). Not surprisingly, a lower initial interest translates to lower engineering enrollment and lower engineer-

TABLE 4.1  Gender differences in engineering interest, enrollment, and degrees

| Measure | African American | White | Latino | Asian | Native American |
|---|---|---|---|---|---|
| Intention to major in engineering (2010) (percent of first-year students) | | | | | |
| Women | 3.2 | 4.0 | 3.2 | 7.2 | 1.6 |
| Men | 14.6 | 18.0 | 18.2 | 22.7 | 11.2 |
| Full-time, first-year undergraduate engineering enrollment (fall 2005) (percent distribution) | | | | | |
| Women | 25.3 | 14.3 | 19.7 | 18.7 | 22.0 |
| Men | 74.7 | 85.7 | 80.3 | 81.3 | 78.0 |
| Engineering bachelor's degrees (2011) (percent distribution) | | | | | |
| Women | 26.3 | 17.1 | 21.2 | 23.0 | 25.2 |
| Men | 73.7 | 82.9 | 78.8 | 77.0 | 74.8 |

SOURCES: National Science Foundation (2007, 2013); US Department of Education (2014a).

ing degree attainment. In 2011, 24.0% of degree-seeking, African American undergraduates enrolled in engineering programs were women; in 2011, women earned 26.3% of engineering bachelor's degrees awarded to African Americans compared to 65.8% of all bachelor's degrees awarded to African Americans (see tables 4.1 and 4.2; National Science Foundation 2013). Though their share of all engineering bachelor's degrees awarded to African Americans has increased markedly since 1977, the female advantage that African American women experience in college enrollment and earned bachelor's degrees does not translate to large numbers in the field of engineering.

There are wide variations in all women's distribution and African American women's distribution across engineering fields. Though African American women are underrepresented in every engineering subfield relative to their share of all bachelor's degrees, African American women exceeded participation levels of women from other racial/ethnic groups in 2012 (see table 4.2). Women earned just 8.2% of all computer engineering bachelor's degrees, whereas African American women earned 20.1% of bachelor's degrees in computer engineering awarded to African Americans (US Department of Education 2014a). Table 4.2 also indicates that African American women

TABLE 4.2   Share of engineering bachelor's degrees earned by women by race and subfield, 2012

| Field | African American (%) | White (%) | Latino (%) | Asian (%) | Native American (%) |
|---|---|---|---|---|---|
| All fields | 65.8 | 56.2 | 60.9 | 54.5 | 61.1 |
| Engineering, total | 27.1 | 17.6 | 22.5 | 23.1 | 23.2 |
| Aerospace, aeronautical and astronautical engineering | 21.1 | 12.5 | 15.7 | 14.3 | 20.0 |
| Biomedical/medical engineering | 49.0 | 41.6 | 40.5 | 33.9 | 27.3 |
| Chemical engineering | 43.8 | 29.7 | 43.7 | 36.4 | 33.3 |
| Civil engineering | 27.3 | 19.2 | 25.3 | 25.6 | 25.0 |
| Computer engineering | 20.1 | 5.5 | 11.0 | 13.1 | 6.3 |
| Electrical, electronics, and communications engineering | 19.2 | 9.2 | 13.5 | 13.5 | 20.5 |
| Mechanical engineering | 19.2 | 11.2 | 14.4 | 16.3 | 10.5 |
| Industrial engineering | 34.9 | 29.7 | 39.4 | 32.1 | 70.0 |
| Other engineering[a] | 34.1 | 22.6 | 28.7 | 29.5 | 32.7 |

SOURCE: US Department of Education (2014a).

[a] The "Other engineering" category includes agricultural engineering; architectural engineering; ceramic sciences and engineering; engineering mechanics; engineering physics; engineering science; environmental/environmental health engineering; materials engineering; metallurgical engineering; mining and mineral engineering; naval architecture and marine engineering; nuclear engineering; ocean engineering; petroleum engineering; systems engineering; textile sciences and engineering; polymer/plastics engineering; construction engineering; forest engineering; manufacturing engineering; operations research; surveying engineering; geological/geophysical engineering; paper science and engineering; electromechanical engineering; mechatronics, robotics, and automaton engineering; biochemical engineering; engineering chemistry; and biological/biosystems engineering.

received 43.8% of chemical engineering bachelor's degrees awarded to African Americans in 2012, while all women earned 32.8% of such degrees (US Department of Education 2014a).

Researchers attribute the lower tendency of women from all racial/ethnic backgrounds to enter engineering to a number of factors, including a lack of exposure to engineering and weak personal identification with engineering—both of which stem from gender stereotypes about math and science ability and "gender-appropriate" careers (Hill, Corbett, and St. Rose 2010). These stereotypes result in women and girls having fewer opportunities to engage in engineering experiences both in and outside of formal learning environments, which, in part, fuels the "interest gap" (Hill, Corbett, and St. Rose 2010). Certainly, all students can benefit from early exposure to and experiences with engineering through formal curricula, or through family or mentors who possess a background in these fields. Indeed, the inclusion of engineering and technology topics and practices in the Next Generation Science Standards for K–12 education can begin to increase engagement with the field among boys and girls. However, it is important to emphasize that all women, including African American women, confront unique gender-related barriers to engineering.

We do not present these data to suggest that African American men are "doing fine" with respect to gaining access to and succeeding in engineering programs. Quite the contrary—just 3.9% of African American male bachelor's degree earners were awarded their degrees in engineering, compared to 8.5% of White men and 13.5% of Asian men in 2012 (US Department of Education 2014a). However, we do note that just 0.8% of African American women bachelor's degree holders earn degrees in engineering (US Department of Education 2014a). If African American women earned bachelor's degrees in engineering at the same rate as their male counterparts, this would yield more than 3,500 additional African American engineering bachelor's degree holders annually (US Department of Education 2014a).

## Community College Pathways to Engineering

African American students are heavily enrolled in the community college and four-year, for-profit sectors (see chap. 12 in this volume). In fall 2012, nearly 40% of African American undergraduates attended public two-year institutions, compared to 35.8% of Whites, 38.1% of Asian Americans, and 44.5% of Latinos (see table 4.3; US Department of Education 2014b). Though African Americans are not the most likely to attend community colleges, their relatively high concentration in these institutions is noteworthy

| Institution Type | African American (%) | White (%) | Latino (%) | Asian (%) | Native American (%) |
|---|---|---|---|---|---|
| Four-year institutions | | | | | |
| Public | 30.1 | 39.7 | 31.0 | 42.7 | 33.3 |
| Private, not-for-profit | 12.5 | 16.4 | 11.6 | 13.7 | 10.1 |
| Private, for-profit | 11.1 | 4.7 | 5.0 | 2.6 | 6.7 |
| Two-year institutions | | | | | |
| Public (community colleges) | 38.6 | 35.8 | 44.5 | 38.1 | 44.2 |
| Private, not-for-profit | 0.5 | 0.2 | 0.2 | 0.2 | 0.6 |
| Private, for-profit | 3.7 | 1.6 | 4.1 | 1.3 | 2.5 |
| Less than two-year institutions | | | | | |
| Public | 0.2 | 0.4 | 0.3 | 0.2 | 1.6 |
| Private, not-for-profit | 0.1 | <0.1 | 0.2 | 0.1 | 0.1 |
| Private, for-profit | 3.2 | 1.1 | 3.0 | 1.0 | 1.0 |

SOURCE: US Department of Education (2014b).

because they are much more geographically dispersed than Latinos. The large Latino population in California and Texas is likely responsible, in part, for the high prevalence of community college attendance among Latinos. For African Americans, however, geography is not as strong of a driver of the patterns of community college attendance. Just around 21% of African Americans in community colleges attend institutions in California and Texas (US Department of Education 2014b). Instead, concerns surrounding college costs, inadequate preparation for four-year institutions, counseling, and college knowledge may be responsible for the rates at which African Americans attend community colleges.

Whatever the reasons, the prevalence of community college attendance among African Americans has important consequences for access to engineering bachelor's degrees. As scholars have noted for quite some time, community college transfer rates are low, with an estimated 20% of community college transfer students who aspire to attain a bachelor's degree actually transferring to a four-year institution (Mullin 2012). Community college transfer pathways to bachelor's degrees are narrow in many science, technology, engineering, and mathematics (STEM) related fields, including engineering. Indeed, data from the 2008 National Survey of Recent College Graduates (NSRCG) show that rates of community college attendance were lowest among engineering graduates compared to other science- and health-related fields (Mooney and Foley 2011). According to the 2008 NSRCG, just over 38% of individuals (of all races/ethnicities) who received a bachelor's or master's degree in engineering in academic years 2005–2006 or 2006–2007 ever attended a community college, compared to slightly above 50% of all science, engineering, and health graduates (Mooney and Foley 2011). In fact, the only field whose graduates were less likely to have attended community college than engineering was physical and related sciences (Mooney and Foley 2011). Unfortunately, large standard errors prevent us from characterizing racial/ethnic differences in community college attendance among engineering bachelor's degree earners. Nonetheless, data do illustrate that, compared to nearly all other fields, community college pathways to engineering remain narrow for all students.

Though the unique barriers that students face on the pathway from community college to engineering bachelor's degrees have not been investigated empirically, we offer some possible factors that ought to be considered. Not all state community college systems have established statewide or system-wide articulation agreements (Ignash and Townsend 2001; US Department of Education 2014b). A lack of curricular alignment and easily transferable courses is a likely barrier to transferring into four-year engineering degree programs, as course sequencing and completion of prerequisites are critically important in the field. It is also possible that the lower prevalence of community college attendance among engineering degree holders may be driven in part by issues related to academic preparation. Decades of research have established the importance of completing advanced coursework in chemistry, physics, and mathematics during high school to engineering degree completion (e.g., Adelman 2006; National Action Council for Minorities in Engineering 2008). If students who opt to attend community colleges do so, in part, for reasons related to inadequate preparation (i.e., not being college ready in terms of mathematics ability), they are less likely to be "engineering-eligible" (National Action Council for Minorities in Engineering 2008; see also Malcom-Piqueux and Malcom 2013). Again, these questions require additional study; however, it is important that changes in policies and practices be made to remove the barriers along community college pathways to engineering degrees.

Community colleges remain an attractive option for a range of students, including African Americans, owing to their lower costs, geographical accessibility, less arduous admissions policies, and more forgiving, remedial, and credit-bearing curriculum. As demand for higher education continues to grow along with rising college costs, community college enrollments will remain high. Thus, attention ought to be paid to smoothing the pathways to engineering through community colleges as a part of a comprehensive approach to broadening participation in engineering.

## For-Profit Institutions

Though African American enrollment in community colleges exceeds that in for-profit colleges, a significant proportion of African Americans attend proprietary institutions. As shown in table 4.3, 11.1% of African American undergraduates are enrolled in four-year proprietary institutions, 3.7% are enrolled in two-year proprietary institutions, and 3.2% attend less-than-two-year, for-profit institutions. African Americans are more likely than any other ethnic group to attend for-profit institutions at all postsecondary levels (US Department of Education 2014b). Though an in-depth discussion of the reasons for the relatively high enrollment rates of African Americans in proprietary institutions is beyond the scope of this analysis, targeted marketing and recruitment, geographic accessibility, prevalence of distance education, and an emphasis on career education have been identified as potential factors (Deming, Goldin, and Katz 2012). The concentration of African Americans in the for-profit sector likely limits access to engineering degrees because these institutions lack program offerings in the field. This is evidenced by recent degree data. In the 2011–2012 academic year, just 0.3% of African American men who earned bachelor's degrees from for-profit institutions were engineering majors, compared to 5.1% in public, four-year institutions and 2.8% in private, not-for-profit, four-year institutions. The pattern is similar among African American women. Just 0.03% of African American women who earned bachelor's degrees from for-profit institutions were engineering majors, compared to 1.0% in public, four-year institutions and 0.7% in private, not-for-profit, four-year institutions (US Department of Education 2014a).

Clearly, for-profit institutions are inefficient pathways to engineering for African Americans, owing to a lack of program offerings, among other factors. We note, however, that for-profit institutions offer a range of coursework in computer science and information technology. Given that the for-profit postsecondary sector is the fastest growing, and with African American enrollments at proprietary institutions steadily increasing, it is important to consider the implications of these trends on African Americans' participation in engineering.

## Institutional Pathways to Engineering:
## The Role of Four-Year HBCUs and PWIs

In the mid-1960s over half of African Americans in higher education were enrolled in HBCUs. By 1973, only 25% of African American students were enrolled in HBCUs (Anderson 1984); a dramatic shift occurred in this intervening period, with more students enrolling in PWIs, especially in the previously segregated state institutions in the South (Allen and Jewell 2002). Though many of these PWIs in the Southern states initially resisted desegregation and were slower to admit and enroll African American students, Black enrollments significantly increased at these institutions in the decades following *Brown* (Allen and Jewell 2002; Gasman 2013). While African American enrollment continued to rise at HBCUs during this period, the numbers at which Blacks entered PWIs outpaced the growth in HBCUs. Since the 1980s, however, the number of African Americans enrolled in four-year HBCUs has remained roughly constant, while their enrollment in four-year PWIs has increased. The shift in African American enrollment to PWIs has continued to the present, such that by fall 2012, HBCUs enrolled just 14.5% of African American undergraduates in four-year institutions (US Department of Education 2014b).

The shift in African American enrollment from HBCUs toward PWIs is of note for many reasons, one of which is the implication for bachelor's degree production among this population in all fields, and in engineering and other STEM fields in particular. HBCUs have long awarded a disproportionately high share of bachelor's degrees to African Americans (Allen 1992; Allen and Jewell 2002) and continue to do so. HBCUs accounted for 16.7% of all bachelor's degrees and 19.5% of engineering bachelor's degrees awarded to African Americans in 2011–2012, despite enrolling just 14.8% of African American undergraduates attending four-year institutions (US Department of Education 2014b). With only 14 universities offering accredited engineering programs among HBCUs, a very small percentage of all engineering programs in US universities produces a disproportionate share of African American engineering bachelor's degree graduates.

The contribution of HBCUs is even more striking when examining the engineering bachelor's degree data disaggregated by gender (see table 4.4). In 2011–2012, 25.1% of all engineering bachelor's degrees awarded to African American women were conferred by HBCUs, while just 14.2% of African American women enrolled in four-year institutions attended HBCUs during the same year (US Department of Education 2014a, 2014b). HBCUs also overperformed among African American men, awarding 17.5%

TABLE 4.4  Engineering bachelor's degrees awarded to African Americans by HBCUs and PWIs by subfield and gender, 2011–2012

| Field | Women | | Men | | Total | |
|---|---|---|---|---|---|---|
| | HBCUs | PWIs | HBCUs | PWIs | HBCUs | PWIs |
| | | | Number | | | |
| Engineering, total | 213 | 637 | 400 | 1,888 | 613 | 2,525 |
| Aerospace, aeronautical, and astronautical engineering | 2 | 13 | 4 | 52 | 6 | 65 |
| Biomedical/medical engineering | 0 | 76 | 0 | 79 | 0 | 155 |
| Chemical engineering | 41 | 94 | 23 | 150 | 64 | 244 |
| Civil engineering | 26 | 99 | 63 | 270 | 89 | 369 |
| Computer engineering | 20 | 38 | 40 | 190 | 60 | 228 |
| Electrical, electronics, and communications engineering | 43 | 92 | 137 | 432 | 180 | 524 |
| Mechanical engineering | 45 | 70 | 68 | 416 | 113 | 486 |
| Industrial engineering | 14 | 51 | 23 | 98 | 37 | 149 |
| Other engineering[a] | 22 | 104 | 42 | 201 | 64 | 305 |

| | Percent Distribution | | | | | |
|---|---|---|---|---|---|---|
| Engineering, total | 25.1 | 74.9 | 17.5 | 82.5 | 19.5 | 80.5 |
| Aerospace, aeronautical, and astronautical engineering | 13.3 | 86.7 | 7.1 | 92.9 | 8.5 | 91.5 |
| Biomedical/medical engineering | 0 | 100 | 0 | 100 | 0 | 100 |
| Chemical engineering | 30.4 | 69.6 | 13.3 | 86.7 | 20.8 | 79.2 |
| Civil engineering | 20.8 | 79.2 | 18.9 | 81.1 | 19.4 | 80.6 |
| Computer engineering | 34.5 | 65.5 | 17.4 | 82.6 | 20.8 | 79.2 |
| Electrical, electronics, and communications engineering | 31.9 | 68.1 | 24.1 | 75.9 | 25.6 | 74.4 |
| Mechanical engineering | 39.1 | 60.9 | 14.0 | 86.0 | 18.9 | 81.1 |
| Industrial engineering | 21.5 | 78.5 | 19.0 | 81.0 | 19.9 | 80.1 |
| Other engineering[a] | 17.5 | 82.5 | 17.3 | 82.7 | 17.3 | 82.7 |

SOURCE: US Department of Education (2014a).

[a] The "Other engineering" category includes agricultural engineering; architectural engineering; ceramic sciences and engineering; engineering mechanics; engineering physics; engineering science; environmental/environmental health engineering; materials engineering; metallurgical engineering; mining and mineral engineering; naval architecture and marine engineering; nuclear engineering; ocean engineering; petroleum engineering; systems engineering; textile sciences and engineering; polymer/plastics engineering; construction engineering; forest engineering; manufacturing engineering; operations research; surveying engineering; geological/geophysical engineering; paper science and engineering; electromechanical engineering; mechatronics, robotics, and automaton engineering; biochemical engineering; engineering chemistry; and biological/biosystems engineering.

TABLE 4.5   Subfield distribution of engineering bachelor's degrees awarded to African American women and all women by institutional pathway, 2011–2012

| | African American Women (%) | | All Women (%) |
|---|---|---|---|
| Field | HBCUs | PWIs | All Institutions |
| Engineering, total | 100 | 100 | 100 |
| Aerospace, aeronautical and astronautical engineering | 0.9 | 2.0 | 2.9 |
| Biomedical/medical engineering | 0 | 11.9 | 11.5 |
| Chemical engineering | 19.2 | 14.8 | 14.9 |
| Civil engineering | 12.2 | 15.5 | 17.9 |
| Computer engineering | 9.4 | 6.0 | 2.6 |
| Electrical, electronics, and communications engineering | 20.2 | 14.4 | 8.3 |
| Mechanical engineering | 21.1 | 11.0 | 16.1 |
| Industrial engineering | 6.6 | 8.0 | 6.8 |
| Other engineering[a] | 10.3 | 16.3 | 18.7 |

SOURCE: US Department of Education (2014a).

[a]The "Other engineering" category includes agricultural engineering; architectural engineering; ceramic sciences and engineering; engineering mechanics; engineering physics; engineering science; environmental/environmental health engineering; materials engineering; metallurgical engineering; mining and mineral engineering; naval architecture and marine engineering; nuclear engineering; ocean engineering; petroleum engineering; systems engineering; textile sciences and engineering; polymer/plastics engineering; construction engineering; forest engineering; manufacturing engineering; operations research; surveying engineering; geological/geophysical engineering; paper science and engineering; electromechanical engineering; mechatronics, robotics, and automaton engineering; biochemical engineering; engineering chemistry; and biological/biosystems engineering.

of engineering bachelor's degrees while enrolling 15.8% of African American undergraduate men at four-year institutions (US Department of Education 2014a, 2014b).

HBCUs' impact is greater in some engineering fields than in others (see table 4.4). They awarded 30.4% of chemical engineering, 34.5% of computer engineering, 31.9% of electrical engineering, and 39.1% of mechanical engineering bachelor's degrees earned by African American women in 2011. The impact of HBCUs is also seen for African American men in electrical engineering, which awarded 24.1% of electrical engineering bachelor's degrees earned by this group in 2011. HBCUs awarded 18.9% of civil and 19.0% of industrial engineering bachelor's degrees received by African American men in 2011.

For all engineering degrees awarded by HBCUs, electrical engineering dominates for both men and women, though to a greater extent for men (25.8% of all engineering degrees awarded by HBCUs for women [see table 4.5] vs. 30.8% of all engineering degrees awarded by HBCUs for men [see table 4.6] in 2011–2012). For African American men, electrical engineering is also the leading field for all engineering degrees awarded by non-HBCUs (25.5%; see table 4.6); for women the leading field for degrees awarded by non-HBCUs is "other engineering" (26.5%; see table 4.5).* African American men are distributed differently across fields of engineering than men and women of all races/ethnicities. Among all men, mechanical engineering dominates among bachelor's degrees; for African American men the dominant field choice is electrical engineering. Among African American women and women from other racial/ethnic groups, the main field choice is "other engineering." Interestingly, electrical engineering is the only degree field option offered by all 14 HBCU accredited programs.

HIGH-PERFORMING INSTITUTIONS

Institutional-level data reveal which HBCUs act as pathways for a significant number of African Americans in engineering (table 4.7; US Department of Education 2014a). Eight of the top 25 bachelor's degree producers are HBCUs, while the remaining 17 are PWIs. Despite the relative disparities in resources—financial, facilities, human, research support, and otherwise—between HBCUs and these large engineering programs at PWIs, HBCUs were 6 of the top 10 producers of engineering bachelor's degree recipients in 2011–2012. Interestingly, among those PWIs that rank in the top 10 producers of African American engineers, all but one have engineering dual-degree programs. These engineering dual-degree programs enable students to complete their engineering degree across two institutions: an

* The "other engineering" category includes agricultural engineering; architectural engineering; ceramic sciences and engineering; engineering mechanics; engineering physics; engineering science; environmental/environmental health engineering; materials engineering; metallurgical engineering; mining and mineral engineering; naval architecture and marine engineering; nuclear engineering; ocean engineering; petroleum engineering; systems engineering; textile sciences and engineering; polymer/plastics engineering; construction engineering; forest engineering; manufacturing engineering; operations research; surveying engineering; geological/geophysical engineering; paper science and engineering; electromechanical engineering; mechatronics, robotics, and automaton engineering; biochemical engineering; engineering chemistry; and biological/biosystems engineering.

TABLE 4.6 Subfield distribution of engineering bachelor's degrees awarded to African American men and all men by institutional pathway, 2011–2012

| Field | African American Men (%) | | All Men (%) |
| | HBCUs | PWIs | All Institutions |
|---|---|---|---|
| Engineering, total | 100 | 100 | 100 |
| Aerospace, aeronautical, and astronautical engineering | 1.0 | 2.6 | 4.5 |
| Biomedical/medical engineering | 0 | 6.1 | 4.2 |
| Chemical engineering | 10.4 | 9.7 | 7.3 |
| Civil engineering | 14.5 | 14.6 | 16.1 |
| Computer engineering | 9.8 | 9.0 | 6.8 |
| Electrical, electronics, and communications engineering | 29.4 | 20.8 | 15.6 |
| Mechanical engineering | 18.4 | 19.2 | 28.1 |
| Industrial engineering | 6.0 | 5.9 | 3.5 |
| Other engineering[a] | 10.4 | 12.1 | 14.0 |

SOURCE: US Department of Education (2014a).

[a]The "Other engineering" category includes agricultural engineering; architectural engineering; ceramic sciences and engineering; engineering mechanics; engineering physics; engineering science; environmental/environmental health engineering; materials engineering; metallurgical engineering; mining and mineral engineering; naval architecture and marine engineering; nuclear engineering; ocean engineering; petroleum engineering; systems engineering; textile sciences and engineering; polymer/plastics engineering; construction engineering; forest engineering; manufacturing engineering; operations research; surveying engineering; geological/geophysical engineering; paper science and engineering; electromechanical engineering; mechatronics, robotics, and automaton engineering; biochemical engineering; engineering chemistry; and biological/biosystems engineering.

HBCU and a PWI. Typically, students fulfill their general education requirements at the HBCU and complete their advanced engineering coursework at the partnering PWI. Though Integrated Postsecondary Education Data System (IPEDS) completion data do not indicate the proportion of graduates who received their degrees by way of dual-degree programs, the distribution of top institutions for production of engineering bachelor's degrees to African Americans is suggestive: Georgia Institute of Technology ranked second among all institutions in award of bachelor's degrees to African Americans in 2011–2012, North Carolina State University ranked sixth, Auburn University ranked seventh, and the University of Florida ranked ninth (see table 4.7). All these institutions have dual-degree engineering programs in states where there is a significant presence of HBCUs.

TABLE 4.7  Top 26 African American engineering bachelor's degree producers, 2011–2012

| Rank | Institution | HBCU | Dual Degree | Number of Engineering Bachelor's Degrees Awarded | | |
| --- | --- | --- | --- | --- | --- | --- |
| | | | | Women | Men | Total |
| 1 | North Carolina A&T University | X | | 55 | 82 | 137 |
| 2 | Georgia Institute of Technology, main campus | | X | 27 | 70 | 97 |
| 3 | Alabama A&M University | X | | 17 | 53 | 70 |
| 4 | Prairie View A&M University | X | | 46 | 22 | 68 |
| 5 | Morgan State University | X | | 13 | 49 | 62 |
| 6 | North Carolina State University at Raleigh | | X | 18 | 41 | 59 |
| 7 | Auburn University | | X | 12 | 36 | 48 |
| 8 | Southern University and A&M College | X | | 9 | 38 | 47 |
| 9 | University of Florida | | X | 14 | 32 | 46 |
| 10 | Howard University | X | | 21 | 22 | 43 |
| 10 | Massachusetts Institute of Technology | | | 16 | 27 | 43 |
| 12 | Rutgers University, New Brunswick | | | 14 | 28 | 42 |
| 12 | University of Maryland, College Park | | X | 7 | 35 | 42 |
| 14 | Tuskegee University | X | | 17 | 21 | 38 |
| 15 | CUNY City College | | | 5 | 31 | 36 |
| 16 | Clemson University | | X | 9 | 24 | 33 |
| 16 | University of Central Florida | | | 2 | 31 | 33 |
| 18 | Ohio State University, main campus | | | 8 | 24 | 32 |
| 19 | Florida A&M University | X | | 10 | 21 | 31 |
| 19 | University of Virginia, main campus | | | 18 | 13 | 31 |
| 21 | University of Minnesota, Twin Cities | | | 5 | 25 | 30 |
| 21 | Columbia University | | X | 12 | 17 | 29 |
| 23 | Purdue University, main campus | | | 10 | 19 | 29 |
| 24 | Missouri University of Science and Technology | | X | 8 | 20 | 28 |
| 25 | New Jersey Institute of Technology | | X | 6 | 21 | 27 |
| 25 | University of Pittsburgh, Pittsburgh campus | | X | 9 | 18 | 27 |

SOURCE: US Department of Education (2014a).
NOTE: Dual-degree program information collected from institutional websites.

We highlight the contribution of PWIs with dual-degree engineering programs to underscore the overall contribution of HBCUs to engineering degree production among African Americans. The prevalence of PWIs that partner with HBCUs to offer dual-degree programs among the top engineering-degree-granting institutions points to a "hidden" HBCU contribution in these numbers (see also chaps. 5 and 13 in this volume). Further, the data suggest that partnerships can be effective in increasing the number of African American engineering degree recipients, as such dual-degree programs simultaneously provide students with access to a supportive HBCU institutional environment on the one hand and engineering program variety on the other.

## HBCUs AS BACCALAUREATE-ORIGIN INSTITUTIONS

HBCUs also play an important role in the education of African Americans who go on to earn a doctorate in engineering. According to a recent *Info-Brief* published by the National Science Foundation (Fiegener and Proudfoot 2013), 5 of the top 10 baccalaureate-origin institutions of African American engineering doctorate recipients who earned the PhD between 2002 and 2011 were HBCUs (see table 4.8).

In table 4.8, we also present the engineering doctorate production rate, which reflects the number of engineering doctorate recipients per 100 bachelor's degrees awarded by the baccalaureate-origin institution eight years earlier. Some (e.g., Fiegener and Proudfoot 2013) have argued that such a measure of "yield" more accurately reflects the contribution of undergraduate institutions to doctoral degree production compared to raw counts of doctorate recipients who attended those institutions as undergraduates. Similar to Fiegener and Proudfoot (2013), we find that the doctorate production rate of African Americans in engineering is lower for HBCUs than for many highly selective research universities. However, we note that HBCUs serve a very different student population than the most selective US research universities. On average, undergraduates at HBCUs enter college with lower SAT scores, are less likely to have taken rigorous college preparatory curricula (e.g., Advanced Placement), and are more commonly from low-income families than the most selective research universities (National Center for Education Statistics 2007). Thus, while raw numbers are an imperfect measure of the role that HBCUs play as baccalaureate-origin institutions for African Americans in engineering, yield ratios are also imperfect because they fail to account for student-level differences and the "value-added" by the undergraduate institution.

TABLE 4.8 Top 10 baccalaureate-origin institutions of African American engineering doctorate recipients, 2002–2011

| Rank | Institution | HBCU | Dual Degree | Number of Doctorate Recipients | Engineering Doctorate Production Rate[a] (%) |
|------|-------------|------|-------------|-------------------------------|---------------------------------------------|
| 1 | North Carolina A&T University | X | | 48 | 2.9 |
| 2 | Florida A&M University | X | | 37 | 4.2 |
| 3 | Morgan State University | X | | 32 | 4.2 |
| 4 | Brown University | | | 28 | 47.5 |
| 5 | Georgia Institute of Technology, main campus | | X | 25 | 1.9 |
| 6 | Howard University | X | | 22 | 4.2 |
| 7 | Tuskegee University | X | | 20 | 2.3 |
| 8 | University of Michigan, Ann Arbor | | X | 18 | 3.6 |
| 9 | University of Maryland, College Park | | X | 16 | 3.9 |
| 9 | University of Maryland, Baltimore County[b] | | | 16 | 18.4 |

SOURCE: Adapted from Fiegener and Proudfoot (2013). Data from NSF Survey of Earned Doctorates.

[a]Engineering doctorate production rate is defined as the number of engineering doctorate recipients per 100 bachelor's degrees awarded in engineering eight years earlier. The time lag of eight years reflects the median time for engineering doctorate completion.

[b]Although UMBC does not have a formal dual-degree program, we note the presence of the Meyerhoff Scholars Program, which is a unique, nationally recognized support program for underrepresented students in STEM which prepares these students to complete STEM doctorates. (For more details, see chap. 14 in this volume.)

The data above clearly illustrate that, despite the declining share of total African American enrollment, HBCUs continue to play an important role in granting engineering bachelor's degrees and by acting as institutional pathways to engineering doctoral programs for African Americans. Many HBCUs outperform a number of predominantly White research universities in terms of the number of engineering bachelor's degrees granted to African Americans, the absolute number of African American engineering bachelor's degree recipients who go on to earn an engineering doctorate, and the engineering production rate among African Americans. For African American women in particular, HBCUs play a critical role in broadening participation in engineering subfields that are traditionally dominated by men.

## HBCUs: Promoting Retention and Success in Engineering

An additional factor that limits access to engineering degrees among African Americans is the failure of higher education institutions to retain students from initial interest in engineering through degree completion and to graduate programs. This is true for students from all racial and ethnic groups; however, lower initial enrollment rates among African Americans make this failure to retain especially damaging to efforts to diversify engineering. Figure 4.1 illustrates the drop-off between the proportion of African American women and men who initially indicate an intention to major in engineering and the proportion of first-time, first-year undergraduates at four-year institutions who actually enroll in engineering programs. There is a subsequent decline in the proportion of all African American female and male undergraduates in engineering programs and, finally, the proportion of African American bachelor's degree recipients who earn their degree in engineering (see fig. 4.1; National Science Foundation 2013; US Department of Education 2014a, 2014b).

As the data we present above illustrate, PWIs do not perform as well as HBCUs in promoting access and success in engineering among African Americans. There are a number of factors that have been identified to explain the success of HBCUs in granting engineering degrees to African Americans. We offer a brief discussion of these below, as the approaches and strategies that HBCUs employ can be instructive to all institutions that wish to promote success among African Americans.

For quite some time, it has been recognized that HBCUs award a disproportionately high number of degrees in engineering and other STEM fields (Wenglinsky 1996; Wolf-Wendel 1998; Allen and Jewell 2002; Fiegener and Proudfoot 2013). HBCUs also rank highly as baccalaureate-origin institutions of STEM doctorates from their target populations (Burrelli and Rapoport 2008; Fiegener and Proudfoot 2013). As a result, researchers and advocates have studied these institutions in an attempt to understand the factors that make them "exemplars" in facilitating success in STEM among African Americans (e.g., Allen 1992; Wenglinsky 1996; Perna et al. 2009). A review of the literature suggests that the effectiveness of HBCUs in facilitating success in engineering and other STEM fields originates from their (1) institutional priorities and mission, (2) effective faculty and institutional practices, (3) supportive campus cultures, and (4) STEM-specific resources.

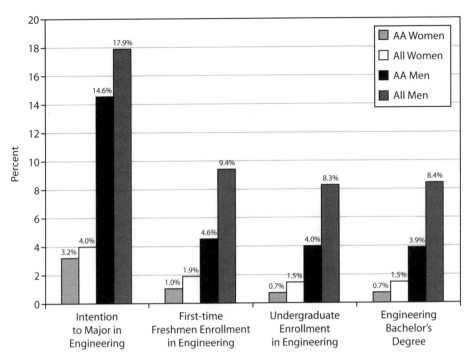

FIGURE 4.1 Intention to major, undergraduate enrollment, and bachelor's degree attainment in engineering among African Americans and all students by sex, 2011
Sources: National Science Foundation (2013); US Department of Education (2014a, 2014b).

INSTITUTIONAL MISSION AND COMMITMENT

Guided by their historical missions, many HBCUs demonstrate a strong commitment to educating and preparing future African American scientists and engineers (Allen 1992). This institutional commitment leads to intentional action to help students overcome barriers that may otherwise hinder degree attainment in STEM fields. For example, in spite of enrolling students who, on average, have lower levels of academic preparation (Allen 1992; Kim 2002) and are more economically disadvantaged (Perna 2001), HBCUs graduate higher proportions of their students in STEM fields than do PWIs (Perna et al. 2009). Further, African American STEM bachelor's degree holders who graduate from HBCUs are more likely to have earned that degree in physical science, engineering, and other math-intensive fields than their counterparts at PWIs (US Department of Education 2014a). HBCUs seem to have prioritized serving African American students with a wide range of academic preparation and facilitating their participation in STEM fields.

## EFFECTIVE FACULTY CAPABILITIES AND PRACTICES

Faculty at HBCUs have also been shown to contribute to these institutions' ability to promote STEM achievement among their target populations. For example, African American faculty, in higher proportions at HBCUs, act as mentors, or mirrors of students' possible selves (Markus and Nurius 1986), and actively encourage their students to pursue STEM fields (Perna et al. 2009). African American STEM majors at HBCUs also benefit from more faculty interaction at these institutions (Allen 1992), which is believed to positively impact student persistence and degree attainment (Astin 1977, 1993). Student-faculty ratios are lower at HBCUs (Kim 2002), which facilitates student-faculty contact. Black students at HBCUs also benefit from more nurturing student-faculty relations at HBCUs (Allen 1987) and report that faculty acknowledge the many psychological, academic, and financial barriers they face on the pathway to a STEM degree (Perna et al. 2009). Particularly for those students majoring in STEM fields, the "role models in excess" among faculty at HBCUs positively impact the educational aspirations and graduate school enrollment of African American students (Brazziel and Brazziel 1997).

## SUPPORTIVE INSTITUTIONAL CULTURE

Many researchers attribute the success of HBCUs in facilitating STEM degree attainment among African Americans to the supportive campus cultures and environments of these institutions. For example, in addition to experiencing less social isolation, dissatisfaction, and racism than their counterparts at PWIs (Perna 2001; Jones, Castellanos, and Cole 2002; Pascarella and Terenzini 2003), African Americans at HBCUs provide a social environment that is more caring, nurturing, and supportive than at non-HBCUs (Fleming 1984; Nettles, Thoeny, and Gosman 1986; Blackwell 1998; Redd 1998; Wagener and Nettles 1998). The culture of STEM programs at HBCUs has also been found to be much more collaborative (Perna et al. 2009), compared to the hypercompetitive culture more common to STEM programs at PWIs.

HBCUs have also been found to facilitate participation and attainment in STEM among their target populations owing to the STEM-specific resources available to their students. For example, chapters of science and engineering professional organizations intended for African Americans (e.g., the National Society of Black Engineers) are more commonly found at HBCUs (Brazziel and Brazziel 1997), thereby increasing opportunities for community building and networking among African American STEM majors. These institutions also offer a variety of forms of academic support to STEM majors, including tutoring and mentoring. They also offer access to undergraduate research opportunities, either on campus or via institutional partnerships, which have been shown to positively impact STEM degree attainment (Perna et al. 2009).

In summary, previous research has identified a number of factors that contribute to the effectiveness of HBCUs in encouraging participation and promoting success in engineering and other STEM fields among African Americans. Though we highlight the success of HBCUs in granting engineering degrees to African Americans, we recognize that there are a number of PWIs that also produce positive outcomes. These PWIs have been able to acquire a critical mass of minority students in engineering by incorporating many of the characteristics and effective practices of HBCUs.

In spite of the success of HBCUs and some PWIs in facilitating access to engineering, the data that we present in this chapter illustrate that African Americans' participation and degree attainment in engineering are not keeping pace with overall participation in higher education. Though we have focused our attention on African Americans in engineering, it is important to note that other minoritized populations—Latinos, Native Americans, Southeast Asians, and Pacific Islanders—experience similar inequitable patterns of participation in engineering. As we have discussed, there are multiple factors that contribute to this disturbing trend, and a comprehensive approach is needed to reverse it and increase African Americans' and other minoritized groups' participation in engineering. Such an approach includes changes in federal, state, and institutional policies; improvements in educational practice; and rigorous research and evaluation to acquire a deeper understanding of the ways in which institutions, programs, and individual faculty members can better promote access and success in engineering. In the following section we offer recommendations in each of these areas.

# Recommendations

Recognizing that currently the overwhelming majority of African American students are enrolled in PWIs, we recommend a comprehensive strategy to support research and development on factors that would need to be incorporated to enable institutions to be "minority-serving." Implementation research would also be needed to determine the necessary and sufficient conditions to effect change under different types of institutional arrangements. In the case of two-year colleges this would mean more attention to conditions that support alignment with four-year transfer institutions, including faculty interactions to ensure course alignment and equivalence. We note that some institutions have established dual-degree programs with HBCUs which provide African American students access to engineering degrees by way of liberal arts programs. Sending institutions, often liberal arts colleges, need support for developing core local infrastructure (e.g., advising systems, core course design) that is key to successful dual-degree relationships. We need to understand the factors that support implementation of successful dual-degree programs, especially in terms of the role of faculty, as well as factors that contribute to the failure of such relationships.

More research is needed on the policies, programs, and practices of institutions (both HBCUs and PWIs) that are successful in attracting and graduating African American students in engineering programs, in developing effective transfer programs, in establishing effective partnerships, in providing early work experiences, and in encouraging students to pursue graduate degrees with a focus on research for future faculty and practice for those who will enter the workforce. A "natural history" of institutional and departmental programs is needed to understand the conditions under which such successes have been achieved. We need to understand, share, and celebrate the practices of faculty who have been recognized as especially effective mentors (e.g., such as through receipt of the AAAS Mentor Awards, which recognize faculty who successfully mentor underrepresented students to a PhD in STEM).

We recommend that research be undertaken to understand the effect of changes in the judicial and legal landscape on the slowing of progress in enrolling, retaining, and graduating African American students in engineering. This would mean an institution-by-institution, department-by-department review of African American student enrollment and completion data, noting critical incidents that might have contributed to discontinuities or downward or upward trends.

In light of the "overperformance" of HBCUs in producing engineering graduates compared with the share of African Americans they enroll, we recommend providing more resources to these institutions to enable them to support more programs and more students in engineering. This would include resources for financial aid, support for partnerships with national labs and other engineering-intensive industries, and more career exploration opportunities, such as internships. This might also include institutional partnerships, two-way faculty and student exchanges with PWIs, and US and international research collaborations. In light of the tremendous resources required to start new programs, the cost-effectiveness of various capacity-building arrangements would need to be assessed.

Many African American students enter PWIs with the intention of majoring in engineering. Some institutions successfully support these students' aspirations through systems of course support, faculty engagement, research and design opportunities, financial aid, and encouragement. However, this is not the case in most institutions, as evidenced by the low output of African American students by many engineering programs. We urge that, through regular processes of internal and external review (e.g., visiting committees), accreditation, and grant consideration (e.g., site visits for large grants), institutions be assessed and held accountable for better performance in recruiting, enrolling, supporting, and retaining these students and mentoring them toward success.

## RECOMMENDATIONS FOR PRACTICE

We recommend support for research-based initiatives aimed at the precollege level to increase the size of the pool of "engineering-eligible" African American students. This includes both school-based and informal education efforts to increase the readiness of students in mathematics, communications, and science. While the focus of this chapter is on college pathways, we recognize that precollege preparation can broaden or narrow the postsecondary options for students, affect time to degree, affect cost of degree (such as when a need exists to take noncredit remedial courses), and thereby affect program persistence and success.

In light of the large percentage of women among college-going African Americans, any initiatives to increase engineering output from African Americans must include strategies to engage women in these fields. This would include research to obtain a better understanding of career choices and decisions regarding college majors for African American women. Clearly, efforts are needed to help these young women (as well as their parents) understand where an engineering connection might be made with their interests and career goals. It is not clear whether an explicit connection to solving problems, global or based in communities, rather than more traditional subfield recruitment, would be more effective as a motivating tool. African American women would also benefit from learning how people used their engineering education, including a focus on individuals who have used engineering to pursue a wide range of careers, including as preprofessional preparation.

Whatever the rate of entering programs, African American women and men are not being retained in engineering and are not successfully completing degrees at the same rate. More information is needed on the "points of loss" across the engineering programs. It is important for faculty to determine whether particular patterns of course taking or particular courses are bottlenecks contributing to these losses. Course redesign by faculty will likely be needed to address such structural issues within the curriculum. Institutional, college, and departmental resources should be provided to introduce faculty to more effective, research-based pedagogical models. This might involve funding of teaching initiatives to assist faculty in implementing such design and teaching improvement models. Such a strategy would pertain to all institutions, HBCUs and PWIs alike, and the outcomes of successful implementation would likely accrue to all students as well.

We recommend that engineering faculty work to build and/or expand professional networks that will support their interest in identifying and recruiting a diverse pool of students and faculty. Establishing relationships of mutual trust might allow for student referrals outside the traditional admissions reliance on standardized test scores, including being able to provide insights as to motivation, resilience, leadership, and creativity.

In spite of the success of many HBCUs in promoting participation in engineering degree programs, the range of program offerings in certain engineering subfields remains limited owing in part to the continued under-resourcing of these institutions. We recommend that HBCUs, perhaps in partnership with industry, professional associations, and other higher education institutions, engage in more experimentation regarding the role of technology in expanding engineering program capacity in HBCUs. For example, HBCUs might consider further using online coursework, massive online open courses (MOOCs), and other forms of educational technology to provide a wider range of courses to its engineering students. In order for this recommendation to be implemented, additional resources would be required to enhance the technological infrastructure at HBCUs.

## Conclusion

In this chapter we explore the higher education pathways to engineering being traveled by African American women and men. Understanding these pathways is necessary to effecting changes in levels of participation in these fields that are so vital to US economic growth and national security. The underperformance of most engineering education programs stands in sharp contrast to the overperformance of HBCUs in engineering degree production. Yet promising practices suggest strategies for improvement of outcomes.

Affecting the participation levels of women is essential to increased production of engineering degrees for African Americans. In light of their predominance among African Americans in higher education, it is critical to improve our knowledge of what works to support their career interest and choice of major in engineering and how this might be expanded.

It is clear that some pathways to engineering for African Americans are more effective and efficient than others. It is also the case that some institutions (and/or departments within institutions) do a better job than others in supporting student success and degree production in engineering. Policy makers, as well as students and their families, need this information to help guide investment.

Institutional leadership at HBCUs and PWIs plays a critical role in creating the conditions for success in engineering for African Americans. Boards of trustees, college and university presidents and provosts, engineering school deans, and departmental chairs shape the values and priorities of their institutions and can continuously reinforce the value of broadening participation in engineering. Further, institutional leadership has the ability to ensure that resources and reward structures are aligned with the goal of increasing success in engineering for African Americans and other underrepresented populations.

Faculty are key actors in supporting success by African American students in engineering. Whether through personal engagement, encouragement, provision of research, and practice opportunities or through course redesign and improvement in teaching, they remain the gatekeepers and facilitators along the pathways to engineering.

# References

Adelman, Clifford. 2006. *The Toolbox Revisited: Paths to Degree Completion from High School through College*. Washington, DC: US Department of Education.

Allen, Walter R. 1987. "Black Colleges vs. White Colleges: The Fork in the Road for Black Students." *Change* 19:28–34.

Allen, Walter R. 1992. "The Color of Success: African-American College Student Outcomes at Predominantly White and Historically Black Public Colleges and Universities." *Harvard Education Review* 62 (1): 26–44.

Allen, Walter R., and Joseph O. Jewell. 2002. "A Backward Glance Forward: Past, Present, and Future Perspectives on Historically Black Colleges and Universities." *Review of Higher Education* 25 (3): 241–261.

Anderson, James D. 1984. "The Schooling and Achievement of Black Children: Before and after *Brown v. Topeka*, 1900–1980." In *The Effects of School Desegregation on Motivation and Achievement*, edited by David E. Bartz and Martin L. Maehr, 103–121. Greenwich, CT: JAI Press.

Anderson, James D. 2002. "Race in American Higher Education: Historical Perspectives on Current Conditions." In *The Racial Crisis in American Higher Education: Continuing Challenges for the Twenty-First Century*, edited by William A. Smith, Philip G. Altbach, and Kofi Lomotey, 3–41. Albany: State University of New York Press.

Astin, Alexander W. 1977. *Four Critical Years*. San Francisco: Jossey-Bass.

Astin, Alexander W. 1993. *What Matters in College? Four Critical Years Revisited*. San Francisco: Jossey-Bass.

Blackwell, James E. 1998. "Faculty Issues: The Impact on Minorities." *Review of Higher Education* 11 (4): 417–434.

Bowen, William G., and Derek Bok. 1998. *The Shape of the River*. Princeton, NJ: Princeton University Press.

Brazziel, William F., and Marian E. Brazziel. 1997. "Distinctives of High Producers of Minority Science and Engineering Doctoral Starts." *Journal of Science Education and Technology* 6 (2): 143–153.

Burrelli, Joan, and Alan Rapoport. 2008. *Role of HBCUs as Baccalaureate-Origin Institutions of Black S&E Doctorate Recipients*. NSF 08-319. Arlington, VA: National Science Foundation.

Deming, David J., Claudia Goldin, and Lawrence F. Katz. 2012. "The For-Profit Postsecondary School Sector: Nimble Critters or Agile Predators?" *Journal of Economic Perspectives* 26 (1): 139–164.

Esters, Lorenzo I., and David C. Mosby. 2007. "Disappearing Acts: The Vanishing Black Male on Community College Campuses." *Diverse: Issues in Higher Education*, August 23. http://diverseeducation.com/article/9184/.

Fiegener, Mark K., and Steven L. Proudfoot. 2013. *Baccalaureate Origins of U.S.-trained S&E Doctorate Recipients*. NSF 13-323. Arlington, VA: National Center for Science and Engineering Statistics, National Science Foundation.

Fleming, Jacqueline. 1984. *Blacks in College*. San Francisco: Jossey-Bass.

Gasman, Marybeth. 2013. *The Changing Face of Historically Black Colleges and Universities*. Philadelphia: Center for Minority Serving Institutions, Graduate School of Education, University of Pennsylvania.

Harper, Shaun R., and Frank Harris III. 2012. *Men of Color: A Role for Policymakers in Improving the Status of Black Male Students in U.S. Higher Education*. Washington, DC: Institute for Higher Education Policy.

Hill, Catherine, Christianne Corbett, and Andresse St. Rose. 2010. *Why So Few? Women in Science, Technology, Engineering and Mathematics*. Washington, DC: American Association of University Women.

Ignash, Jan M., and Barbara K. Townsend. 2001. "Statewide Transfer and Articulation Policies: Current Practices and Emerging Issues." In *Community Colleges: Policy in the Future Context*, edited by Barbara K. Townsend and Susan B. Twombly, 173–192. Westport, CT: Ablex.

Jackson, Jerlando F. L., and James L. Moore III. 2008. "The African American Male Crisis in Education: A Popular Media Infatuation or Needed Public Policy Response." *American Behavioral Scientist* 51 (7): 847–853. doi:10.1177/0002764207311992.

Jones, Lee, Jeanett Castellanos, and Darnell Cole. 2002. "Examining the Ethnic Minority Student Experience at Predominantly White Institutions: A Case Study." *Journal of Hispanic Higher Education* 1:19–39.

Kim, Mikyong. M. 2002. "Historically Black vs. White Institutions: Academic Development among Black Students." *Review of Higher Education* 25 (4): 385–407.

Malcom-Piqueux, Lindsey E., and Shirley M. Malcom. 2013. "Engineering Diversity: Fixing the Educational System to Promote Equity." *Bridge* 43 (1): 24–34.

Markus, Hazel, and Paula Nurius. 1986. "Possible Selves." *American Psychologist* 41 (9): 954–969.

Mooney, Geraldine M., and Daniel J. Foley. 2011. *Community Colleges: Playing an Important Role in the Education of Science, Engineering, and Health Graduates*. NSF 11-317. Arlington, VA: National Center for Science and Engineering Statistics, National Science Foundation.

Mullin, Christopher M. 2012. *Transfer: An Indispensible Part of the Community College Mission*. AACC Policy Brief 2012-03PBL. Washington, DC: American Association of Community Colleges.

National Action Council for Minorities in Engineering. 2008. *Confronting the "New" American Dilemma—Underrepresented Minorities in Engineering: A Data-Based Look at Diversity*, by Lisa M. Frehill, Nicole M. DiFabio, and

Susan T. Hill. Washington, DC: National Action Council for Minorities in Engineering.

National Center for Education Statistics. 2007. *Characteristics of Minority-Serving Institutions and Minority Undergraduates Enrolled in These Institutions*. NCES 2008-156. Washington, DC: US Department of Education.

National Center for Education Statistics. 2012. *Digest of Education Statistics: 2011*. NCES 2012-001. Washington, DC: US Department of Education.

National Center for Education Statistics. 2013. Advance Release of Selected *2013 Digest of Education Statistics* Tables. Washington, DC: US Department of Education. http://nces.ed.gov/programs/digest/2013menu_tables.asp.

National Science Foundation. 2007. *Women, Minorities, and Persons with Disabilities in Science and Engineering: 2007*. NSF 07-315. Arlington, VA: Division of Science Resources Statistics, National Science Foundation.

National Science Foundation. 2013. *Women, Minorities, and Persons with Disabilities in Science and Engineering: 2013*. NSF 13-304. Arlington, VA: National Center for Science and Engineering Statistics, National Science Foundation.

Nettles, Michael T., A. Robert Thoeny, and Erica J. Gosman. 1986. "Comparative and Predictive Analyses of Black and White Students' College Achievement and Experiences." *Journal of Higher Education* 57:289–328.

Pascarella, Ernest T., and Patrick T. Terenzini. 2003. *How College Affects Students: A Third Decade of Research*. San Francisco: Jossey-Bass.

Perna, Laura W. 2001. "The Contribution of Historically Black Colleges and Universities to the Preparation of African Americans for Faculty Careers." *Research in Higher Education* 42 (3): 267–294.

Perna, Laura, Valerie Lundy-Wagner, Noah D. Drezner, Marybeth Gasman, Susan Yoon, Enakshi Bose, and Shannon Gary. 2009. "The Contribution of HBCUs to the Preparation of African American Women for STEM Careers: A Case Study." *Research in Higher Education* 50:1–23.

Redd, Kenneth E. 1998. "Historically Black Colleges and Universities: Making a Comeback." *New Directions for Higher Education* 102:33–43.

US Bureau of the Census. 1968. *Statistical Abstract of the United States: 1968*. Washington, DC: US Department of Commerce.

US Census Bureau. 2011. United States, B01001 Sex by Age [Data]. *2011 American Community Survey 1-Year Estimates*. http://factfinder2.census.gov.

US Department of Education, National Center for Education Statistics. 2014a. *Integrated Postsecondary Education Data System (IPEDS) Completion Survey*. http://nces.ed.gov/ipeds/datacenter/.

US Department of Education, National Center for Education Statistics. 2014b. *Integrated Postsecondary Education Data System (IPEDS) Enrollment Survey*. http://nces.ed.gov/ipeds/datacenter/.

Wagener, Ursula, and Michael T. Nettles. 1998. "It Takes a Community to Educate Students." *Change* 30 (2): 18–25.

Wenglinsky, Harold H. 1996. "The Educational Justification of Historically Black Colleges and Universities: A Policy Response to the U.S. Supreme Court." *Educational Evaluation and Policy Analysis* 18 (1): 91–103.

Wolf-Wendel, Lisa E. 1998. "Models of Excellence: The Baccalaureate Origins of Successful European American Women, African American Women, and Latinas." *Journal of Higher Education* 69 (2): 141–186.

# CHAPTER FIVE

# Clarifying the Contributions of Historically Black Colleges and Universities in Engineering Education

TAFAYA RANSOM

A T JUST 26% OF PARITY with their share of the US resident population, African Americans are the least well represented racial/ethnic group in the engineering workforce (table 5.1). Likewise, African Americans have consistently posted the weakest persistence and bachelor's degree completion rates of all racial/ethnic groups in engineering. The National Action Council for Minorities in Engineering (2012) reported that compared to all other racial/ethnic groups, Black engineering students are less likely to complete their degrees, take longer to complete their degrees, and more often transfer out of bachelor's degree programs into associate's degree or certificate programs. Other research confirms that African American engineering students have graduation rates substantially lower than all other groups (Georges 1999; Brown, Morning, and Watkins 2005; Morse and Babcock 2009). For example, Morse and Babcock (2009) reported a six-year graduation rate of 31% for African American engineering students, compared to 68% for nonminorities and 45% for Hispanics. Recent trends also indicate that African American postsecondary outcomes in engineering are worsening (National Science Board 2012). Therefore, identifying postsecondary institutional levers for improving African American engineering outcomes seems an important avenue for cultivating a stronger, more inclusive engineering workforce. Given their long record of educating African American engineers, historically Black colleges and universities (HBCUs) might be uniquely positioned to offer insights along these lines.

TABLE 5.1 Racial/ethnic distribution (percent) of US residential population, college graduates, S&E degree holders, S&E occupations, and engineering occupations, 2008

| Race/ethnicity | Total US Residential Population | College Degree Holders | S&E Degree Holders | S&E Occupations | Engineering Occupations | Engineering Parity (Occupation/ Population) |
|---|---|---|---|---|---|---|
| Asian | 4.7 | 8.5 | 11.2 | 16.9 | 15.6 | 3.3 |
| American Indian/Alaska Native | 0.7 | 0.3 | 0.4 | 0.3 | 0.2 | 0.29 |
| Black | 11.7 | 7.2 | 5.5 | 3.9 | 3.1 | 0.26 |
| Hispanic | 13.9 | 6.2 | 5.6 | 4.9 | 5.6 | 0.40 |
| White | 67.6 | 76.5 | 75.2 | 71.8 | 73.4 | 1.1 |
| Native Hawaiian/Other Pacific Islander | 0.1 | 0.1 | 0.4 | 0.4 | 0.6 | 6.0 |
| Two or more races | 1.2 | 1.1 | 1.7 | 1.7 | 1.6 | 1.3 |

SOURCE: Tabulations of data provided in the National Science Board's *Science and Engineering Indicators 2012*.
NOTE: S&E = science and engineering.

## HBCUs in STEM Higher Education

Descriptive reports commend HBCUs for a long-standing record of producing African American science, technology, engineering, and mathematics (STEM) baccalaureates, suggesting, "In almost every STEM field, HBCUs lead the nation's larger, better-equipped colleges in producing Black graduates" (Southern Education Foundation 2005, 5). While generalizations like this are debatable and some of the oft-quoted statistics about the role of HBCUs in STEM are certainly outdated,* HBCUs remain key players. HBCUs, which represented 3% of all four-year postsecondary institutions and enrolled roughly 16% of African American students in four-year institutions in 2010, conferred more than 33% of bachelor's degrees to African Americans in agricultural sciences, mathematical sciences, or physical sciences and 20% of bachelor's degrees to African Americans in engineering (NSF 2013).

That HBCUs produce a disproportionate share of STEM graduates is seen as remarkable for at least four reasons. First, HBCUs represent a small segment of STEM degree-granting institutions, for example, HBCUs make up only 17 (4%) of the nearly 400 institutions with ABET-accredited engineering programs (ABET 2013). Second, HBCU students have, on average, lower socioeconomic status backgrounds and lower high school GPAs and college entrance exam scores than their non-HBCU peers (Allen 1992; Kim 2002; Kim and Conrad 2006; Li and Carroll 2007), both of which have been linked to lower rates of STEM persistence and degree attainment (Elliott et al. 1996; Smyth and McArdle 2004). HBCUs also have lower institutional resources (i.e., proportions of faculty with doctorates, average faculty salaries, per-student instructional expenditures, and endowments) and lower STEM resources (e.g., research and development funding and infrastructure) than non-HBCUs (Kim 2002; Suitts 2003; Swail, Redd, and Perna 2003; Kim and Conrad 2006; Bennof 2009; Clewell, de Cohen, and Tsui 2010; Gasman et al. 2010; Matthews 2011).

A small body of research directly examines links between students' attendance at HBCUs and their science and engineering educational outcomes. Available studies suggest that African American students at HBCUs are more likely to choose STEM majors than those at non-HBCUs (Thomas 1987, 1991; Trent 1991; Trent and Hill 1994; Wenglinsky 1997). Georges's (1999) descriptive analysis of Engineering Workforce Commission data indicated

---

*For example, one recent article stated, incorrectly, that HBCUs "[confer] 40% of all STEM degrees and 60% of all engineering degrees earned by Black students" (Goode 2011).

that, on average, the retention rates of Blacks in engineering were higher at HBCUs than the national average for Black students. More recently, researchers affiliated with the University of California, Los Angeles (UCLA), Higher Education Research Institute (HERI) have found that the relationship between institutional selectivity and STEM persistence depended on HBCU status (Chang et al. 2008), Black engineering and computer science majors at HBCUs were less likely than those at non-HBCUs to switch to a non-STEM major (Newman 2011a), HBCU attendance was associated with higher four-year STEM degree completion rates among Black students (Hurtado, Eagan, and Hughes 2012), and HBCU attendance was positively related to graduate and professional degree aspirations among underrepresented minorities (Eagan 2010). Although these studies fall short of identifying the specific factors within HBCUs which might drive improved outcomes in STEM for African American students, the various findings point to the importance of institutional structures and contexts in shaping students' outcomes in STEM.

Additionally, a substantial literature considers the role of HBCUs as the institution of baccalaureate origin for African American STEM doctorate degree recipients (Pearson and Pearson 1985; Solórzano 1995; Leggon and Pearson 1997; Wolf-Wendel 1998; Wolf-Wendel et al. 2000; Burelli and Rapoport 2008; Hubbard and Stage 2010; Sibulkin and Butler 2011). This research consistently indicates that, in absolute terms, HBCUs are the baccalaureate-origin institutions of a disproportionate share of Black STEM doctorate recipients. But the use of different methods to account for the size of an institution's pool of potential doctorate recipients (for a review see Sibulkin and Butler 2011) has led to equivocal findings as to the relative significance of the role of HBCUs along these lines.

Other research examines the ways that HBCU environments might foster success in STEM for Black students. This research takes the form of case studies of HBCUs that demonstrate success in STEM (e.g., Culotta 1992; Brazziel and Brazziel 1997; Southern Education Foundation 2005; Perna et al. 2009) and comparisons between Black students' experiences in STEM at HBCUs and those at non-HBCUs (e.g., Wenglinsky 1997; Suitts 2003; Brown, Morning, and Watkins 2005; Lent et al. 2005; Fries-Britt, Younger, and Hall 2010). By and large, this work indicates that HBCUs provide supportive and affirming STEM environments, with cooperative rather than competitive peer climates (Hurtado et al. 2009; Perna et al. 2009; Fries-Britt, Younger, and Hall 2010). Likewise, this research suggests that HBCU STEM students tend to have more positive perceptions of their educational climates and experiences (Brown, Morning, and Watkins 2005), as well as higher self-efficacy and postbaccalaureate educational aspirations relative to African American STEM students at non-HBCUs (Lent et al. 2005).

Nevertheless, little research examines African American experiences and outcomes specifically in engineering (e.g., Good, Halpin, and Halpin 2002; Moore, Madison-Colmore, and Smith 2003; Brown, Morning, and Watkins 2005; Moore 2006; Slaughter 2009; Newman 2011a, 2011b; Ransom 2013). Even less research considers the role of HBCUs specifically in engineering (see, e.g., Gasman et al. 2010; Ransom 2013; Weinberger, forthcoming). Therefore, to lay the groundwork for developing understanding about the contributions of HBCUs to the engineering education enterprise, a comprehensive descriptive analysis of engineering degree production at HBCUs is warranted. Drawing on degree completion data from the US Department of Education National Center for Education Statistics' Integrated Postsecondary Education Data System (IPEDS) over the period 1990–2012, the chapter specifically addresses four research questions:

1. Over time, what has been the racial/ethnic makeup of engineering bachelor's, master's, and doctorate degree recipients at HBCUs?
2. What share of African American male and female engineering graduates earned their bachelor's, master's, or doctorate degrees from HBCUs over time?
3. To what extent are HBCUs represented among the top (institutional) producers of African American male and female engineers at all degree levels over time?
4. How has the engineering gender gap at HBCUs compared with the engineering gender gap at large over time?

All of the analyses utilize data on US citizens and permanent residents only. Temporary residents (i.e., nonresident aliens) are not included. Before reporting the results, a brief history of engineering education at HBCUs is presented.

## A Historical View of Engineering Education and HBCUs

From the founding of the Military Academy at West Point in 1802 up to the enactment of the first Morrill Act of 1862,* the growth of engineering edu-

---

*The Land Grant Colleges Act (commonly known as the Morrill Act) of 1862 and 1890 served to expand higher education in the United States, specifically funding institutions focused on "agricultural and mechanical arts." The Morrill Act of 1890 led to the establishment of Black land-grant institutions in every Southern state (Thelin 2004).

cation in the United States was slow. In the 10 years after the first Morrill Act, the number of engineering schools increased by 483% (from 12 to 70), curricula were standardized, and new disciplines and professional societies emerged (Grayson 1980). National accreditation for engineering education through ABET began in 1932 and helped to further unify engineering curricular structures and establish the profession. By 1940, accredited engineering schools in the United States numbered 125; by 1976, there were 234 (Grayson 1980).

Engineering education at HBCUs developed more gradually and more modestly. The first HBCU engineering programs were established at Howard University (1914) and North Carolina A&T (1920). Despite that nearly all of the White land-grant colleges operated engineering programs by 1928, Black land-grant colleges generally lacked the resources to do so. Along these lines, researchers have long documented the historic funding disparities between Black and White land-grant institutions—with state legislatures favoring White institutions in both state appropriations and the distribution of federal funds (Jenkins 1942; Craig 1991, 1992; Humphries 1991; Kujovich 1993). By 1963, six HBCUs had established engineering programs, yet only Howard had earned national accreditation (in electrical and civil engineering; Gibbs 1952; McGrath 1965; ABET 2013).

The consistent development of engineering education in the United States writ large is evidence of the steady demand for engineers throughout the twentieth century. Yet Wilburn (1974, 1148) noted that before the enactment of equal employment opportunity (EEO) policies, "no serious attempts [were] made to develop and utilize the scientific and technical talents of racial and ethnic minorities." With the 1970s began a period of rapid growth in Black engineering degree attainment. Weinberger (forthcoming) traced this growth to organized, well-funded efforts that were propelled by a coalition of corporations, professional organizations, and foundations. To expand African American participation in engineering, these engineering education stakeholders specifically leveraged HBCUs by expanding and improving existing engineering programs and by establishing dual-degree partnerships between engineering institutions and HBCUs without engineering programs. By 1979, HBCUs conferred 35% of all engineering bachelor's degrees earned by African Americans, based on tabulations of *IPEDS Completions Survey* data. Today, 17 HBCUs offer ABET-accredited engineering bachelor's degree programs, and 11 offer graduate programs (see table 5.2).

TABLE 5.2 HBCUs with engineering degree programs

| Institution | Location | ABET-Accredited BS Engineering Program | MS Engineering Program | PhD Engineering Program |
|---|---|---|---|---|
| Alabama A&M University* | Normal, AL | X | X | |
| Central State University | Wilberforce, OH | X | | |
| Florida A&M University* | Tallahassee, FL | X | X | X |
| Hampton University | Hampton, VA | X | | |
| Howard University | Washington, DC | X | X | X |
| Jackson State University | Jackson, MS | X | X | |
| Morgan State University | Baltimore, MD | X | X | X |
| Norfolk State University | Norfolk, VA | X | X | |
| North Carolina A&T State University* | Greensboro, NC | X | X | X |
| Prairie View A&M University* | Prairie View, TX | X | | |
| South Carolina State University* | Orangeburg, SC | X | | |
| Southern University and A&M College* | Baton Rouge, LA | X | X | |
| Tennessee State University* | Nashville, TN | X | X | X |
| Tuskegee University* | Tuskegee, AL | X | X | X |
| University of Maryland Eastern Shore* | Princess Anne, MD | X | | |
| University of the District of Colombia | Washington, DC | X | X | |
| Virginia State University* | Petersburg, VA | X | | |

SOURCE: Accredited programs searched via ABET.org and institution websites.
NOTE: An asterisk indicates an 1890 land-grant institution.

## Results of Analysis

This section of the chapter presents the results of an analysis of IPEDS degree completion data to clarify the role of HBCUs in American engineering education. First, I explore the racial/ethnic makeup of HBCU engineering graduates before turning to engineering degree production trends at HBCUs across gender and postsecondary levels. Then, I consider differences in the engineering gender gap at HBCUs relative to all institutions.

RACIAL/ETHNIC DIVERSITY IN HBCU ENGINEERING

Because most HBCUs primarily serve African American students, most HBCU research examines only the outcomes and experiences of Black students—often without explicitly considering whether other racial/ethnic subgroups warrant attention. In line with the overall objective to more fully clarify the contributions of HBCUs in engineering education, this study first offers a preliminary look at the racial/ethnic makeup of HBCU engineering graduates. To be sure, about three decades ago, HBCUs were nearly all Black (Gasman et al. 2010), suggesting no strong rationale for systematic consideration of other groups in research focused on HBCU students. However, researchers have recently begun highlighting the racial/ethnic diversity within HBCUs in general and in public HBCUs in particular (Gasman et al. 2010; Gasman 2013). Efforts to limit program duplication in previously segregated state systems of higher education have led to increasing enrollments of non-Black students at HBCUs, especially at the graduate level. For example, at 10 out of the 40 public, four-year HBCUs, non-Blacks made up 30% or more of graduate degree recipients in 2012. Broader demographic shifts in the US population have also contributed to growing racial/ethnic diversity within HBCUs. As of 2012, non-Blacks accounted for substantial shares of HBCU graduates across all fields and degree levels (see table 5.3).

Figure 5.1 indicates that HBCU engineering is more racially/ethnically diverse at the graduate level. The figure also shows clear gender differences in the racial/ethnic makeup of HBCU engineering cohorts. Particularly at the bachelor's and master's degree levels, African Americans make up larger shares of female than male HBCU engineering graduates. However, for both male and female bachelor's degree cohorts, the percentages of African Americans are trending downward. In the most recent 10 years shown (2002–2012), female engineering baccalaureates at HBCUs went from 98% to 86% Black, male baccalaureates from 90% to 80% Black. During the entire study period, at least one-quarter of male engineering master's degree earners at

TABLE 5.3    Percentage of African Americans in HBCU degree cohorts, all
fields, 2012

|  | All Four-year HBCUs (88 institutions) | Public Four-year HBCUs (40 institutions) |
|---|---|---|
| Bachelor's degrees | | |
| Number of degree recipients | 33,919 | 23,967 |
| % Black | 85.2 | 81.9 |
| Master's degrees | | |
| Number of degree recipients | 7,654 | 6,518 |
| % Black | 72.2 | 70.0 |
| Doctorate degrees | | |
| Number of degree recipients | 2,285 | 1,214 |
| % Black | 64.2 | 55.0 |

SOURCE: US Department of Education, National Center for Education Statistics, IPEDS
Data Center.

HBCUs were non-Black; in 2012 the figure was 39%. The racial/ethnic makeup of engineering doctorate degree cohorts has been more variable. All in all, reflecting patterns in HBCU enrollments across all fields, engineering degree cohorts at HBCUs are increasingly not "all Black," suggesting a need for further research to examine whether the "HBCU experience" similarly impacts students of different racial/ethnic backgrounds.

AFRICAN AMERICAN ENGINEERING DEGREE PRODUCTION AT HBCUs

**Trends in Bachelor's Degree Production**

Addressing the second research question, figure 5.2 shows that HBCUs, which make up 4% of ABET-accredited engineering programs, graduated a substantial share of African American engineering undergraduates throughout the 22-year period from 1990 to 2012. However, figure 5.2 also indicates that after peaking at nearly 30% in 1996, the share of engineering bachelor's degrees conferred to African Americans by HBCUs has been in decline, hovering around 19% in 2012—a 35% decline in the share of degrees awarded.

Also clear from figure 5.2 is that HBCUs have consistently educated greater shares of Black female than Black male engineering undergraduates. In fact, as many as 34% of all Black women engineering baccalaureates earned their degrees at HBCUs in 1997; this figure declined to 25% by

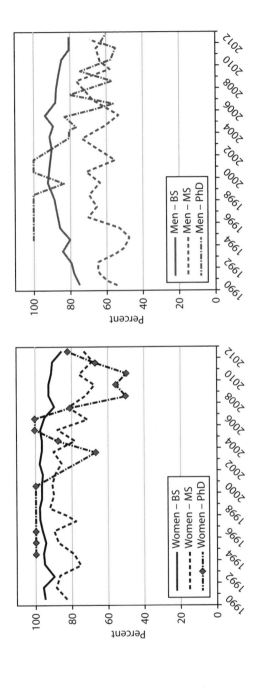

FIGURE 5.1 Percentage of African Americans in HBCU engineering degree cohorts by gender and degree level, 1990–2012. Tabulations based on number of degrees conferred to US citizens and permanent residents.

*Source:* US Department of Education, National Center for Education Statistics, IPEDS Completions Surveys drawn from National Science Foundation, National Center for Science and Engineering Statistics WebCASPAR System.

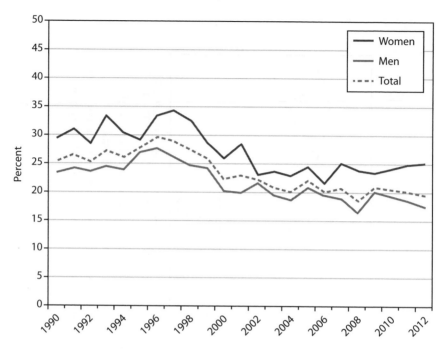

FIGURE 5.2 Share of engineering bachelor's degrees conferred to African Americans by HBCUs, 1990–2012. Tabulations based on number of degrees conferred to African American US citizens and permanent residents.
*Source:* US Department of Education, National Center for Education Statistics, IPEDS Completions Surveys drawn from National Science Foundation, National Center for Science and Engineering Statistics WebCASPAR System.

2012. Over the same period, the share of engineering bachelor's degrees conferred to Black men by HBCUs peaked at 27% in 1996 and fell to 17% by 2012. Despite clear declines, the role of HBCUs remains significant and disproportionate, since HBCUs accounted for just 14% of Black female and 16% of Black male undergraduates enrolled in four-year institutions in 2012 (see chap. 4 in this volume). Nevertheless, identifying the underlying reasons for the regressive trends in engineering degree production at HBCUs is beyond the scope of this descriptive study.

Declining shares of African American engineering undergrads earning their degrees from HBCUs might reflect a number of scenarios: (1) a growing pool of Black engineering undergraduates, of which the absolute number who choose to attend HBCUs has remained relatively flat; (2) a relatively stable pool of Black engineering undergraduates, of which declining numbers

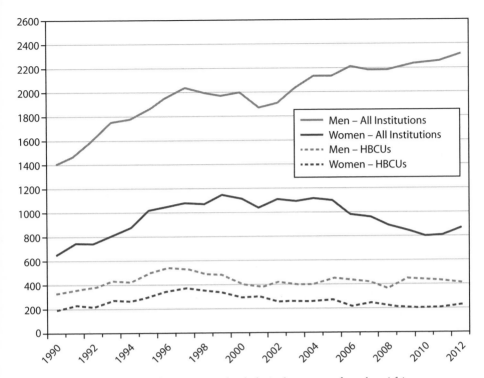

FIGURE 5.3 Number of engineering bachelor's degrees conferred to African Americans by all institutions and HBCUs by gender, 1990–2012. Tabulations based on number of degrees conferred to African American US citizens and permanent residents.
*Source:* US Department of Education, National Center for Education Statistics, IPEDS Completions Surveys drawn from National Science Foundation, National Center for Science and Engineering Statistics WebCASPAR System.

choose to attend HBCUs; (3) little change in enrollments and improved engineering retention at non-HBCUs and/or declining engineering retention at HBCUs; (4) changing demographics or characteristics of Black engineering students which exacerbate scenario 3; or (5) some combination of these.

Figure 5.3 offers additional context for considering the apparent decline in African American engineering degree production at HBCUs. In addition to declining shares of Black undergrads relative to all institutions, the figure shows that the absolute numbers of African American HBCU engineering baccalaureates are also down since the late 1990s. For example, since 1999, the number of engineering bachelor's degrees awarded to Black males declined 16% at HBCUs, even though the same period saw a 17% increase in engineering bachelor's degrees awarded to Black males at all institutions. On the other hand, since 1999, the number of engineering bachelor's degrees earned by Black females declined both at HBCUs (35%) and, though to a lesser extent, at all institutions (25%).

## Top Bachelor's Degree Producers

To further unpack the role of HBCUs in African American engineering bachelor's degree production, the next step in the analysis involved examining the extent to which HBCUs have ranked among the top producers of African American baccalaureate engineers. In particular, this analysis foregrounds the relative role of specific institutions and offers insights about how individual institutional patterns add up to the overall trend observed across institution types. Totaling the number of engineering bachelor's degrees awarded to African Americans by institution and gender over each decade during the study period (1990–2012), tables 5.4–5.6 list institutions that were among the top 10 engineering bachelor's degree producers in at least one decade, collectively and by gender. Table 5.4 indicates that five HBCUs have been fairly consistent top producers of African American engineering baccalaureates on the whole: North Carolina A&T, Prairie View A&M, Tuskegee, Southern, and Morgan State. But along gender lines (tables 5.5 and 5.6), Tuskegee ranked considerably higher in producing Black women engineers (#5) compared to Black men engineers (#13) by 2010–2012. Referring back to table 5.4, three HBCUs dropped substantially in rank of total degree production for African Americans: Howard, Florida A&M, and Tennessee State. But again, gender differences are clear, as Howard ranked much higher in producing Black female engineers (#6) than Black male engineers (#25) by 2010–2012 (tables 5.5 and 5.6). Only one HBCU, Alabama A&M, has risen substantially in rank of total bachelor's degrees awarded to African Americans. Alabama A&M achieved ABET accreditation for its bachelor's degree engineering programs in 1998, which partly explains its dramatic shift in rank (ABET 2013). And by gender, Alabama A&M ranked higher in producing Black male engineers (#5) compared to Black female engineers (#12) during 2010–2012.

While this analysis is helpful for clarifying the patterns evident in institutional degree completion data collected through IPEDS, the data are not rich enough to provide insights into why these patterns emerge. However, these findings could be leveraged to identify key institutions for future qualitative research investigating institutional policies, programs, or other

TABLE 5.4 Top engineering bachelor's degree producers for all African
Americans by decade, 1990–2012

| | | Rank | | |
| | | 1990– | 2000– | 2010– |
| Institution | HBCU | 1999 | 2009 | 2012 |
| --- | --- | --- | --- | --- |
| North Carolina A&T State University | X | 1 | 1 | 1 |
| Georgia Institute of Technology | | 2 | 2 | 2 |
| Prairie View A&M University | X | 3 | 7 | 4 |
| Tuskegee University | X | 4 | 8 | 10 |
| Howard University | X | 5 | 14 | 16 |
| Southern University and A&M College | X | 6 | 6 | 6 |
| North Carolina State University at Raleigh | | 7 | 4 | 3 |
| Morgan State University | X | 8 | 5 | 5 |
| Florida A&M University | X | 9 | 3 | 21 |
| CUNY City College | | 10 | 15 | 15 |
| University of Michigan–Ann Arbor | | 11 | 10 | 15 |
| Tennessee State University | X | 12 | 9 | 32 |
| University of Maryland–College Park | | 16 | 11 | 9 |
| University of Florida | | 21 | 12 | 8 |
| Alabama A&M University[a] | X | 109 | 18 | 7 |

SOURCE: US Department of Education, National Center for Education Statistics, IPEDS
Completions Surveys drawn from National Science Foundation, National Center for
Science and Engineering Statistics WebCASPAR System.
NOTES: Rank tabulations based on total number of degrees conferred during each decade
to African American US citizens and permanent residents. As seen from the table, there
were seven, seven, and six HBCUs ranked in the top 10 in the time periods 1990–1999,
2000–2009, and 2010–2012, respectively.
[a]Achieved ABET accreditation in 1998.

dynamics that might be associated with the changing ranks in engineering
degree production.

### Graduate Degree Production

Compared to the undergraduate level, HBCUs educate a smaller but still
noteworthy share of African American engineering graduate degree recipients. In 2007, HBCUs conferred 13% of engineering master's degrees earned
by African Americans, but this fell to 8% by 2012 (fig. 5.4). Similar to the
undergraduate level, greater shares of Black female (11% in 2012) than
Black male (8% in 2012) engineers earn their master's degrees at HBCUs. But
unlike the undergraduate level, the HBCU share of African American master's degrees has not changed substantially over the study period. Figure 5.5

TABLE 5.5    Top engineering bachelor's degree producers for African American women by decade, 1990–2012

| Institution | HBCU | 1990–1999 | 2000–2009 | 2010–2012 |
|---|---|---|---|---|
| | | *Rank* | | |
| North Carolina A&T State University | X | 1 | 1 | 1 |
| Georgia Institute of Technology | | 2 | 2 | 3 |
| Tuskegee University | X | 3 | 5 | 5 |
| Prairie View A&M University | X | 4 | 8 | 2 |
| Howard University | X | 5 | 11 | 6 |
| North Carolina State University at Raleigh | | 6 | 6 | 4 |
| Southern University and A&M College | X | 7 | 10 | 10 |
| Morgan State University | X | 8 | 3 | 7 |
| Florida A&M University | X | 9 | 4 | 17 |
| Tennessee State University | X | 10 | 9 | 19 |
| University of Michigan–Ann Arbor | | 11 | 7 | 8 |
| Massachusetts Institute of Technology | | 16 | 21 | 9 |
| University of Florida | | 33 | 14 | 7 |

SOURCE: US Department of Education, National Center for Education Statistics, IPEDS Completions Surveys drawn from National Science Foundation, National Center for Science and Engineering Statistics WebCASPAR System.

NOTES: Rank tabulations based on total number of degrees conferred during each decade to African American female US citizens and permanent residents. As seen from the table, there were eight, seven, and six HBCUs ranked in the top 10 in the time periods 1990–1999, 2000–2009, and 2010–2012, respectively.

shows that the numbers of Black men and women earning engineering master's degrees have increased at HBCUs and at all institutions.

Currently, only six HBCUs offer engineering doctorate degrees (table 5.1), but the trending share of African Americans who earn their engineering doctorates at HBCUs (not shown) suggests that their impact at this level is far from negligible. Analysis of IPEDS data indicates that as many as 22% of Black female engineers earned their doctorate degrees at HBCUs in 2004, but only 14% in 2012. Similarly, Black male engineering doctorates from HBCUs peaked at 17% in 2008 and fell slightly to 12% by 2012. This translates to a yearly average of roughly eight Black men and four Black women earning engineering doctorate degrees from HBCUs since 2000.

### Top Graduate Degree Producers
While the data indicate that HBCUs are an important source of engineering graduate education for African Americans, their small numbers and rela-

| Institution | HBCU | Rank | | |
|---|---|---|---|---|
| | | 1990–1999 | 2000–2009 | 2010–2012 |
| North Carolina A&T State University | X | 1 | 1 | 1 |
| Prairie View A&M University | X | 2 | 7 | 7 |
| Tuskegee University | X | 3 | 10 | 13 |
| Georgia Institute of Technology | | 4 | 2 | 2 |
| Southern University and A&M College | X | 5 | 6 | 6 |
| Howard University | X | 6 | 18 | 25 |
| CUNY City College | | 7 | 12 | 10 |
| North Carolina State University at Raleigh | | 8 | 3 | 4 |
| Morgan State University | X | 9 | 5 | 3 |
| Florida A&M University | X | 10 | 4 | 21 |
| University of Maryland–College Park | | 11 | 8 | 9 |
| Tennessee State University | X | 15 | 9 | 34 |
| University of Florida | | 20 | 13 | 8 |
| Alabama A&M University[a] | X | 97 | 16 | 5 |

SOURCE: US Department of Education, National Center for Education Statistics, IPEDS Completions Surveys drawn from National Science Foundation, National Center for Science and Engineering Statistics WebCASPAR System.
NOTES: Rank tabulations based on total number of degrees conferred during each decade to African American male US citizens and permanent residents. As seen from the table, there were seven, seven, and five HBCUs ranked in the top 10 in the time periods 1990–1999, 2000–2009, and 2010–2012, respectively.
[a]Achieved ABET accreditation in 1998.

tively limited disciplinary offerings leave their impact less apparent when comparing their numbers of graduate degrees to the larger engineering institutions. At the master's degree level, only North Carolina A&T and Howard University—the oldest HBCU engineering programs—appeared among the top 10 degree producers for all African Americans during at least one decade of the study period. And Howard was only a top producer during the 1990s, dropping to #38 by 2010–2012 (not shown).

At the doctorate degree level, North Carolina A&T has been a top African American engineering degree producer since at least the early 2000s, and Howard fell out of the top 10 by 2010–2012. However, other HBCUs have become more productive at this level, including Tuskegee, Morgan State, and Florida A&M (table 5.7).

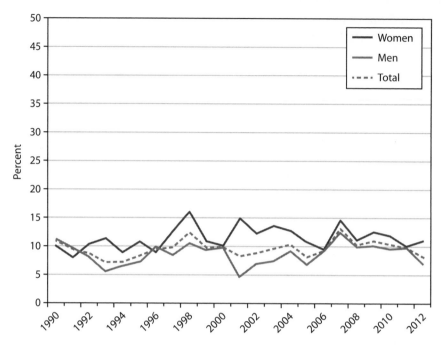

FIGURE 5.4 Share of engineering master's degrees conferred to African Americans by HBCUs, 1990–2012. Tabulations based on number of degrees conferred to African American US citizens and permanent residents.

*Source:* US Department of Education, National Center for Education Statistics, IPEDS Completions Surveys drawn from National Science Foundation, National Center for Science and Engineering Statistics WebCASPAR System.

HBCUs AND THE ENGINEERING GENDER GAP

Data presented above hint at a critical role for HBCUs with respect to the engineering gender gap. In 2011, women made up 12% of employed engineers, an increase from 10% in 2003 but a far cry from the 47% share of all occupations that women held the same year (NSF 2013). The underrepresentation of women in the engineering workforce has persisted even as women now outnumber men in higher education. For example, data compiled by NSF (2013) indicate that women made up 57% of enrollment in four-year institutions, 57% of bachelor's degree earners in all fields, but only 18% of engineering bachelor's degree recipients in 2010. On the other hand, Black women accounted for 63% of Black enrollment in four-year institutions, 66% of bachelor's degrees to Blacks in all fields, but only 26% of engineering bachelor's degrees conferred to Blacks in 2010 (NSF 2013).

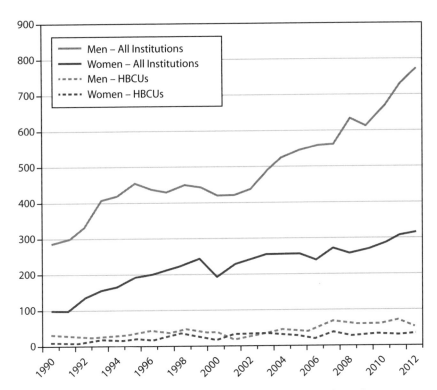

FIGURE 5.5 Number of engineering master's degrees conferred to African Americans by all institutions and HBCUs by gender, 1990–2012. Tabulations based on number of degrees conferred to African American US citizens and permanent residents.
*Source:* US Department of Education, National Center for Education Statistics, IPEDS Completions Surveys drawn from National Science Foundation, National Center for Science and Engineering Statistics WebCASPAR System.

Understanding and closing the persistent engineering gender gap has received much scholarly attention. In fact, every year since 2001 the Society of Women Engineers has published an annual review of research on women in engineering specifically and in STEM broadly, providing an extensive bibliography and critical analysis of selected work (Society of Women Engineers 2013). Yet despite this interest, few researchers have considered that HBCUs appear to be particularly effective at educating women engineers. Figure 5.6 shows that during 1990–2012, HBCU bachelor's degree cohorts have had larger shares of women when compared to all institutions. In 2001, 43% of engineers who earned bachelor's degrees at HBCUs were women.

TABLE 5.7  Top engineering doctorate degree producers for all African Americans by decade, 1990–2012

| | | Rank | | |
| --- | --- | --- | --- | --- |
| Institution | HBCU | 1990–1999 | 2000–2009 | 2010–2012 |
| Georgia Institute of Technology | | 1 | 1 | 1 |
| North Carolina A&T State University | X | 16 | 2 | 3 |
| Massachusetts Institute of Technology | | 2 | 3 | 5 |
| Stanford University | | 8 | 4 | 10 |
| University of California–Berkeley | | 4 | 5 | 7 |
| University of Florida | | 8 | 5 | 5 |
| University of Maryland–College Park | | 11 | 5 | 5 |
| University of Michigan–Ann Arbor | | 5 | 5 | 2 |
| Morgan State University | X | N/A | 6 | 6 |
| Purdue University | | 11 | 7 | 5 |
| North Carolina State University at Raleigh | | 6 | 8 | 10 |
| Northwestern University | | 10 | 9 | 10 |
| Texas A&M University | | 7 | 9 | 8 |
| Florida A&M University | X | 20 | 10 | 9 |
| Virginia Polytechnic Institute and State University | | 13 | 11 | 4 |
| George Washington University | | 12 | 12 | 5 |
| Pennsylvania State University | | 12 | 13 | 10 |
| Howard University | X | 3 | 14 | N/A |
| Auburn University | | 18 | 16 | 6 |
| Tuskegee University | X | 19 | 16 | 10 |
| University of Illinois at Urbana-Champaign | | 10 | 16 | 5 |
| Michigan State University | | 13 | 17 | 9 |
| University of Iowa | | 19 | 18 | 10 |
| Carnegie Mellon University | | 16 | 19 | 8 |
| CUNY Graduate School and University Center | | 9 | 19 | 11 |
| Cornell University | | 9 | 21 | 12 |
| University of Central Florida | | 17 | 22 | 7 |
| University of South Florida | | 14 | 22 | 3 |
| Clemson University | | 17 | 24 | 7 |
| Michigan Technological University | | 20 | 25 | 9 |
| University of Virginia | | N/A | 25 | 8 |
| Dartmouth College | | 9 | 26 | 11 |
| University of Kansas | | 19 | 26 | 10 |

SOURCE: US Department of Education, National Center for Education Statistics, IPEDS Completions Surveys drawn from National Science Foundation, National Center for Science and Engineering Statistics WebCASPAR System.
NOTE: Rank tabulations based on total number of degrees conferred during each decade to African American US citizens and permanent residents.

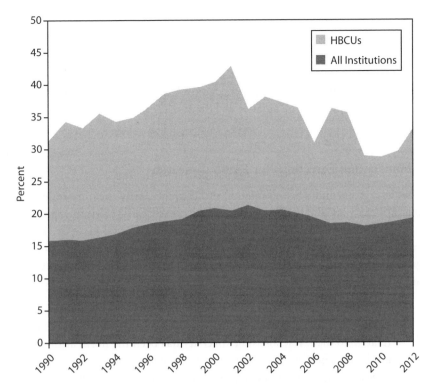

FIGURE 5.6 Share of women among engineering bachelor's degree recipients, 1990–2012. Tabulations based on number of degrees conferred to US citizens and permanent residents.
*Source:* US Department of Education, National Center for Education Statistics, IPEDS Completions Surveys drawn from National Science Foundation, National Center for Science and Engineering Statistics WebCASPAR System.

Nevertheless, HBCUs have not been immune to the broader trending declines in women's participation in engineering (see fig. 5.2); by 2012, 33% of HBCU baccalaureate engineers were women.

HBCUs have also narrowed the engineering gender gap at the graduate degree levels (figures not shown). HBCUs have graduated engineering cohorts that were 50% or more women at the master's degree (in 2001) and doctorate degree (in 2003) levels. Since 1990, women have, on average, made up 32% of HBCU engineering master's degree cohorts, compared to 20% at all institutions. And since 2003, women have averaged 31% of HBCU engineering doctorate degree cohorts, compared to 23% at all institutions.

However, given the small number and relative newness of engineering graduate programs at HBCUs, the gender makeup of graduate cohorts fluctuates much more so than that at the undergraduate level. Also, it is worth noting again that all of the descriptive findings reported here are only suggestive; more research is needed to determine why the engineering gender gap appears to be smaller at HBCUs across postsecondary degree levels.

## Summary and Directions for Future Research

The descriptive findings presented in this chapter clarify the important role that HBCUs play and have played in educating African American engineers. In particular, by providing a longitudinal and genderized view of HBCU engineering, this chapter uncovers some of the nuance behind trends in engineering education for African Americans. That said, several key findings are worth repeating:

1. Undergraduate engineering at HBCUs has grown increasingly racially/ethnically diverse, though at a slower rate among women relative to men. Racial/ethnic diversity has been the norm at the graduate degree level at least since 1990, with about 30%–40% of master's degree cohorts composed of non-Blacks in recent years. (Research Question #1)

2. From 1990 to 2012, HBCU graduates have accounted for substantial shares of African American engineering degree recipients. At the bachelor's and master's degree levels, HBCUs have accounted for larger shares of women than men degree recipients. And since the late 1990s, the HBCU share of African American engineering baccalaureates has trended downward for both men and women. (Research Question #2)

3. Not only is the share of African American engineering baccalaureates from HBCUs in decline for men and women, but the absolute numbers of African American engineering baccalaureates are down at HBCUs since the 1990s (despite that the number of Black male engineering baccalaureates is up at all institutions). (Research Question #2)

4. Compared to the undergraduate level, the HBCU share of African American engineering master's degree recipients did not change substantially between 1999 and 2012, and in recent years, the number of Black men and women earning engineering master's

degrees is up within HBCUs and all institutions. (Research Question #2)

5. In 2012, 14% of female and 12% of male African American engineering doctorates earned their PhDs from one of six HBCUs. (Research Question #2)

6. Seven of the top 10 engineering bachelor's degree producers for African Americans were HBCUs during the 1990s and 2000s, and six have made the top 10 list so far in the 2010s. However, gender differences were evident both in the number of HBCUs that were top producers for men and women and in which HBCUs were top producers. The role of HBCUs at the graduate degree level was less prominent in rankings of top degree producers. (Research Question #3)

7. At every degree level HBCUs have closed the engineering education gender gap relative to all institutions. In 2012, 33% of HBCU baccalaureate engineers were women, compared to 19% at all institutions. (Research Question #4)

All of the descriptive findings presented in this chapter are meant to offer insights into the patterns of African American engineering degree production playing out at HBCUs relative to all institutions. This report, therefore, is intended not to address the underlying causes and consequences of these patterns but to offer potential directions for examining them in the future. With respect to the first research question, the shifting racial/ethnic demographics of engineering degree cohorts at HBCUs suggest that it may be time to consider either the educational outcomes and experiences of non-Black HBCU engineering students (especially at the graduate level) or whether demographic shifts impact the essence of the "HBCU engineering experience." Future qualitative research might attempt to characterize the "HBCU engineering experience" at various degree levels, with attention to variations in the extent of racial/ethnic diversity within and across institutions.

Concerning undergraduate engineering at HBCUs, the findings from this analysis raise several questions. For example, to what extent do downward trends in African American degree production reflect declining engineering enrollments, student retention problems, or resource constraints within HBCUs? That is to say, are HBCUs simply competing less well with non-HBCUs for Black engineering students, performing less well with their engineering students, or both? Ransom (2013), who examined how relationships between engineering faculty predictors and engineering bachelor's degree production differ by student race/ethnicity and institutional contexts, found that HBCUs are not only among the most productive but also among the most technically efficient producers of Black engineers.* Her research indicates that enrollment declines may be the primary driver of regressive degree production trends at HBCUs. More research is needed to take a closer look at engineering enrollment trends across various institutional contexts.

Another line of future research should shine more light on dual-degree engineering partnerships between HBCUs, such as Morehouse College and Spelman College, and non-HBCUs, such as the University of Michigan and Georgia Institute of Technology. Scholars should explore the extent to which African Americans who earn their engineering bachelor's degrees from non-HBCUs matriculate through these arrangements, how the role or impact of dual-degree engineering programs has changed over time, and how students who benefit from these programs might be unmasked.

Future research should also investigate the extent to which the declining number of HBCU engineering baccalaureates might be reflected in African American representation in the engineering workforce. Also, because HBCUs disproportionately serve lower-income students who generally have lower levels of academic preparation (Kim 2002; Kim and Conrad 2006; Li and Carroll 2007; Gasman 2013), does declining engineering degree production at HBCUs mean that fewer low-income or academically underprepared African American students are getting a chance to pursue engineering? And given that HBCUs are among the top baccalaureate-origin institutions for African American engineering doctorate recipients (Burelli and Rapoport 2008; Hubbard and Stage 2010; Sibulkin and Butler 2011) and evidence suggests that underrepresented minority student success in STEM fields is positively related to interactions with same-race faculty (Fries-Britt 1998; Fries-Britt, Younger, and Hall 2010; Price 2010; Newman 2011a; Ransom 2013), might the downward trend in HBCU engineering bachelor's degree production be multiply deleterious in the long run?

## Conclusion

This chapter has offered a broad descriptive analysis of engineering degree production at HBCUs in order to lay the groundwork for developing under-

---

* In the study, Ransom (2013) explored the extent to which engineering schools and colleges maximized bachelor's degree production output with a given set of institutional inputs (i.e., maximized technical efficiency) based on a production model developed in the earlier stages of her analysis.

standing about the contributions of HBCUs to the engineering education enterprise. Using publicly available national degree completion data covering the period from 1990 to 2012, with attention to gender and degree-level variations, we have extended what is known about HBCU engineering. The findings presented echo earlier work that underscored the importance of HBCUs in engineering specifically (Brown, Morning, and Watkins 2005; Gasman et al. 2010; Weinberger, forthcoming) and in STEM more broadly (Southern Education Foundation 2005; Perna et al. 2009). In particular, the study confirms that HBCUs remain a critical lever for broadening African American participation in engineering.

In order for researchers to pursue the lines of inquiry suggested in this study, they must have access to relevant data. For example, existing efforts to collect engineering education data (e.g., through the American Society of Engineering Education and the Engineering Workforce Commission annual surveys) target ABET-accredited engineering schools and colleges and do not collect data on the numbers, race/ethnicity, gender, or partner institutions of students who matriculate through dual-degree, 3/2, or other collaborative arrangements. Including survey questions along these lines or surveying non-ABET institutions that operate such programs would facilitate more complete analyses of the educational pathways of engineering students and offer more instructive research insights. Similarly, identifying what about HBCUs might foster positive engineering outcomes will require data related to the range of institutional characteristics and structures that research suggests contribute to student educational outcomes (see Ransom 2013). Researchers might leverage existing consortia such as the United Negro College Fund or the Multi-Institution Database for Investigating Engineering Longitudinal Development to solicit relevant institutional data and even to carry out study designs subject to institutional review board (IRB) approval in a systematic way.

Ultimately, should future research advance our understanding of how particular aspects of HBCUs contribute to participation and success in engineering for Black students, such understanding could ultimately lead to more strategic use of resources and better-targeted interventions to improve students' outcomes in engineering writ large.

## References

ABET (Accreditation Board for Engineering and Technology). 2013. "Find Accredited Programs." http://main.abet.org/aps/Accreditedprogramsearch.aspx.

Allen, Walter R. 1992. "The Color of Success: African American College Student Outcomes at Predominantly White and Historically Black Colleges and Universities." *Harvard Educational Review* 62 (1): 26–44.

Anderson, Eugene, and Dongbin Kim. 2006. *Increasing the Success of Minority Students in Science and Technology.* Washington, DC: American Council on Education.

Bennof, Richard J. 2009. *Federal S&E Obligations to Three Types of Minority-Serving Institutions Decline in FY 2007 (NSF-09319).* Arlington, VA: National Science Foundation.

Brazziel, William F., and Marian E. Brazziel. 1997. *Distinctives of High Producers of Minority Science and Engineering Doctoral Starts.* Washington, DC: National Science Foundation.

Brown, A. Ramona, Carole Morning, and Charles B. Watkins. 2005. "Influence of African American Engineering Student Perceptions of Campus Climate on Graduation Rates." *Journal of Engineering Education* 94 (2): 263–271.

Burelli, Joan, and Alan Rapoport. 2008. *Role of HBCUs as Baccalaureate-Origin Institutions of Black S&E Doctorate Recipients.* NSF-08-319. Arlington, VA: National Science Foundation.

Chang, Mitchell. J., Oscar Cerna, June Han, and Victor Sàenz. 2008. "The Contradictory Roles of Institution Status in Retaining Underrepresented Minorities in Biomedical and Behavioral Science Majors." *Review of Higher Education* 31 (4): 433–464.

Clewell, Beatriz C., Clemencia C. de Cohen, and Lisa Tsui. 2010. *Capacity Development to Diversify STEM: Realizing the Potential among HBCUs.* Washington, DC: Urban Institute.

Craig, Lee A. 1991. "Constrained Resource Allocation and the Investment in the Education of Black Americans: The 1890 Land-Grant Colleges." *Agricultural History* 65 (2): 73–84.

Craig, Lee A. 1992. "'Raising among Themselves': Black Educational Advancement and the Morrill Act of 1890." *Agriculture and Human Values* 9 (1): 31–37.

Culotta, Elizabeth. 1992. "Black Colleges Cultivate Scientists." *Science, New Series* 258 (5085): 1216–1218.

Eagan, Mark K. 2010. "Moving beyond Frontiers: How Institutional Context Affects Degree Production and Student Aspirations in STEM." PhD diss., University of California, Los Angeles.

Elliott, Rogers A., A. Christopher Strenta, Russell Adair, Michael Matier, and Jannah Scott. 1996. "The Role of Ethnicity in Choosing and Leaving Science in Highly Selective Institutions." *Research in Higher Education* 37 (6): 681–709.

Fries-Britt, Sharon. 1998. "Moving beyond Black Achiever Isolation: Experiences of Gifted Black Collegians." *Journal of Higher Education* 69 (5): 556–576.

Fries-Britt, Sharon L., Toyia K. Younger, and Wendell D. Hall. 2010. "Lessons from High-Achieving Students of Color in Physics." *New Directions for Institutional Research* 148:75–83.

Gasman, Marybeth. 2013. *The Changing Face of Historically Black Colleges and Universities*. Philadelphia: University of Pennsylvania Graduate School of Education, Center for Minority Serving Institutions.

Gasman, Marybeth, Valerie Lundy-Wagner, Tafaya Ransom, and Nelson Bowman. 2010. "Special Issue: Unearthing Promise and Potential—Our Nations Historically Black Colleges and Universities." *ASHE Higher Education Report* 35 (5): 1–134.

Georges, Annie. 1999. *Keeping What We've Got: The Impact of Financial Aid on Minority Retention in Engineering*. New York: National Action Council for Minorities in Engineering.

Gibbs, Warmoth T. 1952. "Engineering Education in Negro Land Grant Colleges." *Journal of Negro Education* 21 (4): 546–550.

Good, Jennifer, Glennelle Halpin, and Gerald Halpin. 2002. "Retaining Black Students in Engineering: Do Minority Programs Have a Longitudinal Impact?" *Journal of College Student Retention* 3:351–364.

Goode, Robin W. 2011. "The HBCU Debate: Are Black Colleges & Universities Still Needed?" *Black Enterprise*, February 15. www.blackenterprise.com/lifestyle /are-hbcus-still-relevant/.

Grayson, Lawrence P. 1980. "A Brief History of Engineering Education in the United States." *Aerospace and Electronic Systems, IEEE Transactions* 16 (3): 373–392.

Hubbard, Steven M., and Frances K. Stage. 2010. "Identifying Comprehensive Public Institutions That Develop Minority Scientists." *New Directions for Institutional Research* 148:53–62.

Humphries, Frederick S. 1991. "1890 Land-Grant Institutions: Their Struggle for Survival and Equality." *Agricultural History* 65 (2): 3–11.

Hurtado, Sylvia, Nolan L. Cabrera, Monica H. Lin, Lucy Arellano, and Lorelle L. Espinosa. 2009. "Diversifying Science: Underrepresented Student Experiences in Structured Research Programs." *Research in Higher Education* 50 (2): 189–214.

Hurtado, Sylvia, Mark K. Eagan, and Bryce Hughes. 2012. "Priming the Pump or the Sieve: Institutional Contexts and URM STEM Degree Attainments." Paper presented at the Annual Forum of the Association for Institutional Research Association, New Orleans, LA.

Jenkins, Martin D. 1942. "The National Survey of Negro Higher Education and Post-war Reconstruction: The Resources of Negro Higher Education." *Journal of Negro Education* 11 (3): 382–390.

Kim, Mikyong M. 2002. "Historically Black vs. White Institutions: Academic Development among Black Students." *Review of Higher Education* 25 (4): 385–407.

Kim, Mikyong M., and Clifton F. Conrad. 2006. "The Impact of Historically Black Colleges and Universities on the Academic Success of African-American Students." *Research in Higher Education* 47 (4): 399–427.

Kujovich, Gil. 1993. "Public Black Colleges: The Long History of Unequal Funding." *Journal of Blacks in Higher Education*, no. 2 (Winter), 73–82.

Leggon, Cheryl B., and Willie Pearson, Jr. 1997. "The Baccalaureate Origins of African American Female Ph.D. Scientists." *Journal of Women and Minorities in Science and Engineering* 3 (4): 213–224.

Lent, Robert W., Steven D. Brown, Hung-Bin Sheu, Janet Schmidt, Bradley R. Brenner, Clay S. Gloster, Gregory Wilkins, Linda C. Schmidt, Heather Lyons, and Dana Treistman. 2005. "Social Cognitive Predictors of Academic Interests and Goals in Engineering: Utility for Women and Students at Historically Black Universities." *Journal of Counseling Psychology* 52 (1): 84–92.

Li, Xiaojie, and C. Dennis Carroll. 2007. *Characteristics of Minority-Serving Institutions and Minority Undergraduates Enrolled in These Institutions: Postsecondary Education Descriptive Analysis Report (NCES 2008-156).* Washington, DC: National Center for Education Statistics.

Matthews, Christine M. 2011. *Federal Research and Development Funding at Historically Black Colleges and Universities.* Washington, DC: Congressional Research Service.

McGrath, Earl J. 1965. *The Predominantly Negro Colleges and Universities in Transition.* New York: Bureau of Publications, Teachers College, Columbia University.

Moore, James L. 2006. "A Qualitative Investigation of African American Males' Career Trajectory in Engineering: Implications for Teachers, School Counselors, and Parents." *Teachers College Record* 108 (2): 246–266.

Moore, James L., Octavia Madison-Colmore, and Dionne M. Smith. 2003. "The Prove-Them-Wrong Syndrome: Voices from Unheard African-American Males in Engineering Disciplines." *Journal of Men's Studies* 12 (1): 61–73.

Morse, Lucy C., and Daniel L. Babcock. *Managing Engineering and Technology: An Introduction to Management for Engineers.* Upper Saddle River, NJ: Prentice Hall, 2009.

National Action Council for Minorities in Engineering. 2012. "African Americans in Engineering." *NACME Research Briefs* 2 (4): 1–2.

National Science Board. 2012. *Science and Engineering Indicators 2012 (NSB 12-01).* Arlington, VA: National Science Foundation.

Newman, Christopher B. 2011a. "Access and Success for African American Engineers and Computer Scientists: A Case Study of Two Predominantly White Public Research Universities." PhD diss., University of California, Los Angeles.

Newman, Christopher B. 2011b. "Engineering Success: The Role of Faculty Relationships with African American Undergraduates." *Journal of Women and Minorities in Science and Engineering* 17 (3): 193–209.

NSF (National Science Foundation). 2013. *Women, Minorities, and Persons with Disabilities in Science and Engineering: 2013 (NSF 13-304).* Arlington, VA: National Science Foundation, National Center for Science and Engineering Statistics. www.nsf.gov/statistics/wmpd/.

Pearson, Willie, Jr., and LaRue C. Pearson. 1985. "Baccalaureate Origins of Black American Scientists: A Cohort Analysis." *Journal of Negro Education* 54 (1): 24–34.

Perna, Laura W., Valerie C. Lundy-Wagner, Noah D. Drezner, Marybeth Gasman, Susan Yoon, Enakshi Bose, and Shannon Gary. 2009. "The Contribution of HBCUs to the Preparation of African American Women for STEM Careers: A Case Study." *Research in Higher Education* 50 (1): 1–23.

Price, Joshua. 2010. "The Effect of Instructor Race and Gender on Student Persistence in STEM Fields." *Economics of Education Review* 29:901–910.

Ransom, Tafaya. 2013. "When Do Faculty Inputs Matter? A Panel Study of Racial Ethnic Differences in Engineering Bachelor's Degree Production." PhD diss., University of Pennsylvania.

Sibulkin, Amy E., and J. S. Butler. 2011. "Diverse Colleges of Origin of African American Doctoral Recipients, 2001–2005: Historically Black Colleges and Universities and Beyond." *Research in Higher Education* 52 (8): 830–852.

Slaughter, John B. 2009. "African American Males in Engineering: Past, Present, and Future of Opportunity." *Diversity in Higher Education* 7:193–208.

Smyth, Frederick L., and John J. McArdle. 2004. "Ethnic and Gender Differences in Science Graduation at Selective Colleges with Implications for Admission Policy and College Choice." *Research in Higher Education* 45:353–381.

Society of Women Engineers. 2013. *A Compendium of the SWE Annual Literature Reviews on Women in Engineering.* Chicago: Society of Women Engineers. www.nxtbook.com/nxtbooks/swe/litreview2012/#/o.

Solórzano, Daniel G. 1995. "The Doctorate Production and Baccalaureate Origins of African Americans in the Sciences and Engineering." *Journal of Negro Education* 64 (1): 15–32.

Southern Education Foundation. 2005. *Igniting Potential: Historically Black Colleges and Universities in Science, Technology, Engineering and Mathematics.* Atlanta: Southern Education Foundation.

Suitts, Steve. 2003. "Fueling Education Reform: Historically Black Colleges Are Meeting a National Science Imperative." *Cell Biology Education* 2:205–206.

Swail, Watson S., Kenneth E. Redd, and Laura W. Perna. 2003. "Retaining Minority Students in Higher Education: A Framework for Success." *ASHE-ERIC Higher Education Report* 30 (2).

Thelin, John R. 2004. *A History of American Higher Education.* Baltimore: Johns Hopkins University Press.

Thomas, Gail E. 1987. "African-American College Students and Their Major Field Choice." In *In Pursuit of Equality in Higher Education,* edited by Anne S. Pruitt, 105–115. Dix Hills, NY: General Hall.

Thomas, Gail E. 1991. "Assessing the College Major Selection Process for African-American Students." In *College in Black and White: African American Students in Predominantly White and Historically Black Universities,* edited by

Walter R. Allen, Edgar G. Epps, and Nesha Z. Hanniff, 75–91. Albany: State University of New York Press.

Trent, William T. 1991. "Focus on Equity: Race and Gender Differences in Degree Attainment, 1975–76, 1980–81." In *College in Black and White: African American Students in Predominantly White and Historically Black Universities*, edited by Walter R. Allen, Edgar G. Epps, and Nesha Z. Hanniff, 41–60. Albany: State University of New York Press.

Trent, William T., and J. Hill. 1994. "The Contributions of Historically Black Colleges and Universities to the Production of African American Scientists and Engineers." In *Who Will Do Science? Educating the Next Generation*, edited by Willie Pearson, Jr., and Alan Fetcher, 68–80. Baltimore: Johns Hopkins University Press.

Weinberger, Catherine. Forthcoming. "Engineering Educational Opportunity: Impacts of 1980s Policies to Increase the Share of Black College Graduates with a Major in Engineering or Computer Science." In *U.S. Engineering in the Global Economy*, edited by Richard Freeman and Hal Salzman. Chicago: National Bureau of Economic Research/University of Chicago Press.

Wenglinsky, Harold. 1997. *Students at Historically Black Colleges and Universities: Their Aspirations & Accomplishments*. Princeton, NJ: Educational Testing Service, Policy Information Center.

Wilburn, Adolph Y. 1974. "Careers in Science and Engineering for Black Americans." *Science* 184 (4142): 1148–1154.

Wolf-Wendel, Lisa. 1998. "Models of Excellence: The Baccalaureate Origins of Successful European American Women, African American Women, and Latinas." *Journal of Higher Education* 69 (2): 141–186.

Wolf-Wendel, Lisa E., Bruce D. Baker, and Christopher C. Morphew. 2000. "Dollars and $ense: Institutional Resources and the Baccalaureate Origins of Women Doctorates." *Journal of Higher Education* 71 (2): 165–186.

# CHAPTER SIX

# Beyond the Black-White Minority Experience

## Undergraduate Engineering Trends among African Americans

SYBRINA Y. ATWATERS, JOHN D. LEONARD II,
and WILLIE PEARSON, JR.

PRIOR TO THE 1970S, approaches to diversifying the engineering land-scape were structured in binary terms of Black and White. Many of the minority initiatives were targeted to increase access to college education for African Americans, who had been denied equal access for centuries in America. Fiss (1991, 261) argued that antidiscrimination laws, which build on the Fourteenth Amendment and the significant court cases that followed, "became the principal legal instrument for achieving equality." Eventually, antidiscrimination laws significantly impacted the enrollment of minorities at predominately White institutions (PWIs) across America (Gavins 2009).

As recently as the late 1980s, a shift in the minority engineering population can be observed. Between 1987 and 1997, Asian enrollment increased from 8.4% to 10.8% and Hispanic enrollment almost doubled, from 4.7% to 8.4% (National Science Foundation 2000). For the same period, African American enrollment increased from 4.9% to 6.8%. In response, many diversity and minority development offices at higher education institutions were charged with attending to the needs of the changing college population. It is questionable whether these new initiatives were accompanied by additional resources to meet the expanding and multidimensional needs of a more diverse student population. Soon scholars realized that antidiscrimination laws governed the selection process by which a pool of applicants was constituted but did not address issues of isolation and injustice embedded within the institutional climate which impeded the successful matriculation of many minority students (Fiss 1991). It soon became clear that access-focused laws were necessary but not sufficient, because long-term policies, programs, and permanent infrastructure were also needed.

As the US population continues to transform, with projections of 54% of the population in 2060 consisting of Asians, African Americans, and Hispanics (US Census Bureau 2012), researchers, policy makers, and institutional leaders find themselves struggling to prepare for the changing demographics. Science and engineering (S&E) have been at the forefront of diversity conversations, warning institutional leaders and policy makers of the urgency for a more diverse S&E talent pool (Chubin, May, and Babco 2005; National Research Council 2007; National Science Foundation 2007; Page 2007; Leggon and Pearson 2009; National Academies 2010, 2011). Overall undergraduate enrollment in S&E fields has increased as much as 15% between 2006 and 2009 (National Science Board 2012). Engineering has been impacted by the growing college-age population, reaching a noticeable 5.3% increase in enrollment in 2010 (Gibbons 2010; National Science Board 2012) and an approximately 8% increase between 2012 and 2013 (Yoder 2013). Yet, in spite of projections for a growing minority population and a push for initiatives to increase minority talent within S&E, recent data demonstrate that African Americans remain significantly underrepresented in engineering and are experiencing a decline in their percentage of undergraduate engineering enrollment and degrees awarded. Such disparities at the undergraduate level leave little reason to expect significant changes in the future engineering workforce, in which African Americans currently have a lower percentage (4%) of representation than in all other S&E occupations (National Science Foundation 2013). Leggon and McNeely (2012) emphasize that underrepresentation in science, technology, engineering, and mathematics (STEM) fields has become a global issue of economic and national development rather than merely a reflection of social injustice and inequality.

In the midst of global competitiveness, as other countries invest as much as 47% of their college-age human capital into S&E for economic development and sustainability (National Academies 2010; Ong et al. 2011), the United States must produce engineers who provide culturally diverse approaches and insights to solving complex local, national, and global problems. Innovative strategies and disaggregated data are essential to analyze the multidimensional variables involved in recruiting and retaining a diverse engineering talent pool. In response, this chapter examines trends in undergraduate engineering for African American citizens and permanent residents from 2005 to 2011; the most consistent race and gender data were collected during this time period. Specifically, the chapter focuses on gender differences in enrollment, field choice, persistence and retention, degrees awarded, and institutional baccalaureate origin. Statistical data are derived primarily from three sources: American Society for Engineering Education (ASEE),

National Center for Education Statistics (NCES), and National Science Foundation (NSF)/National Center for Science and Engineering Statistics (NCSES). Using the Georgia Institute of Technology (Georgia Tech) as a case study, enrollment, degree awards, retention, and performance trends among select engineering disciplines are also examined during the 2007–2013 time period. Relevant literature regarding racial and gender differences in degrees awarded, persistence, barriers, and perception of institutional climate is reviewed briefly for the study period.

## Review of Relevant Literature

Until recently, few academic studies focused on minorities in engineering (Bond and LeBold 1977; Morrison and Williams 1993). Many publications on the topics were produced by federal agencies (NSF), professional organizations such as the National Action Council for Minorities in Engineering (NACME), and honorific institutions such as the National Academy of Engineering (NAE). Most of the reports and earlier academic studies analyzed demographic trends and disparities impacting African Americans' participation in higher education across S&E as a whole, rather than focusing on individual disciplines. Once data were disaggregated, engineering demonstrated slower progress than other S&E fields in achieving a more racially and gender-diverse talent pool (National Science Foundation 2013). Issues of recruitment, persistence and retention, and institutional type and climate emerged in recent studies as factors that vary in their impact by race, ethnicity, and gender on engineering education.

RECRUITMENT

Established during the early 1970s, Minority Engineering Programs (MEPs) were aimed at increasing the recruitment of minorities. With appropriate resources, several MEPs were successful in increasing recruitment, retention, and performance of African Americans in engineering (Morrison and Williams 1993). Ohland and Zhang (2002) found that participants in the Florida A&M University–Florida State (FAMU-FSU) Engineering Concepts Institute were 25% more likely to graduate in engineering, even when controlling for high school GPA. Interestingly, Ohland and Zhang (2002) reported that the program was discontinued. Recent literature regarding the evaluation of diversity programs established in the past decade is sparse. Leggon and Pearson (2009) note that programs focusing on broadening the participation of underrepresented groups in STEM need to be informed by evaluation literature as much as research literature. The lack of rigorous program evaluation may lead to erroneous assessments of programs' success and utility (Leonard et al. 2013).

The latest findings by Malcom-Piqueux and Malcom (2013) validate that focusing on pre-engineering practices and curriculum, as was the practice of several MEPs, is crucial for recruitment of African Americans to engineering. The authors attributed the lack of diversity in the engineering workforce to such precollege factors as lack of access to high-quality teaching and science infrastructure and to career information and counseling. These gaps in access at the precollege level contribute to inadequate academic preparation and lack of awareness of engineering as a profession (Pearson and Miller 2012). In addition to academic preparedness, financial concerns also significantly impact the recruitment of African Americans into engineering (Fleming, Engerman, and Griffin 2005). Academic preparation, awareness, and financing are key to the recruitment of African Americans in engineering. Institutions may find that collaborations with high schools and state education boards may assist in addressing issues of academic preparation and awareness, while more targeted and competitive funding programs may be necessary to address economic barriers to enhancing the participation of African Americans in engineering. Nonetheless, retention seems to present an even greater challenge to the production of African American engineers.

RETENTION

African American males' persistence in higher education as a whole is declining, with college graduates decreasing from 18.7% in 2008 to 17.7% in 2010 (US Census Bureau 2012). These dynamics impact African American women's gender representation (when comparing gender only), since at the same time African American women college graduates increased from 20.4% to 21.4% (US Census Bureau 2012). Yet, African American women's enrollment and retention in undergraduate engineering reflect troubling levels of underrepresentation.

In a study of undergraduate engineering students at nine public universities in the southeastern United States, Lord et al. (2009) report that women persist in engineering at similar rates to men, 53.6% versus 55.1%, respectively. When disaggregated by race/ethnicity and gender, it is revealed that African American males and females have lower rates of persistence in engineering than White, Asian, and Hispanic males and females. African American males persist at a rate of 48.4%, lower than all other groups, including their female counterparts, who persist at a rate of 51.1%. African American women's STEM average GPA was 2.34, compared to African American men's 2.10. Consequently, academic performance does not explain gender disparities in engineering within the same racial/ethnic group. According to Lord et al., the gender disparities observed in engineering can be attributed to disparities in enrollment rather than persistence since the data show that women persist at similar rates to men, once enrolled in engineering disciplines. Noticeably, women who do not persist in engineering are more likely than men to switch majors rather than leave the university altogether (Lord et al. 2009). Approximately 5%–8% more women switched their major to a field outside of engineering as opposed to leaving the institution. African American women were the exception, reflecting similar migration patterns as their male counterparts by leaving the university at a rate more consistent with their male counterparts rather than switching majors at rates comparable with other female groups (Lord et al. 2009). For African American women in engineering both gender and racially unique patterns are observed in the Lord et al. findings.

Concannon and Barrow (2010) conducted a quantitative study of 493 undergraduate engineering students' intentions to persist in engineering. While self-efficacy factors that impact men and women engineering students' intention to persist differed, the researchers concluded that persistence is best predicted by career outcome expectations. In their study, Byars-Winston et al. (2010) found that a combination of self-efficacy, career outcome expectations, and perceived campus climate impacted African American, Latino, Southeast Asian, and Native American students' interest and retention in engineering. DeFreitas (2012) argues that although a relationship between self-efficacy and academic performance has been well established, examining persistence in terms of self-efficacy might not be the best approach to understanding persistence in African Americans' academic performance. For African Americans, negative outcome expectations (beliefs about the results of one's academic efforts) are often perceived as the result of external factors (i.e., racism or discrimination) rather than individual factors related to self-efficacy. It is noteworthy that African American participants with more negative outcome expectations have higher GPAs than African Americans with more positive outcome expectations. DeFreitas attributes this dynamic to the impact of preparedness for racial microaggressions on performance. Thus, African Americans with negative outcome expectations seek out ways to compensate for external factors. McGee and Martin (2011) reported that high-performing African American engineering students learn to develop coping mechanisms to guard against racial stereotypes and assaults. More research is needed to explore the impact of outcome expectations on academic performance in engineering by race and gender, as well as to unearth key factors that impact academic performance among African Americans in engineering and how performance relates to retention rates and matriculation patterns.

Owing to low rates of student retention, both the structure and delivery of engineering have become an increasing concern to the engineering community. Despite rigorous admission screening practices, only 56% of engineering undergraduates are retained (Case and Jawitz 2003; Fortenberry et al. 2007). Shehab et al. (2007) point out that despite experiencing some of the same struggles and using some of the same coping strategies as majority students, racial and ethnic minority students tend to experience these struggles more deeply. Racial and ethnic minority students cited faculty as a major cause of academic problems and reported that some faculty did not know how to teach; other faculty, in spite of having knowledge of how to teach, were poor in their execution of teaching skills; and still other faculty seemed to delight in making courses difficult (Shehab et al. 2007). Although poor teaching negatively impacts all students, studies suggest that it has a greater impact on students from racial and ethnic groups that are underrepresented in engineering (Leonard et al. 2013). According to a 2011 NSF report, the racial diversity of full-time professors in S&E remained relatively unchanged from 1979 (with approximately 2% minorities—Asian, African American, Hispanic, and Pacific Islanders) to 2008 (6%). While some African American students persist to graduation, many who are haunted by the experiences within these institutions, as well as perceptions of these experiences, leave the S&E pipeline relinquishing their aspirations toward engineering careers. Minorities represent less than 30% of the S&E workforce. Over half of the minority S&E workforce is Asian. African American males compose 3% and African American females 2% of the S&E workforce.

Chubin, May, and Babco (2005) assert that particular attention is needed at critical transitional points along the S&E pathway to ensure that the United States has a diverse (culturally competent) engineering workforce. Programs that address research, mentoring, and funding challenges for minorities are imperative to successfully recruit underrepresented minorities into engineering faculty. These researchers argue that industry and government research institutions would enhance their value by employing intellectual talent that reflects the diversity of the US population. Since industry and government institutions often work in collaboration with academia to develop or establish directives for research agendas, they have the advantage of establishing programs and initiatives for minority engineering recruitment and development which work across academic institutions.

The type of institution awarding S&E degrees to African Americans is chang-ing. Simultaneously, colleges' and universities' administrations are changing their diversity recruitment and retention practices. Since 2009, Georgia Tech has awarded the most bachelor's degrees in engineering. It has also awarded the most engineering bachelor's degrees to women (Yoder 2011). Yet, North Carolina A&T State University has been continually noted as awarding the most engineering bachelor's degrees to African American students (Chubin, May, and Babco 2005; Gibbons 2010; Yoder 2011). In 2011, historically Black colleges and universities (HBCUs) were 5 of the top 10 baccalaureate-origin institutions for African American PhD engineering recipients (Fiegener and Proudfoot 2013). Consequently, HBCUs continue to play a pivotal role in the production of African American engineers. However, there has been an overall decline in the number of African Americans earning engineering bachelor's degrees from HBCUs (National Science Foundation 2011; for more details, see chap. 5 in this volume).

HBCUs demonstrate that long-term programmatic initiatives that target the recruitment and retention of African American students in STEM fields are critical factors for improving production of African American engineers (Perna et al. 2009). Not only are there significant changes among HBCUs and non-HBCUs in the production of African American engineers, but commu-nity colleges can also be an important resource for African Americans with engineering aspirations (Chubin, May, and Babco 2005; Freeman and Hug-gans 2009). According to Freeman and Huggans (2009), community (two-year) colleges play a significant role in educating approximately 46% of America's undergraduate population. Further, PWIs such as Louisiana State University's (LSU) Howard Hughes Medical Institute (HHMI) have taken the long-term, multiple-year, programmatic approach to the retention of African American students, demonstrating 55% graduation rates in com-parison with 20% graduation rates for nonparticipants at LSU and 25% graduation rates for African American students in STEM nationwide (Wil-son et al. 2012).

Long-term targeted programmatic initiatives, such as dual-degree programs, may prove to be effective in recruiting and retaining African Americans in engineering fields. However, much more evaluation research is required to understand the impact of such programs over time (Leonard et al. 2013). As studies move beyond examining individual background and self-efficacy to reviewing institutional climate as a significant factor in under-standing trends in African Americans' engineering participation, findings regarding the type of university successful in producing African Americans in engineering reveal new opportunities for interinstitutional collaborations in programmatic efforts, curriculum and research development, and funding. Review of the most recent literature exposes gaps in knowledge and data regarding the decline in African American engineering representation and the development of policies, programs, and initiatives to reverse current trends and enhance the effectiveness of future projections. The next section describes current trends of African Americans in engineering at the baccalaureate level.

CURRENT TRENDS

African American representation among engineering bachelor's degree recipients peaked at 5.6% in 2000 after a quarter-century climb from just 3% in 1977 (NACME 2011). Trends are complex when disaggregated by gender and race.* Although the proportion of African American women as a percentage of total African American enrollment is higher than is the proportion of women enrolled with any other racial group, African American women persist in engineering at lower rates than White, Asian, and Hispanic women (Lord et al. 2009). The demarcation of race and gender in previous initiatives to broaden participation of underrepresented populations in STEM fields resulted in programs that largely benefited White women or minority men (Ong et al. 2011) in overall engineering representation but had an interesting impact within racial groups as well. Assessment of current data trends shows that African American gender patterns are unique to their racial group and that African American racial patterns are unique within gender categories.

## Data and Methods

The primary source of data for this analysis was a multiyear, annual survey of engineering colleges and universities administered by the ASEE. ASEE annually collects data from nearly 350 engineering colleges and schools across North America and publishes these data on its website (http://profiles.asee.org).

---

* Few studies disaggregate race/ethnicity by gender. However, such disaggregation is crucial because multiple layers of difference are not mutually exclusive. Race/ethnicity impacts how one experiences being male or female, and gender impacts how one experiences belonging to a particular race and/or ethnic group (Hrabowski and Pearson 1993; Leggon 2006). Attention to the nuances in current trends revealed by disaggregated data is highlighted throughout this section.

Data for this study were downloaded from the ASEE Profiles site during June 2014. ASEE data collection generally lags by one year; thus, data for this study represent the period ending June 30, 2013.

Data for this analysis include degrees awarded and fall enrollments from 2007 to 2013. Degrees awarded are reported for the entire year (for example, summer, fall, and spring semesters), while enrollments are reported for full-time headcounts during the fall term.* Each measure (degrees awarded or enrollments) is disaggregated by ethnicity and gender. Data are accepted at the level of disaggregation provided by the survey participant. For example, schools with small numbers of degrees awarded might not report disaggregated ethnicity data, but rather only submit their counts as a single aggregate.

ASEE enrollment and degrees awarded data were merged with basic institutional data available from the Integrated Postsecondary Education Data System (IPEDS) (National Center for Education Statistics, http://nces.ed.gov /ipeds/datacenter/Default.aspx). Attributes merged with the ASEE data include Carnegie Classification of the institution and an HBCU indicator available from the institutional data file.

Figures and tables were created, based on queries used to extract appropriate variables from ASEE and IPEDS databases. Data were disaggregated by gender and race/ethnicity across four institutional categories: HBCUs, non-HBCUs, research universities (RUs), and master's colleges and universities (MCUs). Data were also delineated into four categories for analysis: (1) trends in undergraduate engineering enrollment, (2) trends in field choice, (3) trends in bachelor's degrees awarded, and (4) trends in institutional type.

Further, performance, discipline, and retention data were collected from Georgia Tech's Office of Institutional Research and Planning internal database. Transcript records for 54,180 students enrolled in the College of Engineering from fall 2008 through fall 2013 were examined. Final grades for each course completed were recorded. Student records were categorized based on academic year (fall and spring semesters), degree level (UG—undergraduate, MS—master's, DR—doctorate), home college and department (based on their discipline major), and ethnicity-gender pair.

---

*By analyzing enrollment trends based on enrollment data that reflect full-time headcounts, we recognize the limitations involved with the lack of available data regarding part-time student enrollment. Underrepresented minorities who enroll in master's institutions on a part-time basis will not be reflected in enrollment trends in tables 6.1 and 6.2 and figs. 6.1–6.4.

## Trends in Enrollment

Since 2007, undergraduate engineering enrollment for African Americans has remained below 5.7% (table 6.1). While the number of African American students enrolled in engineering increased from 23,694 in 2007 to 28,731 in 2013 (see the appendix, table 6A.1), the overall percentage of African Americans' undergraduate engineering enrollment declined from 5.6% to 4.9%. In 2013, there were a total of 583,816 students enrolled in engineering education at the baccalaureate level (see the appendix, table 6A.1). African Americans (13% of the US population) are significantly underrepresented in undergraduate engineering enrollment when compared with other racial/ethnic groups. Asian Americans compose 5.1% of the US population and 10.9% of undergraduate engineering students. Hispanics are 17% of the US population versus 11.3% of undergraduate engineering students (table 6.1).

Figure 6.1 further disaggregates undergraduate African American engineering enrollments by gender. While the number of African American males enrolled in engineering increased from 17,787 to 21,842 between 2007 and 2013, the number of African American females enrolled in engineering has remained relatively flat, ranging from 5,907 in 2007 to 6,889 in 2013. As a proportion of all undergraduate engineering students, there were declines in percentages of both male and female African American undergraduate engineers, from 4.2% in 2007 to 3.7% in 2013 for African Ameri-

TABLE 6.1    Baccalaureate engineering enrollment percentages by ethnicity, 2007–2013

| Ethnicity | FY07 | FY08 | FY09 | FY10 | FY11 | FY12 | FY13 |
|---|---|---|---|---|---|---|---|
| White | 60.9 | 60.6 | 58.8 | 59.1 | 58.9 | 57.9 | 56.6 |
| Asian | 10.7 | 10.9 | 10.8 | 10.5 | 10.6 | 10.8 | 10.9 |
| Black | 5.6 | 5.7 | 5.4 | 5.4 | 5.3 | 5.2 | 4.9 |
| Hispanic | 9.5 | 10.0 | 10.2 | 10.4 | 10.8 | 11.1 | 11.3 |
| Native | 0.6 | 0.6 | 0.6 | 0.7 | 0.7 | 0.7 | 0.6 |
| Other | 7.3 | 8.3 | 8.3 | 7.3 | 6.7 | 6.1 | 7.1 |
| Foreign | 5.3 | 5.9 | 5.9 | 6.5 | 7.0 | 8.3 | 8.5 |
| All | 100.0 | 100.0 | 100.0 | 100.0 | 100.0 | 100.0 | 100.0 |

NOTES: ASEE ethnicity data represent US domestic students, i.e., US citizens and resident aliens. The "Foreign" category is not an ethnicity per se, but rather represents the remaining pool of international students (e.g., nonresident aliens attending US schools on visas) aggregated across all ethnicities segregated from the US domestic students. The "Native" category aggregates Native American and Native Hawaiian ASEE survey categories. The "Other" category aggregates the "Two or more," "Unknown," and "Other" ASEE survey categories.

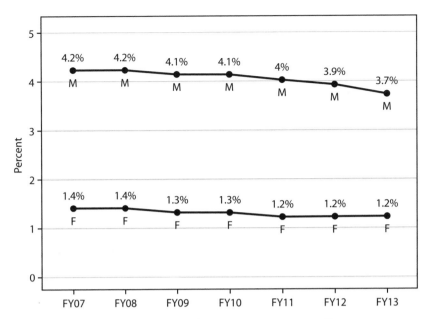

FIGURE 6.1  African American baccalaureate engineering enrollment percentages by sex, 2007–2013
*Source:* Data from ASEE.

can males, and from 1.4% in 2007 to 1.2% in 2013 for African American females (fig. 6.1).

Disaggregating the data by gender and race reveals other interesting trends (table 6.2). The percentage of African American males has changed slightly over time within the group of all male undergraduate engineering students, from 5.1% of all males in 2007 to 4.7% in 2013. The proportion of African American females has dropped from 8.2% of all female undergraduate engineers in 2007 to 6.0% in 2013. Other ethnic groups do not exhibit as large a range of changes among females. However, White males show a significant decline, while Hispanic males show a significant increase.

Other notable trends are revealed by comparing the proportions of African American males and females within their respective gender groups. In 2013, African American females represented 6.0% (6,889) of the total females (114,383) enrolled, while African American males represented 4.7% (21,842) of the total males (469,433) enrolled in engineering at the baccalaureate level. In contrast, White females represent 52.2% (59,741) of the total females enrolled and White males represent 57.7% (270,990) of the total males enrolled in engineering at the baccalaureate level (see the appendix,

| Ethnicity | FY07 | FY08 | FY09 | FY10 | FY11 | FY12 | FY13 |
|---|---|---|---|---|---|---|---|
| | | | | Male | | | |
| White | 62.4 | 62.2 | 60.3 | 60.6 | 60.1 | 59.1 | 57.7 |
| Asian | 10.2 | 10.4 | 10.3 | 10.1 | 10.2 | 10.3 | 10.4 |
| Black | 5.1 | 5.1 | 5.0 | 5.1 | 4.9 | 4.9 | 4.7 |
| Hispanic | 9.1 | 9.6 | 9.9 | 10.0 | 10.6 | 10.9 | 11.1 |
| Native | 0.6 | 0.6 | 0.6 | 0.6 | 0.7 | 0.6 | 0.6 |
| Other | 7.4 | 6.7 | 8.1 | 7.2 | 6.6 | 6.0 | 6.9 |
| Foreign | 5.2 | 5.4 | 5.8 | 6.4 | 6.9 | 8.2 | 8.6 |
| All | 100.0 | 100.0 | 100.0 | 100.0 | 100.0 | 100.0 | 100.0 |
| | | | | Female | | | |
| White | 53.4 | 53.4 | 52.1 | 52.7 | 53.5 | 52.9 | 52.2 |
| Asian | 13.1 | 13.0 | 12.7 | 12.4 | 12.5 | 12.8 | 13.0 |
| Black | 8.2 | 8.1 | 7.4 | 7.1 | 6.7 | 6.5 | 6.0 |
| Hispanic | 11.7 | 11.9 | 11.7 | 12.2 | 12.0 | 11.9 | 11.8 |
| Native | 0.8 | 0.8 | 0.8 | 0.8 | 0.8 | 0.8 | 0.7 |
| Other | 7.2 | 7.0 | 8.9 | 7.8 | 7.3 | 6.7 | 7.9 |
| Foreign | 5.6 | 6.0 | 6.3 | 7.0 | 7.3 | 8.4 | 8.3 |
| All | 100.0 | 100.0 | 100.0 | 100.0 | 100.0 | 100.0 | 100.0 |

SOURCE: ASEE.

table 6A.1). Asian females are 13% (14,911) and Hispanic females are 11.8% (13,528), while Asian males are 10.4% (48,969) and Hispanic males are 11.1% (52,214) of enrollment. Across all years, minority females represent larger proportions of their gender than the males in their respective ethnic groups.

Within ethnic groups, African American females represent 24% of all African American undergraduate engineering students, the largest proportion across all ethnic groups, while White females represent the lowest proportion at 18.1% of all White undergraduate engineering students (fig. 6.2). Enrollment challenges persist for women in general—and African American females specifically—in undergraduate engineering, suggesting that a different approach for recruitment is needed. Moreover, the results of comparing African American males' and African American females' engineering enrollment reveal that African American females' enrollment trends are more closely aligned to their male counterparts' enrollment trends than those of their gender counterparts. Both findings demonstrate that race and gender continue to play a major role in engineering enrollment disparities for African Americans.

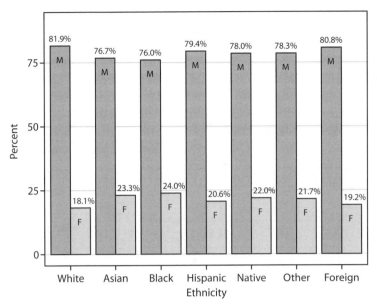

FIGURE 6.2 Baccalaureate enrollment percentages within race/ethnicity by sex, 2013
*Source:* Data from ASEE.

Similarity in discipline choices at the national level by African American undergraduate engineers can be seen in figure 6.3. Out of 10 selected engineering majors, 13% of African American females were enrolled in electrical engineering in 2007, with 11% enrolled in each of mechanical, chemical, and civil engineering. By 2013, 12% of African American engineering females were enrolled in mechanical or chemical engineering, and 11%, 9%, and 8% were enrolled in civil, electrical, and biomedical engineering majors, respectively. Electrical engineering and mechanical engineering are also the top two field choices for African American males, with 21% and 18% of African American males enrolled in these fields in 2007 and in 2013. However, African American males choose civil engineering and computer engineering as their next highest field choices, while African American women have fluctuated between civil and chemical engineering over the past six years.

Examining electrical engineering enrollment trends in greater detail is illuminating. African Americans' enrollment reflects a percentage increase comparable to the percentage of African Americans enrolled in undergraduate engineering overall. African American males represent 6.5% of all males

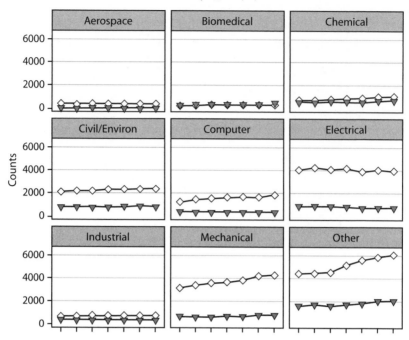

FIGURE 6.3 African American baccalaureate enrollments by discipline and sex, 2007–2013. The "Other" category is an aggregate of nine small enrollment boutique engineering majors surveyed by ASEE, including computer science (inside engineering); metallurgical and materials engineering; engineering (general); biological engineering and agricultural engineering; engineering science and engineering physics; engineering management; mining engineering; petroleum engineering; and other engineering disciplines. Georgia Tech offers only one of the majors in the "other" category—materials science and engineering. *Source:* Data from ASEE.

selecting electrical engineering, and African American females represent 9.1% of all females selecting electrical engineering (fig. 6.4A). African American females represent 16.2% of all African Americans enrolled in electrical engineering, a comparable decrease to African American female representation in undergraduate engineering overall (fig. 6.4B).

Enrollment trends at the discipline level reveal racial and gender disparities in undergraduate engineering for African Americans, often demonstrating that African Americans (apart from Native Americans) have come to compose the lowest percentage of minority engineering students. Several

A.

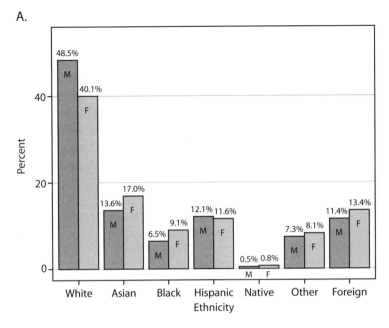

B.

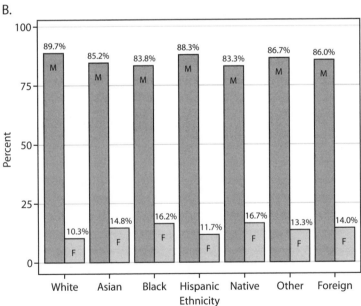

FIGURE 6.4 (A) Baccalaureate electrical engineering enrollments by race/ethnicity and sex, 2013. (B) Baccalaureate electrical engineering enrollments within race/ethnicity by sex, 2013.
*Source:* Data from ASEE.

factors are perceived to have a major impact on recent trends in African American student engineering enrollment. As mentioned earlier, financing, pre-engineering academic preparedness, and awareness impact the recruitment and enrollment of African Americans particularly. Many of the efforts of the late 1970s and early 1980s led to a slight increase in African American student engineering enrollment by the early 1990s. However, by the mid-1990s demographics on many college campuses began to change. As higher education moved beyond the Black-White dichotomy into multiracial diversity initiatives, to meet the growing demands of the new college demographics, funding and recruitment initiatives were adjusted.

## Trends in Undergraduate Degree Awards

In 2013 a total of 92,809 baccalaureate degrees in engineering were awarded across all schools reporting to ASEE, up from 73,380 baccalaureate engineering degrees awarded in 2007 (see the appendix, table 6A.2). The percentage of total bachelor's degrees awarded in engineering to African Americans dropped from 4.5% (3,283) in 2007 to 3.9% (3,591) in 2013 (table 6.3). The percentage of White students receiving engineering degrees decreased from 61.9% (45,407) to 60.7% (56,436) during the same time period. The percentage of Hispanic students receiving degrees increased from 7.0% (5,160) to 8.5% (7,917), and the percentage of bachelor's degrees awarded to Asian students was 12.3% (9,000) in 2007 and 11.9% (11,056) in 2013.

Disaggregation by gender reveals analogous patterns in degree awards for African American males and females. The percentage of total bachelor's degrees awarded in engineering to African Americans declined from 3.2%

TABLE 6.3    Baccalaureate engineering degree percentages by race/ethnicity, 2007–2013

| Ethnicity | FY07 | FY08 | FY09 | FY10 | FY11 | FY12 | FY13 |
|-----------|------|------|------|------|------|------|------|
| White | 61.9 | 61.9 | 60.5 | 62.8 | 62.2 | 61.2 | 60.7 |
| Asian | 12.3 | 12.0 | 11.5 | 11.2 | 11.4 | 11.3 | 11.9 |
| Black | 4.5 | 4.4 | 4.3 | 4.1 | 3.9 | 3.8 | 3.9 |
| Hispanic | 7.0 | 7.3 | 7.4 | 7.7 | 7.9 | 8.3 | 8.5 |
| Native | 0.5 | 0.5 | 0.5 | 0.6 | 0.6 | 0.6 | 0.7 |
| Other | 7.2 | 7.8 | 9.7 | 7.5 | 7.4 | 7.2 | 6.6 |
| Foreign | 6.7 | 6.1 | 6.1 | 6.1 | 6.8 | 7.5 | 7.8 |
| All | 100.0 | 100.0 | 100.0 | 100.0 | 100.0 | 100.0 | 100.0 |

SOURCE: ASEE.

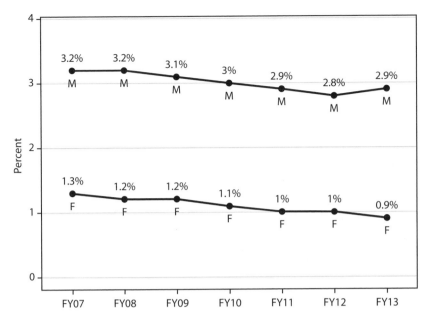

FIGURE 6.5 Baccalaureate degrees awarded to African Americans by sex, 2007–2013
*Source:* Data from ASEE.

for African American males and 1.3% for African American females in 2007 to 2.9% for African American males and 0.9% for African American females in 2013 (fig. 6.5).

In 2013, African American females received 4.8% of bachelor's degrees awarded to women in engineering, while African American males received 3.6% of all engineering bachelor's degrees awarded to males (fig. 6.6A). Similar to enrollment patterns in 2013, African American females represent the largest female gender percentage (23.7%) of engineering degrees awarded within an ethnic group, while White females (17.4%) represent the lowest female gender percentage within an ethnic group (fig. 6.6B). Overall, African American females receive around 1% of the total engineering degrees awarded at all levels of engineering education (Leonard et al. 2013).

While there was a decline in enrollment of African Americans in engineering between 2007 and 2013, the decline in degrees awarded to African Americans is more disconcerting. Funding and academic preparedness not only may impact African American's student enrollment but also may be exacerbated over time and have a greater impact on African American student

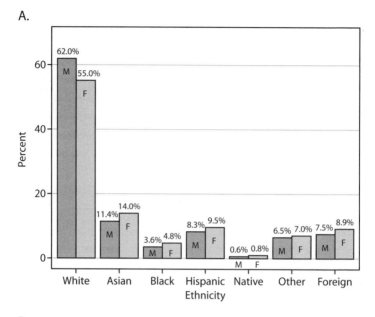

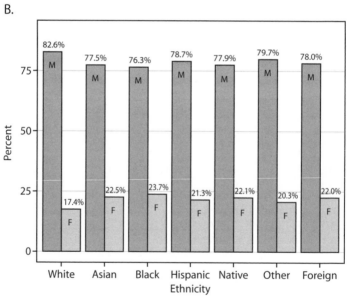

FIGURE 6.6 (A) Baccalaureate degree awards by race/ethnicity and sex, 2013. (B) Baccalaureate degree awards within race/ethnicity and sex, 2013. *Source:* Data from ASEE.

retention in engineering. Further, the decline in degrees awarded to African Americans during the same time period indicates that changes in post-enrollment factors affect African American students' persistance significantly. Key studies have shown that self-efficacy and expected outcomes affect African Americans' academic performance and retention rates. Yet, for African Americans, self-efficacy and expected outcomes are based more on external factors such as racism, faculty expectations, and behavior than on such internal factors as ability or skill. The declining trends in the production of African American engineers may suggest that there are recent changes in racial attitudes and faculty-to-student demographics within the institutional climate which adversely affect African Americans. Moreover, the coping mechanisms of African Americans for a dual-racial environment may not be adequate to sustain them in the changing multiracial environment of higher education. More longitudinal research is required to understand the profiles of high-achieving African American engineering students in the midst of institutional change in terms of ethnicity and gender. Trends in female representation within racial/ethnic groups among engineering degree recipients at the baccalaureate level highlight that gender disparities are not easily explained. When controlling for race, African American females increasingly exhibit interests in engineering fields and an ability to persist in engineering to completion, analogous to those of their male counterparts.

## Production Trends by Institution Type

Significantly, the types of institutions producing African American engineers continue to vary by gender. The number of bachelor's degrees awarded in engineering to African American females declined across HBCUs, non-HBCUs, RUs, and MCUs.* The number of bachelor's degrees in engineering awarded to African American females by non-HBCUs decreased by 5% from 2007 (710) to 2013 (663) (fig. 6.7). During the same time period, the number of bachelor's degrees in engineering awarded to African American males by non-HBCUs increased by 16% (1,809 to 2,108). HBCU and non-HBCU MCUs have remained relatively close in terms of degree production (fig. 6.7). However, HBCUs classified as RUs have not remained on par with non-HBCUs in the production of African American engineers. This may be due to the small number of HBCUs classified as RUs. Comparison of HBCUs

---

* Although HBCUs and non-HBCUs are also research and master's institutions, in order to compare HBCUs with non-HBCUs, as well as RUs and MUs, each institutional type is reflected in fig. 6.7.

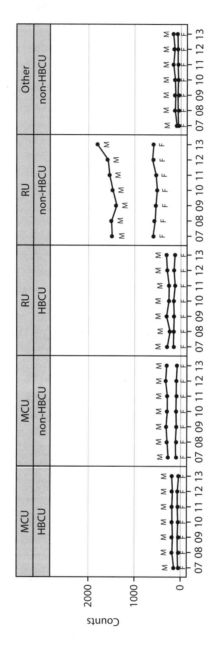

FIGURE 6.7 Baccalaureate degrees awarded to African Americans by sex and institution type, 2007–2013

*Source:* Data from ASEE.

and non-HBCUs by yield per thousand students may show similar yield rates among HBCU and non-HBCU RUs. The decline in engineering bachelor's degrees awarded to African American males by research HBCUs may also be associated with the increase in number of bachelor's degrees awarded to African American males by non-HBCU RUs. The same pattern is not observed for African American females. (For more details on HBCUs, see chap. 4 in this volume.)

The list of the top 25 colleges and univerisities awarding engineering bachelor's degrees to African Americans also varies by gender. The differences between HBCUs and non-HBCUs in the production of African American engineers are more telling when research degree conferring status is controlled for. There are three HBCU and three non-HBCU MCUs among the top 25 producers of African American female and male engineers. Comparatively, there are six HBCU and approximately 12 non-HBCU RUs among the top 25 producers of African American female and male engineers. In fact, approximately 70% of the top producers of African American engineers at the baccalaureate level are RUs. Some of the salient aspects of HBCUs are the mentoring and supporting environments, when coupled with adequate resources for research. While only a few HBCUs are classified as RUs, strategic partnerships between HBCU MCUs and non-HBCU RUs (such as dual-degree programs) can be used to enhance the production of African American engineers.

Academic leaders and policy makers should not only focus on recruiting African Americans into RUs but also capitalize on opportunities to acknowledge and cultivate the talent pool matriculating at HBCU MCUs into the engineering pipeline at greater levels than observed in the past. HBCUs' role in the production of African American engineering remains significant and can be strengthened through institutional partnerships and program initiatives. Studies that evaluate the impact of dual-degree programs on African American engineering enrollment and degree awards at some of the top-producing institutions of African Americans may reveal more tangible insights.

## Institutional Case Study: Georgia Institute of Technology

Georgia Tech was selected as the institutional case study for microlevel analysis. Georgia Tech has been noted as the top producer of minority and women engineers for the past five years (Yoder 2011). It is also the second-highest producer of African American male and female graduates at the baccalaureate level. Matthews (2006) identified Georgia Tech as the "envy of its peers" for its ranking in production of African American engineers at all levels of engineering education, employment of African American faculty, and economic viability and vitality. Matthews attributes Georgia Tech's successful rankings to important factors such as its close proximity to the Atlanta University System, its dual-degree program, and its willingness to establish partnerships with HBCUs and community colleges. In light of its continual success, Georgia Tech is positioned as a distinctive case study for understanding experiences and trends of African Americans in undergraduate engineering education.

Consistent with the trend analyses conducted in other sections of this chapter, enrollment, degree awards, and field choice were investigated. Additionally, retention and performance data were examined by race, gender, and field to obtain disaggregated data at the microlevel which could inform future studies, policies, and program initiatives.

ENROLLMENT

African Americans represented 6.5% of total engineering enrollment at Georgia Tech in 2007 and 6.0% in 2013 (fig. 6.8). These percentages are slightly higher than African Americans' representation at the national level.

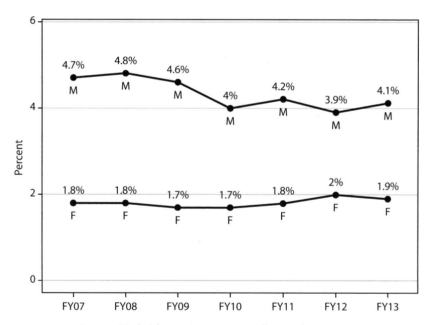

FIGURE 6.8 Georgia Tech African American enrollments by sex, 2007–2013
*Source:* Data from ASEE.

TABLE 6.4    Georgia Tech student-to-faculty ratios by ethnicity, 2007–2013

| Ethnicity | FY07 | FY08 | FY09 | FY10 | FY11 | FY12 | FY13 |
|-----------|------|------|------|------|------|------|------|
| White     | 14.6 | 14.6 | 14.9 | 16.2 | 15.9 | 17.2 | 17.2 |
| Asian     | 13.8 | 13.8 | 14.0 | 14.9 | 14.4 | 16.0 | 13.6 |
| Black     | 28.1 | 25.9 | 24.9 | 24.3 | 29.2 | 28.1 | 34.8 |
| Hispanic  | 40.2 | 39.3 | 36.1 | 39.0 | 41.3 | 49.2 | 48.2 |
| All       | 16.6 | 16.8 | 17.6 | 18.7 | 19.2 | 21.3 | 21.3 |

SOURCE: ASEE.

Similar to national patterns, the percentage of African Americans in engineering has declined over the six-year period. The number of African American males enrolled increased from 342 in 2007 to 377 in 2013, but the percentage decreased from 4.7% in 2007 to 4.1% in 2013. The drop in percentages is attributable to an increase in enrollments among other ethnicities. African American female enrollments increased from 135 to 179, but their percentage remained relatively flat at 1.8% in 2007 and 1.9% in 2013.

When enrollment numbers are analyzed in relation to faculty, a significant ratio imbalance is revealed for faculty of color. The ratio of African American students to African American faculty is almost twice as great as the ratio of White and Asian students to White and Asian faculty (see table 6.4). The slow increase in the number of African American student enrollments between 2007 and 2013 has had an ironic effect on student-to-faculty ratio, from 28:1 to 34:8, amid a changing number of African American engineering faculty. Notably, Hispanic student-to-faculty ratios are almost three times as great as White and Asian student-to-faculty ratios.

Sedlacek et al. (2007) argue that this imbalance has adverse effects on mentoring, as it requires nonminority faculty to garner skills that enable them to mentor minority and nonminority students with equal effectiveness. Further, the lack of faculty of color and the reluctance of some professors to enter into a cross-race mentoring relationship create environments of isolation for students of color, as well as additional cultural demands for faculty of color (Sedlacek et al. 2007).

FIELD CHOICE

Data for field choice by gender disclose similar patterns by engineering discipline, regardless of race/ethnicity, at Georgia Tech. Women are a higher percentage of biomedical engineering students than any of the other selected

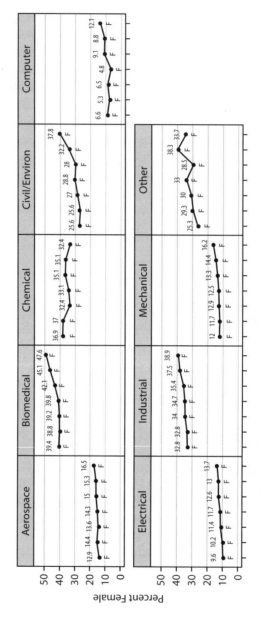

FIGURE 6.9 Georgia Tech female baccalaureate enrollment percentages by discipline, 2007–2013
*Source:* Data from ASEE.

engineering majors (fig. 6.9). Bowman (2011) highlights that the creation of biomedical engineering programs at many American universities has increased female percentage in biomedical engineering, while simultaneously decreasing female participation in electrical, mechanical, chemical, and civil engineering. The redistribution of female engineering talent, rather than the recruitment of new engineering talent, by the creation of biomedical engineering programs is presented as a cautionary conclusion (Bowman 2011).

Gender patterns, by discipline, not only are similar among female students at Georgia Tech but also align with gender patterns observed among faculty. Biomedical engineering also had the highest percentage of female faculty between 2008 and 2012 (table 6.5). In 2013, chemical and civil engineering had roughly 22% female faculty and 25%–37% female student enrollment. Electrical, mechanical, and aerospace engineering disciplines have the lowest percentages of female faculty and are among those with the lowest percentages of female students at Georgia Tech.

Gender patterns for African Americans by engineering discipline reveal that several trends in discipline enrollment at Georgia Tech align around gender rather than race (a pattern not observed at the national level among African Americans). Between 2007 and 2013 African American males were over 75% of African American students majoring in electrical and mechanical engineering at Georgia Tech. African American females chose biomedical and industrial engineering as their top two choices in 2007 (fig. 6.10). By 2013, 49.3% of biomedical African American engineers and 48.3% of chemical African American engineers enrolled at Georgia Tech were female. Since 2011, the number of African American females in computer engineering has doubled. Yet, less than 8% of African American female engineering students major in computer engineering (fig. 6.10). National trends show

TABLE 6.5   Georgia Tech female engineering faculty percentages by discipline, 2008–2013

| Discipline | FY08 | FY09 | FY10 | FY11 | FY12 | FY13 |
|---|---|---|---|---|---|---|
| Aerospace | 10.0 | 7.9 | 7.9 | 9.4 | 8.3 | 8.1 |
| Biomedical | 25.0 | 25.9 | 26.9 | 26.3 | 25.9 | 18.5 |
| Chemical | 21.2 | 25.0 | 25.0 | 26.5 | 25.7 | 22.9 |
| Civil | 13.3 | 13.3 | 15.8 | 18.0 | 18.5 | 22.6 |
| Electrical | 10.6 | 10.5 | 11.1 | 10.9 | 13.5 | 12.7 |
| Industrial | 20.4 | 20.4 | 20.8 | 17.4 | 19.6 | 23.5 |
| Mechanical | 8.6 | 8.4 | 8.4 | 10.1 | 10.6 | 12.2 |

SOURCE: ASEE.

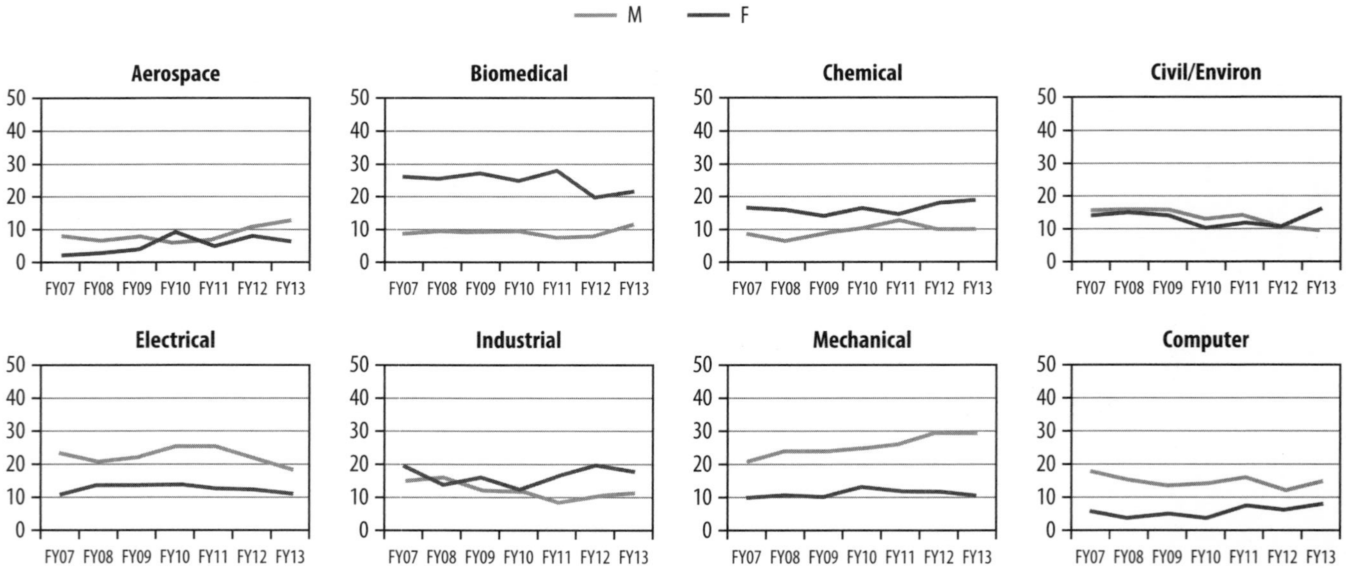

FIGURE 6.10 Georgia Tech African American baccalaureate enrollment by discipline, 2007–2013

*Source:* Data from ASEE.

that electrical and mechanical engineering disciplines continue to garner the highest percentage of African American females, although it is declining (see fig. 6.3).

Georgia Tech presents an illuminating case regarding gender and racial dynamics when analyzing enrollment by discipline compared to national trends. Similarity in African American males' and females' enrollment in electrical and mechanical engineering at the national level may be due to the longevity and access of electrical and mechanical engineering programs at several institutions. However, at Georgia Tech disparities in enrollment between African American females and males in electrical and mechanical engineering are observed. Further, African American females' increasing enrollment in industrial and chemical engineering suggests that gender variances in engineering enrollment are not consistent across disciplines regardless of longevity and access at an institution. Biomedical engineering has started to attract more African American females at noteworthy rates. However, it is difficult to verify or refute Bowman's (2011) claim that this pattern represents a redistribution of African American engineering talent rather than recruitment of new talent, in the midst of declining African American engineering enrollment at Georgia Tech. Trends in degree awards for African Americans at Georgia Tech also present complex and unique patterns.

DEGREES AWARDED

In 2007, African Americans earned 6.5% (95) of bachelor's degrees awarded in engineering at Georgia Tech. By 2013, they earned 6.0% (109) of total degrees awarded. The number of degrees awarded to African American females has increased slightly since 2010 (fig. 6.11). The percentage of engineering degrees awarded to African American males has declined, while the number of degrees awarded increased over the six-year period, with the exception of a dip in degree awards in 2012. It is important to note that these shifts are greatly influenced by the relatively small numbers of students.

The percentage of African American representation among engineering degree award recipients at Georgia Tech exceeds the percentage of African American representation at the national level, comparatively by 2%–3% during the 2007–2013 time frame. Nonetheless, African Americans continue to be underrepresented in engineering enrollment and degree awards at the baccalaureate level at Georgia Tech. Retention and performance data indicate some of the areas and fields that present the most barriers to African Americans at Georgia Tech.

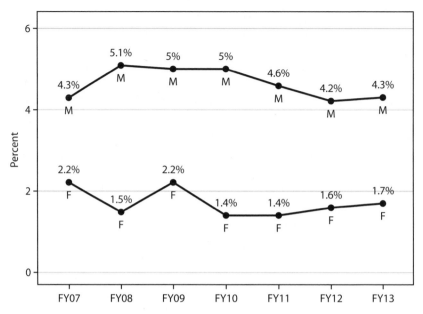

FIGURE 6.11 Georgia Tech engineering baccalaureate degrees to African Americans by sex, 2007–2013
*Source:* Data from ASEE.

RETENTION AND PERFORMANCE

High school GPA, first-semester GPA, and participation in a freshman orientation course were noted as significantly impacting second-year retention for all cohorts entering Georgia Tech since 2005. Although standardized testing is still used as a key factor in the admission's review process, SAT math and verbal scores did not prove to be a significant factor in second-year retention (Georgia Institute of Technology 2012a).

African Americans' five-year retention rates have improved at Georgia Tech, since dipping in 2005. For the 2007 freshman cohort, African Americans' six-year retention rates increase to 80.3%. African Americans' five-year retention rates for the 2008 freshman cohort reach a historic high of 84.4%.

The five-year retention rates for 2005, 2006, and 2007 cohorts in the College of Engineering were approximately 82%, 83%, and 85%, respectively. In the College of Engineering, African Americans' five-year retention rates reflect a change of over 10% between 2005 cohorts and 2007 cohorts (Georgia Institute of Technology 2012a).

The average GPA of all students in select majors within the College of Engineering between 2008 and 2013 ranged from 3.01 to 3.15 (table 6.6). The lowest average GPA in 2008 was among mechanical engineering students, while the lowest average GPA in 2013 was among civil/environmental engineering students. In 2013, electrical engineering students achieved a noted high GPA of 3.18, the highest of all select engineering majors.

Data in table 6.6 also reveal that between 2008 and 2013, African Americans had relatively low GPAs compared to the average Georgia Tech GPA in each engineering discipline. In 2008, African American females' overall GPA was 2.70, compared to an overall GPA of 3.11 for all females in engineering. African American males' average overall GPA was 2.57, compared to the overall male engineering GPA of 2.99. By 2013, African American females' overall GPA was 2.78, compared to an overall female average engineering GPA of 3.19. African American males sustained a 2.72 GPA, compared to an overall male engineering GPA of 3.13.

Additional changes in performance trends are observed when reviewing African Americans' performance at the micro level within select engineering disciplines. In 2008, African American males within the field of chemical engineering demonstrated the lowest average GPA among the select engineering disciplines. Conversely, African American males in civil engineering acquired the highest average GPA of African American male majors in 2008, and electrical engineers acquired the highest average GPA in 2013. African American males' GPA improved for each select engineering discipline in 2013 (with the exception of civil/environmental) compared to 2008. While electrical engineering and mechanical engineering are the least two selected majors for African American females (see fig. 6.11), they are the two majors in which African American females have the highest average GPA.* African American females' GPA of 2.54 in civil/environmental engineering reflects their lowest average GPA among the select engineering disciplines during 2008 and 2013. Consistent with the Lord et al. (2009) findings, academic performance does not explain gender disparities among African Americans in engineering at Georgia Tech. Apart from civil/environmental engineering, African American females outperformed their male counterparts in engineering in 2008 and 2013.

Retention and performance trends for African Americans in engineering at Georgia Tech expose the need for focused systemic initiatives that attend

---

*Aerospace engineering was not considered in performance trends owing to a small *N* for African American females.

TABLE 6.6  Performance (GPA) by discipline and sex, 2008 and 2013

| | | African American Students | | | | All Students | | | |
| | | AY2008 | | AY2013 | | AY2008 | | AY2013 | |
| Major | Sex | N | GPA | N | GPA | N | GPA | N | GPA |
| --- | --- | --- | --- | --- | --- | --- | --- | --- | --- |
| Biomedical | Male | | 2.28 | | 2.62 | 547 | 2.98 | 748 | 3.13 |
| | Female | | 2.64 | | 2.75 | 355 | 3.09 | 617 | 3.19 |
| | All | | 2.49 | | 2.70 | 902 | 3.02 | 1,365 | 3.16 |
| Civil/environmental | Male | | 2.73 | | 2.56 | 670 | 3.00 | 587 | 3.10 |
| | Female | | 2.54 | | 2.65 | 230 | 3.10 | 290 | 3.16 |
| | All | | 2.68 | | 2.59 | 900 | 3.03 | 877 | 3.12 |
| Chemical | Male | | 2.24 | | 2.73 | 382 | 2.95 | 621 | 3.14 |
| | Female | | 2.69 | | 2.77 | 220 | 3.15 | 332 | 3.12 |
| | All | | 2.46 | | 2.75 | 602 | 3.03 | 953 | 3.14 |
| Electrical | Male | | 2.64 | | 2.80 | 1,278 | 2.98 | 1,392 | 3.18 |
| | Female | | 2.82 | | 2.82 | 118 | 3.09 | 181 | 3.20 |
| | All | | 2.67 | | 2.80 | 1,396 | 2.99 | 1,573 | 3.18 |
| Industrial | Male | | 2.46 | | 2.53 | 804 | 3.03 | 1,029 | 3.11 |
| | Female | | 2.79 | | 2.67 | 394 | 3.16 | 624 | 3.20 |
| | All | | 2.59 | | 2.60 | 1,198 | 3.07 | 1,653 | 3.14 |
| Mechanical | Male | | 2.63 | | 2.76 | 1,652 | 2.97 | 2,048 | 3.12 |
| | Female | | 2.71 | | 3.09 | 225 | 3.08 | 347 | 3.24 |
| | All | | 2.64 | | 2.82 | 1,877 | 2.98 | 2,395 | 3.13 |
| All fields | Male | 352 | 2.57 | 363 | 2.72 | 5,333 | 2.99 | 6,425 | 3.13 |
| | Female | 132 | 2.70 | 170 | 2.78 | 1,542 | 3.11 | 2,391 | 3.19 |
| | All | 484 | 2.61 | 533 | 2.74 | 6,875 | 3.01 | 8,816 | 3.15 |

SOURCE: Georgia Tech.

to race and gender rather than race or gender. Diversity initiatives that focus on gender are often developed in response to gender dynamics observed when race is not equally considered. As a result, women become the focus of gender-based initiatives (often to the exclusion of minority men). African American males may have access to racially focused initiatives but are often denied access to gender-focused initiatives.* African American women may have access to gender-focused and/or racially focused initiatives that enhance their persistence and performance in engineering when compared to their African American male counterparts. However, these programs often divide African American females' engineering experiences into categories of gender or race, which does not attend to the double bind effect of their experiences. Additionally, the aggregation of race in diversity initiatives may adversely affect African Americans. US engineering policy makers and institutional leadership have the difficult task of generating diverse globally competitive science and engineering talent, amid a changing college-age population, using limited resources. Trends in baccalaureate engineering enrollment, degree awards, retention, and performance at Georgia Tech highlight both the successes in and challenges to meeting this demand in the twenty-first century.

## Conclusion

Until recently, much of the research on the engineering education and workforce participation of African Americans has been embedded in studies and reports on STEM. Failure to separate engineering from the sciences results in obscuring important distinctions. Similarly, failure to disaggregate data by race/ethnicity and gender masks significant intragroup differences. In both cases, failure to disaggregate data appropriately is likely to result in policies, programs, and practices that are flawed at best and counterproductive at worst.

Recently, the NSF has begun aggregating underrepresented minorities into one collective group in its reports. The results of this chapter suggest that this policy and practice are likely to be detrimental to enhancing scholars' knowledge regarding successful recruitment and retention of African Americans in engineering, especially in the midst of a growing college-age

---

* Georgia Tech's Office of Minority Education and Development has recently established a new initiative, GT PRIME, in alignment with the Board of Regent's African American Male Initiative (AAMI), aimed to enhance black male enrollment and performance in higher education (Georgia Institute of Technology 2012b). The initiative focuses on retention and leadership through mentoring, workshops, tutoring, and leadership training (Board of Regents of the University System of Georgia 2011).

minority population. Funding and recruitment initiatives that target minorities more generally, rather than African Americans specifically, might overlook the cultural, generational, and demographic challenges that impact African Americans differently than other minorities. Aligning trends in enrollment decline highlight the significance of race in the recruitment of African Americans in engineering. However, trends by discipline denote the importance of attending to gender as well as race for recruitment of African Americans in engineering.

Unique gender patterns are observed when African American female representation is analyzed in reference to their African American male counterparts and when African American males and females are examined in relation to engineering students of other racial/ethnic backgrounds. Therefore, addressing the declining trends in African American student enrollment in engineering cannot focus on race or gender alone. It was beyond the scope of this chapter to explore whether changes in funding, recruitment, programming, and admission practices targeting underrepresented groups mirror declining trends in African American student enrollment engineering.

The findings of this exploratory study suggest directions for further research to enhance understanding of dynamics and processes, including the following:

- Impacts of expectations on academic performance in engineering—differences by race/ethnicity and gender.
- Key factors that affect African Americans' academic performance in engineering, and how performance relates to retention rates and persistence.
- The extent to which institutional collaborations such as dual-degree programs influence the academic performance, retention, and graduation rates of African American engineering students.
- The extent to which the presence of African American faculty in engineering programs affects African American students' performance and persistence to degree.
- The extent to which African American engineering students' GPAs limit both their opportunities to pursue graduate studies and options in the workforce.
- Profiles of high-achieving (GPAs of B or higher) African American engineering students by gender.
- The implications of the continuing underrepresentation of African Americans in engineering for the twenty-first-century workforce.

# Appendix

TABLE 6A.1  Engineering enrollment by race/ethnicity and sex, 2007–2013

| Race/Ethnicity | FY07 | FY08 | FY09 | FY10 | FY11 | FY12 | FY13 |
|---|---|---|---|---|---|---|---|
| Caucasian | | | | | | | |
| Male | 217,662 | 224,446 | 229,187 | 242,516 | 251,401 | 260,330 | 270,990 |
| Female | 38,843 | 41,417 | 43,395 | 47,192 | 50,714 | 54,544 | 59,741 |
| Total | 256,505 | 265,863 | 272,582 | 289,708 | 302,115 | 314,874 | 330,731 |
| African American | | | | | | | |
| Male | 17,787 | 18,448 | 18,974 | 20,189 | 20,601 | 21,412 | 21,842 |
| Female | 5,907 | 6,262 | 6,159 | 6,338 | 6,315 | 6,698 | 6,889 |
| Total | 23,694 | 24,710 | 25,133 | 26,527 | 26,916 | 28,110 | 28,731 |
| Asian American | | | | | | | |
| Male | 35,657 | 37,540 | 39,198 | 40,256 | 42,591 | 45,292 | 48,969 |
| Female | 9,539 | 10,056 | 10,575 | 11,132 | 11,836 | 13,158 | 14,911 |
| Total | 45,196 | 47,596 | 49,773 | 51,388 | 54,427 | 58,450 | 63,880 |
| Hispanic | | | | | | | |
| Male | 31,730 | 34,931 | 37,803 | 40,151 | 44,326 | 48,014 | 52,214 |
| Female | 8,526 | 9,271 | 9,813 | 10,983 | 11,392 | 12,245 | 13,528 |
| Total | 40,256 | 44,202 | 47,616 | 51,134 | 55,718 | 60,259 | 65,742 |
| Native American/Hawaiian | | | | | | | |
| Male | 2,146 | 2,128 | 2,294 | 2,620 | 2,936 | 1,882 | 1,850 |
| Female | 609 | 627 | 684 | 783 | 760 | 541 | 528 |
| Total | 2,755 | 2,755 | 2,978 | 3,403 | 3,696 | 2,423 | 2,378 |
| Other/two or more | | | | | | | |
| Male | 25,572 | 24,133 | 30,732 | 4,774 | 7,346 | 9,976 | 11,591 |
| Female | 5,229 | 5,404 | 7,350 | 1,386 | 2,266 | 3,159 | 3,829 |
| Total | 30,801 | 29,537 | 38,082 | 6,160 | 9,612 | 13,135 | 15,420 |

(continued)

TABLE 6A.1 *continued*

| Race/Ethnicity | FY07 | FY08 | FY09 | FY10 | FY11 | FY12 | FY13 |
|---|---|---|---|---|---|---|---|
| Unknown | | | | | | | |
| Male | 0 | 0 | 0 | 24,102 | 20,135 | 16,262 | 20,991 |
| Female | 0 | 0 | 0 | 5,564 | 4,576 | 3,756 | 5,189 |
| Total | 0 | 0 | 0 | 29,666 | 24,711 | 20,018 | 26,180 |
| Foreign | | | | | | | |
| Male | 18,074 | 19,484 | 22,080 | 25,758 | 28,854 | 36,212 | 40,178 |
| Female | 4,079 | 4,621 | 5,253 | 6,233 | 6,908 | 8,681 | 9,546 |
| Total | 22,153 | 24,105 | 27,333 | 31,991 | 35,762 | 44,893 | 49,724 |
| Male total | 348,628 | 361,110 | 380,268 | 400,366 | 418,190 | 440,343 | 469,433 |
| Female total | 72,732 | 77,658 | 83,229 | 89,611 | 94,767 | 103,017 | 114,383 |
| Total | 421,360 | 438,768 | 463,497 | 489,977 | 512,957 | 543,360 | 583,816 |

SOURCE: Data from ASEE and IPEDS national database.

TABLE 6A.2   Bachelor degrees awarded by race/ethnicity and sex, 2007–2013

| Race/Ethnicity | FY07 | FY08 | FY09 | FY10 | FY11 | FY12 | FY13 |
|---|---|---|---|---|---|---|---|
| Caucasian | | | | | | | |
| Male | 38,329 | 38,712 | 37,944 | 41,383 | 42,944 | 44,722 | 46,631 |
| Female | 7,078 | 7,173 | 6,989 | 7,794 | 8,495 | 9,066 | 9,805 |
| Total | 45,407 | 45,885 | 44,933 | 49,177 | 51,439 | 53,788 | 56,436 |
| African American | | | | | | | |
| Male | 2,325 | 2,344 | 2,302 | 2,366 | 2,414 | 2,481 | 2,741 |
| Female | 958 | 909 | 877 | 826 | 821 | 895 | 850 |
| Total | 3,283 | 3,253 | 3,179 | 3,192 | 3,235 | 3,376 | 3,591 |
| Asian American | | | | | | | |
| Male | 6,960 | 6,942 | 6,677 | 6,784 | 7,252 | 7,664 | 8,565 |
| Female | 2,040 | 1,985 | 1,886 | 2,013 | 2,177 | 2,268 | 2,491 |
| Total | 9,000 | 8,927 | 8,563 | 8,797 | 9,429 | 9,932 | 11,056 |
| Hispanic | | | | | | | |
| Male | 3,974 | 4,167 | 4,270 | 4,700 | 5,120 | 5,696 | 6,227 |
| Female | 1,186 | 1,265 | 1,237 | 1,341 | 1,387 | 1,595 | 1,690 |
| Total | 5,160 | 5,432 | 5,507 | 6,041 | 6,507 | 7,291 | 7,917 |
| Native American | | | | | | | |
| Male | 254 | 295 | 271 | 297 | 277 | 275 | 315 |
| Female | 77 | 61 | 73 | 80 | 81 | 76 | 92 |
| Total | 331 | 356 | 344 | 377 | 358 | 351 | 407 |
| Other/two or more | | | | | | | |
| Male | 4,381 | 4,689 | 5,983 | 337 | 650 | 852 | 1,100 |
| Female | 906 | 1,071 | 1,248 | 101 | 208 | 260 | 303 |
| Total | 5,287 | 5,760 | 7,231 | 438 | 858 | 1,112 | 1,403 |

(continued)

TABLE 6A.2 *continued*

| Race/Ethnicity | FY07 | FY08 | FY09 | FY10 | FY11 | FY12 | FY13 |
|---|---|---|---|---|---|---|---|
| Unknown | | | | | | | |
| Male | 0 | 0 | 0 | 4,483 | 4,336 | 4,207 | 3,807 |
| Female | 0 | 0 | 0 | 993 | 888 | 1,036 | 948 |
| Total | 0 | 0 | 0 | 5,476 | 5,224 | 5,243 | 4,755 |
| Foreign | | | | | | | |
| Male | 3,887 | 3,625 | 3,619 | 3,783 | 4,412 | 5,225 | 5,650 |
| Female | 1,025 | 891 | 924 | 999 | 1,176 | 1,403 | 1,594 |
| Total | 4,912 | 4,516 | 4,543 | 4,782 | 5,588 | 6,628 | 7,244 |
| Male total | 60,110 | 60,774 | 61,066 | 64,133 | 67,405 | 71,122 | 75,036 |
| Female total | 13,270 | 13,355 | 13,234 | 14,147 | 15,233 | 16,599 | 17,773 |
| Total | 73,380 | 74,129 | 74,300 | 78,280 | 82,638 | 87,721 | 92,809 |

SOURCE: Data from ASEE and IPEDS national database.

# References

Board of Regents of the University System of Georgia. 2011. *The University System of Georgia's African-American Male Initiative: Creating a More Educated Georgia*. Atlanta: University System of Georgia. www.usg.edu/aami/AAMI_Brochure _2011.pdf.

Bond, Author, and William K. LeBold. 1977. "Factors Associated with Attracting and Retaining African Americans in Engineering." In *1977 Frontiers in Education Conference: Proceedings*, edited by Lawrence P. Grayson and Joseph M. Biedenbach, 123–130. New York: Education Group [Institute of Electrical and Electronics Engineers].

Bowman, Keith J. 2011. "Potential Impacts of Creating Biomedical Engineering Programs on Gender Distribution within Leading Engineering Colleges." *Journal of Diversity in Higher Education* 4:51–64.

Byars-Winston, Angela, Yannine Estrada, Christina Howard, Dalelia Davis, and Juan Zalapa. 2010. "Influence of Social Cognitive and Ethnic Variables on Academic Goals of Underrepresented Students in Science and Engineering: A Multiple-Groups Analysis." *Journal of Counseling Psychology* 57:205–218.

Case, J., & Jawitz, J. 2003. "Educational Paradigms and Engineering Educators in South Africa." *Higher Education* 45:251–256.

Chubin, Daryl E., Gary S. May, and Eleanor L. Babco. 2005. "Diversifying the Engineering Workforce." *Journal of Engineering Education* 94:73–86.

Concannon, James, and Lloyd Barrow. 2010. "Men's and Women's Intentions to Persist in Undergraduate Engineering Degree Programs." *Journal of Science, Education, and Technology* 19:133–145.

DeFreitas, Stacie C. 2012. "Differences between African American and European American First-Year College Students in the Relationship between Self-efficacy, Outcome Expectation, and Academic Achievement." *Social Psychology of Education* 15:109–123.

Fiegener, Mark, and Steven Proudfoot. 2013. *Baccalaureate Origins of U.S.-Trained S&E Doctorate Recipients*. Arlington, VA: National Science Foundation.

Fiss, Owen M. 1991. "An Uncertain Inheritance." In *The Outer Circle: Women in the Scientific Community*, edited by Harriet Zuckerman, Jonathan R. Cole, and John T. Bruer, 259–276. New Haven, CT: Yale University Press.

Fleming, Lorraine, Kimarie Engerman, and Ashley Griffin. 2005. "Persistence in Engineering Education: Experiences of First Year Students at a Historically Black University." In *Proceedings from the 2005 American Society for Engineering Education Annual Conference and Exposition*. Portland, OR: American Society for Engineering Education.

Fortenberry, Norman L., Jacquelyn F. Sullivan, Peter N. Jordan, and Daniel W. Knight. 2007. "Engineering Education Research Aids Instruction." *Science* 31:1175–1176.

Freeman, Terrance, and Marcus Huggans. 2009. "Persistence of African American Male Community College Students in Engineering." In *African American Males in Higher Education: Diminishing Proportions*, edited by Henry T. Frierson, Willie Pearson, and James H. Wyche, 229–252. Bingley, UK: Emerald Group.

Gavins, Raymond. 2009. "An Historical Overview of Barriers Faced by African American Males in Pursuit of Higher Education." In *African American Males in Higher Education: Diminishing Proportions*, edited by Henry T. Frierson, Willie Pearson, and James H. Wyche, 13–30. Bingley, UK: Emerald Group.

Georgia Institute of Technology. 2012a. *Annual First-Time Freshman Retention Study*. Atlanta: Georgia Institute of Technology, Office of Institutional Research and Planning. www.irp.gatech.edu/wp-content/uploads/2011/11/First-Time-Freshman-Retention-Study_Fall-2012_05072013.pdf.

Georgia Institute of Technology. 2012b. *Complete College Georgia Plan*. Atlanta: Georgia Institute of Technology, Office of Institutional Research and Planning. www.irp.gatech.edu/wp-content/uploads/2012/12/Georgia-Tech-Complete-College-Georgia-Plan_August2012.pdf.

Gibbons, Michael T. 2010. *Engineering by the Numbers*. American Society for Engineering Education. www.asee.org.

Hrabowski, Freeman, III, and Willie Pearson, Jr. 1993. "Recruiting and Retaining African American Males in College Science and Engineering." *Journal College Science Teaching* 22:234–238.

Leggon, Cheryl B. 2006. "Women in Science: Racial and Ethnic Differences and the Differences They Make." *Journal of Technology Transfer* 31 (3): 325–333.

Leggon, Cheryl, and Connie McNeely. 2012. "Promising Policies." In *Blueprint for the Future: Framing the Issues of Women in Science in a Global Context*, rapporteurs Catherine Didion, Lisa Frehill, and Willie Pearson, Jr., 109–115. Washington, DC: National Research Council of the National Academies.

Leggon, Cheryl B., and Willie Pearson, Jr. 2009. "Assessing Programs to Improve Minority Participation in the STEM Fields: What We Know and What We Need to Know." In *Doctoral Education and the Faculty of the Future*, edited by Ronald Ehrenberg and Charlotte Kuh, 160–171. Ithaca, NY: Cornell University Press, 2009.

Leonard, J., S. Atwaters, C. Leggon, W. Pearson, Jr., and M. Gaines. 2013. "Blacks, Historical Disadvantage, and Engineering Education: Lessons Learned from the U.S. of America." *South African Society for Engineering Education (SASEE) Biennial Meeting Conference Proceedings*. Cape Town, South Africa: South African Society for Engineering Education.

Lord, Susan M., Michelle Camacho, Richard Layton, Russell Long, Matthew Ohland, and Mara Washburn. 2009. "Who's Persisting in Engineering? A Comparative Analysis of Female and Male Asian, African American, Hispanic, Native American, and White Students." *Journal of Women and Minorities in Science and Engineering* 15:167–190.

Malcom-Piqueux, Lindsey, and Shirley Malcom. 2013. "Engineering Diversity: Fixing the Educational System to Promote Equity." *Bridge: Linking Engineering and Society* 43:24–34.

Matthews, Frank. 2006. "The Envy of Its Peers?" *Diverse: Issues in Higher Education* 23:26–31.

McGee, Ebony, and Danny Martin. 2011. "'You Would Not Believe What I Have to Go Through to Prove My Intellectual Value!' Stereotype Management among Academically Successful African American Mathematics and Engineering Students." *American Educational Research Journal* 48:1347–1389.

Morrison, Catherine, and Lea Williams. 1993. "Minority Engineering Programs: A Case-Study for Institutional Support." *NACME Research Letter* 4:10.

NACME (National Action Council for Minorities in Engineering). 2011. "African Americans in Engineering." *NACME Research and Policy Brief* 1 (4).

National Academies. 2010. *Rising above the Gathering Storm, Revisited: Rapidly Approaching Category 5*. Washington, DC: National Academies Press.

National Academies. 2011. *Expanding Minority Participation: America's Science and Technology Talent at the Crossroads*. Washington, DC: National Academies Press.

National Research Council. 2007. *Beyond Bias and Barriers: Fulfilling the Potential of Women in Academic Science and Engineering*. Washington, DC: National Academies Press.

National Science Board. 2012. *Science and Engineering Indicators 2012*. Arlington, VA: National Science Foundation. www.nsf.gov/statistics/seind12/.

National Science Foundation. 2000. *Women, Minorities, and Persons with Disabilities in S&E*. Special Report NSF 00-327. Arlington, VA: National Science Foundation, 2000. www.nsf.gov/statistics/women/.

National Science Foundation. 2007. *Women, Minorities, and Persons with Disabilities in Science and Engineering*. Arlington, VA: National Science Foundation.

National Science Foundation. 2011. *Women, Minorities, and Persons with Disabilities in Science and Engineering*. Special Report NSF 11-309. Arlington, VA: National Science Foundation. www.nsf.gov/statistics/women/.

National Science Foundation. 2013. *Women, Minorities, and Persons with Disabilities in Science and Engineering*. Special Report NSF 13-304. Arlington, VA: National Science Foundation. www.nsf.gov/statistics/women/.

Ohland, Matthew, and Guili Zhang. "A Study of the Impact of Minority Engineering Programs at the FAMU-FSU College of Engineering." *Journal of Engineering Education* 91:435–440.

Ong, Maria, Carol Wright, Lorelle Espinosa, and Gary Orfield. 2011. "Inside the Double Bind: A Synthesis of Empirical Research on Undergraduate and Graduate Women of Color in Science, Technology, Engineering, and Mathematics." *Harvard Education Review* 81:172–208.

Page, Scott. 2007. *The Difference: How the Power of Diversity Creates Better Groups, Firms, Schools, and Societies*. Princeton, NJ: Princeton University Press.

Pearson, Willie, Jr., and Jon Miller. 2012. "Pathways to an Engineering Career." *Peabody Journal of Education* 87:46–61.

Perna, Laura, Valerie Lundy-Wagnar, Noah Drezner, Marybeth Gasman, Susan Yoon, Enaksh Bose, and Shannon Gary. 2009. "The Contribution of HBCUs to the Preparation of African American Women for STEM Careers: A Case Study." *Research in Higher Education* 50:1–23.

Sedlacek, William, Eric Benjamin, Lewis Schlosser, and Hung-Bin Sheu. 2007. "Mentoring in Academia: Considerations for Diverse Populations." In *The Blackwell Handbook of Mentoring: A Multiple Prospective Approach*, edited by Tammy Allen and Lillian Eby, 259–280. Malden, MA: Blackwell.

Shehab, Randa, Teri Murphy, Jeanette Davidson, Teri Rhoades, Deborah Trytten, and Susan Walden. 2007. "Academic Struggles and Strategies: How Minority Students Persist." Paper presented at 2007 American Society for Engineering Education Annual Conference and Exhibition, Honolulu, Hawaii.

US Census Bureau. 2012. *2012 Statistical Abstract: The National Data Book*. Washington, DC: US Department of Commerce. www.census.gov/compendia /statab/.

Wilson, Zakiya S., Lakenya Holmes, Karin Degravelles, Monica R. Sylvain, Lisa Batiste, Misty Johnson, Saundra Y. McGuire, Su Seng Pang, and Isiah M. Warner. 2012. "Hierarchical Mentoring: A Transformative Strategy for Improving Diversity and Retention in Undergraduate STEM Disciplines." *Journal of Science, Education, and Technology* 21:148–156.

Yoder, B. 2011. Engineering by the Numbers. American Society for Engineering Education. www.asee.org/papers-and-publications/publications/college-profiles /2011-profile-engineering-statistics.pdf.

Yoder, B. 2013. *Engineering by the Numbers*. American Society for Engineering Education. www.asee.org/papers-and-publications/publications/14_11-47.pdf.

# Workforce Participation

The four chapters in part III discuss the participation of African Americans in the engineering workforce in the United States. In chapter 7, Carl S. Person and Lenell Allen profile outstanding African American engineering professionals at the National Aeronautics and Space administration (NASA). These individuals, who come from the 10 NASA centers and NASA Headquarters, were selected based on several criteria, including their tenure with the agency and their broad engineering and leadership skills. The authors highlight the contributions of each individual not only to NASA but also to the broader engineering community.

In chapter 8, Rodney Adkins and Garland L. Thompson highlight the contributions of African American engineers to business and industry. They focus on the gains for corporations that have adopted diversity practices, as well as the contributions African American engineers have made as a result of these efforts.

In chapter 9, Cheryl B. Leggon and Gilda A. Barabino present a case study of African American women in tenure-track positions on engineering faculties who participated in a three-year professional initiative. The primary goal of the initiative is to enhance academic engineering career success through a series of research-based activities. In addition, this program aims at enriching participants' understanding of the rules governing academic engineering in general and in the context of their institutional settings in particular.

In chapter 10, Tyrone D. Taborn and Lango Deen emphasize the importance of stories and the role of recognition as critical tools to motivate and inspire African Americans—especially youth. The authors highlight the stories of African American engineers, inventors, and scientists who have received significant awards and recognition. In addition, they discuss the roles of Black-oriented media and professional organizations in developing and institutionalizing systemic honorific recognitions of African Americans' substantial contributions to engineering and science.

# CHAPTER SEVEN

# Profiles of Distinguished
# African American Engineers at NASA

CARL S. PERSON *and* LENELL ALLEN

T HE PEOPLE OF THE National Aeronautics and Space Administration (NASA) have the privilege of doing things that have never been done before, of enabling people to see things that have never been seen before. The men and women of NASA have been responsible for incredible achievements, such as landing human explorers on the moon, sending robots to scout the solar system, and making transformative discoveries about the nature of our universe. The world has benefited enormously from NASA's communications, weather, and earth monitoring satellites and the agency's leadership in aeronautics research, for example, fly-by-wire technology and reduced drag fuselage configurations. As a result of the new technologies derived from NASA research, our lives are immeasurably better.

African American engineers at NASA have made and are making substantial contributions to the technical and scientific achievements of the agency. NASA, like other science and technology agencies, is striving to increase the diversity of its science and technology workforce. Accordingly, NASA has identified one of its strategic goals and outcomes to address the issue of diversity and inclusion: *Strategic Goal 5: Enable program and institutional capabilities to conduct NASA's aeronautics and space activities. Outcome 5.1: Identify, cultivate, and sustain a diverse workforce and inclusive work environment that is needed to conduct NASA's missions* (NASA 2011). This chapter highlights some of the previous and current African American engineers who have contributed and are contributing to the legacy of NASA and the nation as we continue to explore the unknown.

In the following sections, we will talk about African American pathfinders or individuals who helped to open the door and set the standard for careers in engineering at NASA. Finally, we will present the profiles of an array of successful African American engineers at NASA.

## Reflections on NASA

In October 2008, Neil deGrasse Tyson, world-renowned astrophysicist and director of the Hayden Planetarium, wrote a Happy Fiftieth Birthday letter to NASA. It reads in part as follows:

> During the 1960s, the Civil Rights movement was more real to me than it surely was to you. In fact, it took a directive from President Johnson in 1963 to force you to hire black engineers at your prestigious Marshall Space Flight Center in Huntsville, Ala. I found the correspondence in your archives. Do you remember? James Webb, then head of NASA, wrote to German rocket pioneer Wernher von Braun, who headed the center and who was the chief engineer of the entire manned space program. The letter boldly and bluntly directs von Braun to address the "lack of equal employment opportunity for Negroes" in the region, and to collaborate with the area colleges Alabama A&M and Tuskegee Institute to identify, train and recruit qualified Negro engineers into the NASA Huntsville family. . . .
>
> Over the decades that followed you've come a long way, including, most recently, a presidentially initiated, congressionally endorsed vision statement that finally gets us back out of low earth orbit. Whoever does not yet recognize the value of this adventure to our nation's future soon will, as the rest of the developed and developing world passes us by in every measure of technological and economic strength. Not only that, today you look much more like America—from your senior-level managers to your most decorated astronauts. Congratulations. You now belong to the entire citizenry. Examples of this abound, but I especially remember when the public took ownership of the Hubble Space Telescope, your most beloved unmanned mission. They all spoke loudly back in 2004, ultimately reversing the threat that the telescope might not be serviced a fourth time, extending its life for another decade. Hubble's transcendent images of the cosmos had spoken to us all, as did the personal profiles of the space shuttle astronauts who deployed and serviced the telescope and the scientists who benefited from its data stream.
>
> Not only that, I've even joined the ranks of your most trusted, as I served dutifully on your prestigious Advisory Council. I came to recognize that when you're at your best, nothing in this world can inspire the dreams of a nation the way you can—dreams fueled by a pipeline of ambitious students eager to become scientists, engineers and technologists in the service of the

greatest quest there ever was. You have come to represent a fundamental part of America's identity, not only to itself but to the world. (Tyson 2008)

In the early years of NASA, the presence of African Americans in science, technology, engineering, and mathematics (STEM) careers was minimal at best. The individuals highlighted below represent some true trailblazers in the history of African Americans at NASA.

## African American Pathfinders at NASA

Katherine Johnson, a mathematician and computer scientist, was born on August 26, 1918, in White Sulphur Springs, West Virginia. Her mother was a teacher, and her father was a farmer and janitor. From a young age, Johnson enjoyed mathematics and could easily solve aeronautical trajectory equations. Her father moved their family to Institute, West Virginia, which was 125 miles away from the family home, so that Johnson and her siblings could attend school. She attended West Virginia State High School and graduated at age 14. Johnson received her BS in French and mathematics in 1932 from West Virginia State University (formerly West Virginia State College). At that time, Dr. W. W. Schiefflin Claytor, the third African American to earn a PhD degree in mathematics, created a special course in analytic geometry specifically for Johnson. In 1940, she attended West Virginia University to obtain a graduate degree. Johnson was one of the first African Americans to enroll in the mathematics program. However, family issues kept her from completing the required courses.

After college, Johnson began teaching in elementary and high schools in Virginia and West Virginia. In 1953, she joined Langley Research Center (LaRC) as a research mathematician for the National Advisory Committee for Aeronautics (NACA). Johnson was assigned to the all-male flight research division. Her knowledge made her invaluable to her superiors, and her assertiveness won her a spot in previously all-male meetings. NACA became the National Aeronautics and Space Administration (NASA) in 1958. Upon leaving the Flight Mechanics Branch, Johnson went on to join the Spacecraft Controls Branch, where she calculated the flight trajectory for Alan Shepard, the first American to go into space, in 1959. Johnson also verified the mathematics behind John Glenn's orbit around the earth in 1962 and calculated the flight trajectory for Apollo 11's flight to the moon in 1969. She retired from NASA in 1986 (HistoryMakers 2012).

Harriet G. Jenkins was born on July 26, 1926, in Forth Worth, Texas. She earned her BA in mathematics from Fisk University. Jenkins began her teaching career in the Berkeley, California, school system. She quickly rose through the ranks, serving as the city's first Black female vice principal. In 1957, she earned her master's degree in education from the University of California at Berkeley. She went on to become a school principal, director of elementary education, and assistant superintendent of schools.

From 1974 until 1992, Jenkins worked as the assistant administrator for equal opportunity programs at NASA. At that time, NASA, like other federal agencies, had not been very successful in integrating its workforce. During her tenure at NASA, Jenkins was responsible for implementing several programs that assisted minorities and women, including recruiting some of the agency's first African American astronauts. NASA was successful in bringing about change because it applied to this human resource challenge the same managerial concern and diligence that had been applied to its technical, scientific, and engineering programs and projects. Under her leadership, research centers were established among competitively selected minority universities; and because women and minorities were historically scarce in science and engineering disciplines and resource pools, the underrepresented groups were encouraged to apply for and participate in numerous NASA programs and initiatives. In 1984, she earned her law degree from Georgetown University (HistoryMakers 2004).

During her tenure at NASA, Jenkins received numerous agency honors, including its Distinguished and Outstanding Leadership medals, as well as the Federal Presidential Meritorious and Distinguished Service medals. She left the agency in 1992 and directed the Office of Senate Fair Employment Practices for the United States Senate until 1997. In 2000, NASA honored Jenkins by establishing the Harriett G. Jenkins Pre-Doctoral Fellowship Program, which competitively selects 20 outstanding college graduates each year for assistance with their predoctoral education for up to three years. Since its beginning, 35 fellows have received doctoral degrees and 33 fellows have received master's degrees (NASA 2008).

**Edward Dwight,** the first African American to be trained as an astronaut and the sculptor of major monuments, was born on the outskirts of Kansas City, Kansas, on September 9, 1933. Dwight joined the United States Air Force (USAF) in 1953, pursuing his dream of flying jet aircraft. He became a USAF test pilot, and in 1961 he earned a BS in aeronautical engineering from Arizona State University. At the suggestion of the National Urban League's Whitney M. Young Jr., the Kennedy administration chose Capt. Ed Dwight as the first Negro astronaut trainee in 1962. Catapulted to instant fame, he was featured on the cover of *Ebony, Jet,* and *Sepia* and in news magazines around the world. He resigned in 1966, never having gone into space. Dwight's talents then led him to work as an engineer, in real estate, and for IBM (HistoryMakers 2002).

As a pilot and engineer, Dwight served in a variety of capacities: a jet pilot instructor at Williams Air Force Base, Arizona; a B57 bomber pilot in Japan; a chief of collateral training at Travis Air Force Base, California; a graduate of the test pilot training program and candidate for astronaut training at the Aerospace Research Pilot School at Edwards Air Force Base, California; and the deputy for flight tests in the Aeronautical Systems Division at Wright-Patterson Air Force Base in Ohio.

Dwight represented the top African Americans of his day when he was selected by the Kennedy administration to train as an astronaut. As both the space program and civil rights activism were emerging, Dwight served as a pathfinder for other African American astronauts and aerospace engineers.

**Guion Bluford Jr.** was born on November 22, 1942, in Philadelphia, Pennsylvania. Bluford became the first African American astronaut to fly in space in 1983, as a member of the crew of the space shuttle *Challenger* on mission STS-8. He received a BS in aerospace engineering from the Pennsylvania State University, an MS in aerospace engineering from the Air Force Institute of Technology (AFIT), a PhD in aerospace engineering with a minor in laser physics from AFIT, and an MBA from the University of Houston–Clear Lake.

Upon receiving his undergraduate degree, Bluford began his career in the air force. He earned his pilot wings and flew 144 combat missions, of which more than 60 were over North Vietnam. After Bluford logged more than 5,200 hours of jet flight time, it was time for him to reach even higher. Bluford's NASA career began in 1979, when he was picked to be one of 35 individuals to become space shuttle astronauts (out of 10,000 who applied to the program). (It is significant to note that two other African American astronauts were in Bluford's cohort: Dr. Ronald McNair and Col. Frederick Gregory.) During Bluford's career with NASA, he flew into space four times. In 1983, he was aboard the space shuttle *Challenger* for his first trip to space. This was the first mission to have a night launch and a night landing. In 1985, Bluford went on his second space shuttle mission. Seven other crew members joined Bluford in this flight, making it the largest crew to fly into space. Bluford made his third and fourth trips to space in 1991 and 1992, respectively. In 1993, Bluford retired from NASA and the USAF as a colonel. When asked about being an astronaut, Bluford responded, "The job is so fantastic, you don't need a hobby. The hobby is going to work" (Sylvia 2008).

**Janez Yvonne Lawson Bordeaux**, the first African American to earn a technical position at the Jet Propulsion Laboratory (JPL) in Pasadena, California, was born on February 22, 1930, to Mr. Hilliard L. Lawson and Mrs. Bernice Stout Lawson in Santa Monica, California. From an early age, Mrs. Bordeaux had shown an aptitude for mathematics and subsequently won a scholarship from the Delta Sigma Theta sorority. She graduated from Belmont High School in 1948 and enrolled in the chemical engineering program at the University of California, Los Angeles (UCLA). She was very active in her sorority, including serving as the president during the 1949–1950 academic year ("Delta Mothers Honored Guests at Reception Sunday Afternoon" 1950).

Janez joined JPL in March 1952 as a "computer," one of a team of women hired to carry out the critical mathematical calculations (done by hand) required to plot satellite trajectories. She was selected to attend an IBM training school to "program problems" for JPL's new IBM computer in February 1953 and was promoted to mathematician assistant in 1954—10 years before the Civil Rights Act of 1964. She married Theodore Austin Bordeaux on August 29, 1954, in Santa Monica, California ("Film Series Begins Sunday; Mutiny on the Bounty" 1954; Lab-Oratory 1954). They had two children: Mrs. Janine Yvette Bordeaux Smith, who currently resides in Nipomo, California, and Mr. Todd Austin Bordeaux. Janez preceded her husband and son in death on November 24, 1999 (Allen 2013).

Although it is unknown when Janez left JPL, a 1957 American Mathematical Society conference proceedings listed her employment with the Ramo-Wooldridge Corporation (American Mathematical Society 1957). It is interesting to note that the third JPL director, Dr. Louis G. Dunn, left JPL in August 1954 to take over the Atlas missile project for the Ramo-Wooldridge Corporation. Thompson Products merged with Ramo-Wooldridge in October 1958 and became TRW. In addition, Dr. Louis Dunn became the president of Space Technology Laboratories, a subsidiary of TRW, which built *Pioneer 1*—NASA's first satellite (Portree 1998; Hromas 2005; Christyne Lawson [Janez's sister], personal communication with L. Allen, September 17, 2012).

**Mae Jemison** was born on October 17, 1956, in Decatur, Alabama, but considers Chicago, Illinois, her hometown since her family moved there when she was 3. She became the first African American woman to travel in space when she went into orbit aboard the space shuttle *Endeavour* on September 12, 1992. Jemison earned a BS in chemical engineering and fulfilled the requirements for a BA in African and Afro-American Studies from Stanford University. Jemison also earned her MD from Cornell University. After completing her medical training, Jemison joined the staff of the Peace Corps and served as a medical officer from 1983 to 1985, responsible for the health of Peace Corps volunteers serving in Liberia and Sierra Leone. Jemison also served as a general practitioner prior to becoming an astronaut. Following her return to the United States, Jemison decided to pursue a childhood dream. After the historic flight of Sally Ride, the first American woman in space, Jemison applied to NASA's astronaut program, feeling that more opportunities had opened up. Jemison's inspiration for joining NASA was African American actress Nichelle Nichols, who portrayed Lieutenant Uhura on *Star Trek*. Jemison was turned down on her first application to NASA, but in 1987 she was accepted on her second application. Her work with NASA before her shuttle launch included launch support activities at the Kennedy Space Center in Florida and verification of shuttle computer software in the Shuttle Avionics Integration Laboratory. Jemison flew her only space mission from September 12 to 20, 1992, as a mission specialist on STS-47. Jemison resigned from NASA in March 1993 (NASA 1993). As a tireless advocate for science education and technology, Jemison is a role model for African American girls and boys to pursue their dreams.

## African American Leadership at NASA

Charles Bolden Jr. was selected as the best example of African American leadership at NASA in his capacity as the agency's Administrator. His extraordinary career as an astronaut, Marine Corps general, and NASA Administrator was built on the foundation of a BS in electrical engineering and an MS in systems management.

Bolden, a retired Marine Corps major general, was nominated by President Barack Obama, confirmed by the US Senate, and began his duties as the 12th Administrator of NASA on July 17, 2009. As Administrator, he leads the NASA team and manages its resources to advance the agency's missions and goals. Mr. Bolden is the first African American to serve as the Administrator of NASA.

Bolden's confirmation marks the beginning of his second stint with the nation's space agency. His 34-year career with the Marine Corps included 14 years as a member of NASA's Astronaut Office. After joining the office in 1980, he traveled into orbit four times between 1986 and 1994, commanding two of the missions. His flights included deployment of the *Hubble Space Telescope* and the first joint US-Russian shuttle mission, which featured a cosmonaut as a member of his crew.

Bolden was born on August 19, 1946, in Columbia, South Carolina. He graduated from C. A. Johnson High School in 1964 and received an appointment to the US Naval Academy. Bolden earned a BS in electrical science in

1968 and was commissioned as a second lieutenant in the Marine Corps. After completing flight training in 1970, he became a naval aviator. Bolden flew more than 100 combat missions in North and South Vietnam, Laos, and Cambodia, while stationed in Namphong, Thailand, from 1972 to 1973.

After returning to the United States, Bolden served in a variety of positions in the Marine Corps in California and earned an MS in systems management from the University of Southern California in 1977. Following graduation, he was assigned to the Naval Test Pilot School at Patuxent River, Maryland, and completed his training in 1979. While working at the Naval Air Test Center's Systems Engineering and Strike Aircraft Test Directorates, he tested a variety of ground attack aircraft, until his selection as an astronaut candidate in 1980.

Bolden's NASA astronaut career included technical assignments as the Astronaut Office safety officer; technical assistant to the director of Flight Crew Operations; special assistant to the director of the Johnson Space Center; chief of the Safety Division at Johnson (overseeing safety efforts for the return to flight after the 1986 *Challenger* accident); lead astronaut for vehicle test and checkout at the Kennedy Space Center; and assistant deputy administrator at NASA Headquarters. After his final space shuttle flight in 1994, he left the agency to return to active duty with the operating forces in the Marine Corps as the deputy commandant of midshipmen at the US Naval Academy.

Bolden was assigned as the deputy commanding general of the 1st Marine Expeditionary Force in the Pacific in 1997. During the first half of 1998, he served as commanding general of the 1st Marine Expeditionary Force Forward in support of Operation Desert Thunder in Kuwait. Bolden was promoted to his final rank of major general in July 1998 and named deputy commander of US Forces in Japan. He later served as the commanding general of the 3rd Marine Aircraft Wing at Marine Corps Air Station Miramar in San Diego, California, from 2000 until 2002, before retiring from the Marine Corps in 2003. Bolden's many military decorations include the Defense Superior Service Medal and the Distinguished Flying Cross. He was inducted into the US Astronaut Hall of Fame in May 2006 (NASA 2009).

## Profiles of African American Engineers at NASA

If your image of a NASA engineer or scientist is that of a White male in a crisp white shirt with black clip-on tie and pocket protector, think again. NASA has evolved, and so has its workforce. Drawing on the talents of individuals from all nationalities and cultural backgrounds, NASA is looking to acquire the best of what humanity has to offer.

No one builds a rocket or makes a discovery in space alone. Hundreds and sometimes thousands of people may be involved in a single project. For a mission to succeed, NASA scientists and engineers must share certain qualities despite their inherent differences, "qualities like patience, dedication, optimism, faith in colleagues, a willingness to take informed risks, and the capacity to be a team player," according to former JPL director Edward C. Stone, also the *Voyager* project scientist. Only together can scientists and engineers do the work of NASA, and it has been that way from the start. NASA is blessed with a wealth of talented African American engineers.

The individuals selected below all consented to be interviewed in person or by phone and/or to complete a survey from April 2012 to August 2013. The number of participants and the gender mix are both representative of the proportion of African American engineers at NASA. We selected individuals who represented a broad spectrum of educational pathways (public and private historically Black colleges and universities [HBCUs] and predominantly White institutions [PWIs]) and NASA responsibilities. The African American engineers highlighted below are outstanding examples of the diversity of African American engineers currently employed by NASA. Although not all of the individuals cited below are currently doing engineering, the foundational platform they used to succeed in their current roles was engineering. Besides their engineering prowess, the key thread that connects the engineers cited below is not only their belief in but their actions to "pay it forward."

**Mark Branch.** Born: March 12, 1969, in Louisville, Kentucky. Team Lead in the Environmental Test and Integration Branch at NASA's Goddard Space Flight Center (GSFC) in Greenbelt, Maryland. BS in aerospace engineering and physics from Morgan State University in Baltimore, Maryland.

INTERVIEW QUESTIONS

*1. How early did you know you wanted to be an engineer?*

I wanted to be an engineer since I was a young child, maybe 7 or 8 years of age. As I tell kids during my educational outreach endeavors, it all started when I took apart a toy that had broken . . . and I tried to make it work again. That's what engineers do—we solve problems and we figure out how to make things work.

2. *Who and what influenced you to become an engineer?*

When I was around 6 or 7 years old, my mom—an educator—gave me a book about the evolution of flight. It showed early conceptions of aircraft from Leonardo da Vinci to the Wright brothers. Then it showed how aircraft would evolve from World War I through the modern-day space shuttle. Then it gave a futuristic glimpse of post-shuttle spacecraft leading up to the famous starship *Enterprise* (from *Star Trek*). One thing that was common to all of the pictures from the space shuttle onward was the NASA logo on the side of the vehicle. Those images made me want to work for NASA when I grew up! Also, my high school physics teacher, Mrs. Viele, was instrumental in my desire to become an engineer. She made science fun, and I always wanted to learn more because of her.

3. *What excites you about engineering?*

The thing that excites me most about engineering is taking a real-world, seemingly complex problem and breaking it down until you come to the root cause of the problem. Whenever you can break something down to its root cause, then you are more likely to solve the problem. I *love* solving problems! In many of my experiences, complex (or seemingly complex) problems have very simple solutions.

4. *What is your job at NASA?*

I work in the Environmental Test and Integration Branch at NASA/GSFC. Here, we individually test the instruments and components of a satellite and then integrate them into a full satellite system (a.k.a. an observatory). Accordingly, we perform both subsystem- and system-level testing prior to launching the full observatory into space.

There are different groups within my branch that test different aspects of the space environment. I am the leader of a group that tests the electromagnetic environment of space. As we know the path that the satellite system is likely to travel in space, we use one of my world-class test facilities to simulate the electromagnetic environment that the observatory is likely to encounter during its mission. If the subsystems and fully integrated systems pass the tests that they endure in my lab, then we (the scientists and engineers who built and test those systems) have a good feeling that they will likely survive the real-life environment when launched into space.

5. *How has engineering changed?*

Automation has changed engineering in my 22 plus years as an engineer at NASA. Things that we used to do by hand are now done more precisely, more efficiently, and often much faster than in years prior to automation.

I also see a lot more African American faces in all facets of engineering today versus when I started. Beginning my career at NASA in 1991, I was not only the youngest face at the meetings I attended, but also my face was much darker than other folks in the room. I'm thankful to see various shades of intelligence nowadays throughout my workplace. I pray that trend continues to grow into the future.

6. *What do you see as the future of engineering?*

I see more automation coming into play as we increase our expertise in robotics in a wide range of engineering applications. I also see more minorities becoming involved in engineering. So, the complexion of this field of study is steadily changing as we move forward.

7. *What have been the keys to your success in school and on the job?*

The number one key to success for me has been maintaining a high level of confidence in my knowledge, skills, and abilities in performing my job. As I came from a relatively small school (Morgan State University), I was somewhat intimidated when I first started working at NASA because many of my colleagues were from larger and more prominent schools such as MIT, Cal Berkley, Stanford, and many others. However, I soon realized that my HBCU—Morgan State—prepared me to compete with anyone. So, after realizing 22 years ago that I could compete at a high level with the best of the best—despite the color of my skin and despite few people outside of Maryland knowing about my school—I am still here because I believed in myself.

The other key to my success in school and at work is maintaining a healthy balance between those two areas and other things that I love to do. I love music (deejaying specifically), and I love my job as an aerospace engineer. I have maintained a healthy balance in my life for over two decades by focusing on the things that I *love* to do. As such, my life at NASA and my life outside of NASA have been very rewarding for me.

8. *What are you most proud of as a NASA engineer?*

Besides working on things that actually travel into space (which is very exciting), I am most proud that NASA has allowed me to reach out for the past 22 plus years and encourage others—namely, African American youths—about the joys of STEM and how rewarding it can be for those who choose to pursue a career in one or more of those disciplines. I have talked to and I have encouraged literally thousands of our youths—Black,

White, Yellow, and Brown—to become the next generation of explorers. For that, I am extremely proud—not of myself but of the agency for which I serve because it gave me a platform to give back to my community.

9. *How do you respond when someone asks, do you believe in "paying it forward?"*

I respond by saying that I do this every day when I am asked to talk to kids. I do this for free, as I have never received an honorarium for my public outreach. I know that my investment of time will pay big dividends down the road when the child that I inspire to become an engineer or scientist actually becomes an engineer or scientist, as well as a productive citizen. That makes me feel good! It makes me feel even better when I challenge kids to do the same thing that I'm doing once they've achieved their goals, and they accept that challenge. If they give back and try to uplift the generation that comes after them, then the cycle of giving will continue. Thus, the impact that I have on my community will last many lifetimes!

**Elaine Flowers Duncan.** Born: January 7, 1955, in Montgomery, Alabama. Manager of Aerospace Flight System Engineering at NASA's Marshall Space Flight Center in Huntsville, Alabama. BS in Mathematics from Alabama State University in Montgomery, Alabama, and MS in systems engineering from Howard University in Washington, DC.

INTERVIEW QUESTIONS

1. *How early did you know you wanted to be an engineer?*

As a student, since elementary school, mathematics and science were always my favorite subjects. I always thought I wanted to be a mathematician or computer programmer. It was not until I had completed undergraduate school with a BS in mathematics and a minor in computer science that I realized I wanted to be an engineer.

2. *Who and what influenced you to become an engineer?*

During my senior year at Alabama State University, I participated in an internship as a mathematician/computer programmer at the Brookhaven National Laboratory (BNL). I served on a team of statisticians and simulation engineers, tasked to perform a technical assessment of the Refuse Collection Systems of New York City. We developed numerical representations, using techniques of linear programming to simulate the problem and to find solutions to minimize the amount of refuse left on the streets. My advisor, who was a senior statistician at BNL, told me that I was quick at seeing the "big picture" required to solve complex problems, and that I would be an excellent systems engineer. Although at the moment I had no idea what he was talking about, I believed him and I was anxious to learn what systems engineering was all about. Two years after working as a computer programmer, I applied for a graduate fellowship in urban systems engineering funded by the Departments of Transportation and Energy for studies at Howard University; I was awarded the fellowship with full tuition in addition to a monthly stipend for two years and graduated (summa cum laude) with a masters of science in urban systems engineering.

3. *What excites you about engineering?*
 The field of engineering excites me because it fulfills a need. Systems engineering, which is unlike the typical single discipline engineering fields such as electrical engineering, mechanical engineering, etc., is a multidiscipline engineering field. It excites me because it requires "big picture" system thinking and reasoning. It emphasizes the importance of interfaces, dependencies, interdependencies, and commonalities that are required to make things work in harmony, for the good of the whole. Systems engineering enthralls me because it considers not only the technical aspects of what is to be engineered, to fulfill a need; it also considers the human element as a vital component. It stresses the importance and necessity of teamwork for engineering excellence.

4. *What is your job at NASA?*
 Presently, I am serving as an aerospace flight system engineering manager, where I lead and supervise a group of engineers supporting the design, development, test, and evaluation (DDT&E) of our country's new Space Launch System (SLS), which is to replace the retired space shuttle. We perform systems engineering in the area of operations engineering to ensure the operability and sustainability of the newly designed launch vehicle that will provide an entirely new capability for human exploration beyond earth orbit. It also will back up commercial and international partners' transportation services to the International Space Station. We provide engineering functions to define and develop concepts of operations and design requirements, leading to a design that will be flexible for crew or cargo missions. More specifically, my organization defines technical requirements and processes for ground and flight operations of the SLS for launch readiness up to ascent operations of the integrated system. Our systems engineering efforts ensure the SLS will be safe, affordable, and sustainable.

*5. How has engineering changed?*

Not only has the field of engineering changed in that there are more disciplines, like material, structural, manufacturing, logistics, materiel, environmental, industrial, etc.; the faces of engineers have changed. When I first entered the field of engineering, there was not much diversity in the faces of engineers. The majority of engineers were White males, with hardly any females and minorities. I am now seeing more female engineers. We still need more African American engineers to complete the increasing demands for engineering jobs and other opportunities. Our diverse cultures and experiences are needed for more innovative thinking among teams for greater technical excellence.

*6. What do you see as the future of engineering?*

The demand for engineers is increasing. Advanced technologies require a sustained workforce of engineers with innovative and diverse ideas. The future of engineering in the United States is at risk, due to the decreasing number of STEM college graduates. There is a need for more STEM educational outreach during the formative years of students, prior to exiting middle school. All sectors of the community must be inspired to study STEM subjects. Further, improved STEM retention programs are needed to fulfill the growing demands for engineers.

*7. What have been the keys to your success in school and on the job?*

I give all the credit to my parents, for my success in school and on the job. My parents always had high expectations and a strong determination to succeed, and they always encouraged my siblings and me to study and work hard, even when it presented as a struggle. Although neither of my parents completed high school, they always emphasized the importance of education and its role in bringing people out of poverty. They often reminded us there is "no such thing as a free and/or easy ride" and that we had to get an education as armor to defend ourselves and those who were less privileged than we were. I always had exceptional mathematics and science teachers, and they all encouraged me to be a math teacher. Even though I have always had a great appreciation and admiration for teachers, I never considered being one. My dad encouraged me to dream big and to not be afraid of being different, as long as I was true to myself. My success on the job is due to a willingness to step outside of the box, strong perseverance, steady persistence with a purpose and a goal. I continuously seek opportunities, in spite of difficulties, setbacks, and discouragements. Accordingly, I have a strong belief that I have not failed until I stop trying. I strive to succeed, so I can bless someone else; and for the cycle of success to continue.

*8. What are you most proud of as a NASA engineer?*

As a NASA engineer, I am most proud of the opportunity I had to support the first Spacelab mission, STS-9, to low earth orbit in the space shuttle *Columbia*. Spacelab 1 was a joint NASA and European Space Agency (ESA) program designed to demonstrate the ability to conduct advanced scientific research and operations in space. Astronauts worked in the Spacelab module and coordinated with scientists and engineers in the Payload Operations Control Center (POCC) at the Johnson Space Center. I had the opportunity to build experiment simulators used to train payload specialists, who were astronauts operating the experiments on orbit in the space shuttle. I had the opportunity to work side by side with engineers from ESA and developed lasting relationships. I also had the opportunity to develop experiment flight software for command and control of payloads operating in orbit in the space shuttle's payload bay during the 10-day mission. I worked with scientific principle investigators in the POCC, to develop and uplink commands to the Onboard Experiment Computer during real-time operations. These efforts of teamwork led to a successful mission, with 72 experiments collecting and transmitting scientific data back to earth.

*9. How do you respond when someone asks, do you believe in "paying it forward?"*

"Paying it forward" is a responsibility that we all have, especially those who have more than others—"To whom much is given, much is expected." It does not require a lot of resources, just a desire to give someone an opportunity to prove him- or herself. It could be as simple as encouraging someone to dream and to pursue his/her dreams. "Paying it forward" for me is a personal commitment and provides me with the opportunity to teach and/or give children, especially those who are considered "at-risk," something I wish I knew at an early age. "Paying it forward" is critical to sustain our communities both locally and globally.

**Dr. Robert Howard Jr.** Born: August 3, 1971, in Xenia, OH. Manager of the NASA Habitability Design Center at the Johnson Space Center in Houston, Texas. BS in general science from Morehouse College in Atlanta, Georgia, BS in aerospace engineering from Georgia Tech in Atlanta, Georgia, MS in industrial engineering from North Carolina A&T State University in Greensboro, North Carolina, and PhD in aerospace engineering from the University of Tennessee Space Institute in Tullahoma, Tennessee.

*1. How early did you know you wanted to be an engineer?*

I grew up in a city with very little engineering industry, and as a result I didn't know what engineering was, but I always knew that I wanted to design spaceships. I used to go to the library and read all of the space-related books I could find, and in my spare time I liked to draw different ideas for space vehicles that I had come up with. In the eighth grade, I created a concept for an asteroid mining colony, and during my senior year in high school, I created a concept for a reconfigurable space base that could operate in orbit as a space station, travel in deep space as an interplanetary spacecraft, and land on a lunar or planetary surface as an outpost. I used the reconfigurable base to win a scholarship to attend Space Camp, and I redressed my asteroid colony and won second place in the citywide science fair with what was essentially an eighth grade idea.

*2. Who and what influenced you to become an engineer?*

I was influenced to become an engineer by astronaut Fred Gregory. Since I didn't know what engineering was growing up, I chose to major in computer science when I enrolled at Morehouse College. I had a computer at home and knew that computer science was something I could do very well. But I was on a scholarship that included summer internships at NASA Johnson Space Center. I took advantage of my internships to talk with several of the astronauts, and Fred shared with me that engineering was what really matched my interests the most. So the following fall I changed my major, and by the time I was finished with school I had graduated with a bachelor's in general science from Morehouse, a bachelor's of aerospace engineering from Georgia Tech, a master's in industrial engineering from North Carolina A&T State University, and a PhD in aerospace engineering from the University of Tennessee Space Institute.

*3. What excites you about engineering?*

I most enjoy being able to develop new ideas and nurture them into reality. Design is my favorite part of engineering because I can combine both my creative and analytical sides. I enjoy thinking through a new concept and refining the details of how the spacecraft would look, operate, and be assembled.

*4. What is your job at NASA?*

I manage the NASA Habitability Design Center. This involves leading a diverse team of engineers and interior designers in support of future spacecraft concepts. So far I have had the opportunity to work on the

Orion Multi-Purpose Crew Vehicle, the Altair Lunar Lander, the Multi-Mission Space Exploration Vehicle, and a variety of planetary and deep-space habitats.

*5. How has engineering changed?*

One thing that I think has the potential to drastically change engineering in the next few years is the increase in accessible fabrication tooling. For instance, 3D printers are becoming more and more capable of creating parts that years ago would have required sending drawings to external vendors for fabrication. Right now most 3D printers only make plastic parts, but there are a growing number of metal 3D printers. More machining tools are also accessible from hardware stores at reasonable prices. It is becoming possible for the average engineer to establish a machine shop in his or her own home/garage. This will truly unlock an engineer's creativity, to be able to personally envision a product, build it in computer-aided design (CAD), and then make a physical instance of the design. In the workplace, this may also lead to smaller, more agile design teams.

*6. What do you see as the future of engineering?*

I think there is a major challenge ahead in the field of engineering, and how we respond will directly impact the future success of the United States. Since World War II, the cost of producing vehicles, particularly in aerospace, has risen astronomically. Any major vehicle program tends to cost billions of dollars, and it is clear that the country cannot continue to pay these costs. Worse, these programs take so long to develop that they become obsolete shortly after they enter service. If we do not find a way to reduce both the cost and the development cycle, we will lose ground to others. However, I think it is possible for us to become much leaner, and I think that is the future of engineering. I think individual engineers will become much more responsible for understanding not only the theory of design but also the fabrication, integration, and testing of designed hardware and software. While this will be a more technically challenging career, I also believe it will be much more rewarding.

*7. What have been the keys to your success in school and on the job?*

The primary key to success is to not give up. Sometimes it will seem like the job is insurmountable or for whatever reason you cannot do the task in front of you. It is easy to give up at those moments, but I have found many times over that if I continue to work my hardest even when I see no possible way to succeed, success has a way of emerging to surprise you. When I was in eighth grade, the late astronaut Dr. Ron McNair was speaking to my school and said, "If something looks a bit difficult, go ahead and do it anyway." That philosophy has taken me far, and I believe it will do the same for all others who adopt it.

8. *What are you most proud of as a NASA engineer?*

I think my most enjoyable moment was when a group of Apollo astronauts were touring a mockup for the Altair lunar lander. Hearing their satisfaction with the work we had done made me feel like we had done a very good job in the eyes of some of the only people on the planet who had ever walked on the moon.

9. *How do you respond when someone asks, do you believe in "paying it forward?"*

I think this is very important. It is basically the idea that the recipient of a good deed is to repay the deed by doing a good deed to others. For me, I see Harriet Tubman, Martin Luther King Jr., Cinque, and other Black leaders in history—who sacrificed, led, and even died for our freedom—standing before me and saying that they gave their lives in order for me to be where I am now as a Black aerospace engineer. That motivates me to feel that it is not enough for me to simply do a good job at work and bring home a comfortable salary. I have to find a way to use the skills, talents, and opportunities I have to benefit the Black community, the United States, and the world. That leads to my involvement in the National Society of Black Engineers, where I lead a Space Special Interest Group. We try to find ways as a group to invent new technologies that can help the nation's space programs, as well as to create programs and opportunities to engage the Black community in the benefits of spaceflight. I encourage everyone to "pay it forward" in whatever venues or opportunities are present in your life.

**Dawn Davis.** Born: October 9, 1970, in New Orleans, Louisiana. Lead for the Electrical Design Group in the Design and Analysis Division of the Engineering and Test Directorate at NASA's Stennis Space Center in Mississippi. BS in electrical engineering from Tulane University in New Orleans, Louisiana, and MS in electrical engineering from Georgia Tech in Atlanta, Georgia.

INTERVIEW QUESTIONS

1. *How early did you know you wanted to be an engineer?*

I grew up wanting to become a doctor or a scientist since these were the "science" careers that were familiar to me. I did not know anything about engineers until high school, when I attended an event with a friend sponsored by the Society of Women Engineers at a local college.

2. *Who and what influenced you to become an engineer?*
I always knew that I would pursue a career in a science/technology related field.

3. *What excites you about engineering?*
Engineering allows me to use my creative side in solving real-world problems. Although we rely on a lot of what we learned in school, engineering allows me to continue learning. The problems that we face on the job are often similar, but there is always a unique nuance that allows me to learn something new.

4. *What is your job at NASA?*
I am the team leader for the Electrical Design Group in the Design and Analysis Division of Engineering and Test Directorate at Stennis Space Center. Stennis Space Center is NASA's main center for propulsion testing. We test rocket engines and components for NASA and commercial companies. We have several test facilities with various capabilities that are used to support this mission. The Electrical Design Group is responsible for ensuring that the test facilities' electrical systems meet the customer's requirements. We are responsible for the design of any new facility or modifications to the existing facilities to meet these requirements. Systems that we are responsible for include: low-speed and high-speed data acquisition (we sample data at rates from 250 sps to 100,000 sps), video, communication, hydrogen fire and gas detection systems, power, network, and control systems.

5. *How has engineering changed?*
Our engineering environment has become more dependent on computers and computer skills. At Stennis, all of our tests are automated or controlled by computers. When I first arrived at Stennis, the testing of the space shuttle main engines was performed using hardwired switch gear that required several people to operate for a test. Today most of that has been replaced with computers and software. Computers have enabled us to improve the quality and quantity of data. Analysis tools have enabled us to better predict test results, therefore reducing the number of tests required to prove systems operational. Additionally, technology has enabled engineering teams to work more collaboratively across geographical space. This brings a lot more diversity and experience to teams.

*6. What do you see as the future of engineering?*

Engineers will be faced with new problems that have more global implications.

*7. What have been the keys to your success in school and on the job?*

My aunt, who is a retired high school math teacher, has always been a strong supporter of mine. She always encouraged me to take the challenging classes or assignments. In school, I always surrounded myself with a support team where we shared our strengths to ensure each person's success. At work, I have been fortunate to have several people serve as mentors, both formally and informally. My mentors provided me with guidance and support in my technical and career development.

*8. What are you most proud of as a NASA engineer?*

I am most proud of the fact that I am contributing to the country's space program. I enjoy testing, especially all of the fire and smoke that we make! It is extremely exciting to see hardware that you have worked on perform perfectly and safely on its mission.

*9. How do you respond when someone asks, do you believe in "paying it forward?"*

There are so many people who have contributed to the success that I enjoy as an engineer. It is my responsibility to ensure the next generation of engineers has that same opportunity and that I serve as a vehicle to help others succeed.

**Ray Gilstrap.** Born: March 1, 1974, in Rowlett, Texas. Network engineer and chief technology officer for information technology at the NASA Ames Research Center in Mountain View, California. BS in electrical engineering from Florida A&M University in Tallahassee, Florida, MS in electrical engineering from the University of California, Berkeley, and pursuing PhD in electrical engineering at Stanford University in Palo Alto, California.

*1. How early did you know you wanted to be an engineer?*

I wanted to be an engineer even before I knew what an engineer was. All throughout my childhood, I would disassemble and reassemble the toys that my parents got me in order to learn how they worked. From there, I became interested in designing and building new toys, including spaceships, cars, and Transformers. Later on as a teenager, I found out that people called "engineers" did that for a living, and that's when I decided on that as a career choice.

*2. Who and what influenced you to become an engineer?*

I had a lot of influences, including my parents, who encouraged my curiosity; some great science teachers in middle school and high school; and Mr. Wizard, whose television show I watched religiously.

*3. What excites you about engineering?*

I am excited by the creative aspects of engineering—combining the technical skills that are taught in school with a healthy dose of imagination to create solutions to real problems that people face in their daily lives.

*4. What is your job at NASA?*

I am a network engineer at NASA Ames Research Center, and I work on a number of projects to increase the performance of NASA's computer networks, enable scientists who conduct research in remote field locations to have network access, and protect NASA's networks from hackers. I am also currently acting as the chief technology officer for information technology at Ames; in this capacity I help NASA take advantage of emerging technologies and trends such as virtual collaboration tools and ubiquitous mobile computing.

*5. How has engineering changed?*

Engineering has always been a team effort, especially on larger projects. But one thing that has changed is the geographic distribution of the members of those teams. Once, teams were localized to a single office, but now teams can have members from all over the world. This leads to the inclusion of different perspectives on the problems at hand, generally resulting in better solutions.

*6. What do you see as the future of engineering?*

I see the trend toward globalization continuing, including a much-needed increase in participation from regions of the world that are just now developing native engineering talent.

*7. What have been the keys to your success in school and on the job?*

I would say that the most important thing has been to maintain a positive attitude. Like any other field of endeavor, success in engineering careers depends on working well with others, and staying positive is key to successfully interacting with colleagues. Before the career stage, success in an engineering degree program also depends on that same attitude: a willingness to ask questions when something isn't clear and old-fashioned hard work in engineering courses. It's also important to be well-rounded as engineering doesn't happen in a vacuum, and being a great engineer requires an understanding of the human context in the work as it's taking place.

8. *What are you most proud of as a NASA engineer?*

I am most proud of the accomplishments of the summer students I have mentored over the years. The eight students I have worked with have so far gone on to collectively earn nine bachelor's degrees, four master's degrees, and two PhDs, all in electrical engineering or computer science. I'm looking forward to seeing all of the great things they'll accomplish in their careers.

9. *How do you respond when someone asks, do you believe in "paying it forward?"*

Absolutely! I've had a lot of support over the years myself, and I think it's essential to keep our momentum going by helping the next generation reach their goals.

**Dr. Christyl Johnson.** Born in North Carolina. Deputy center director for science and technology at the NASA GSFC in Greenbelt, Maryland. BS in physics from Lincoln University in Pennsylvania, MS in electrical engineering from Pennsylvania State University, and PhD in systems engineering from George Washington University in Washington, DC.

INTERVIEW QUESTIONS

*1. How early did you know you wanted to be an engineer?*

I wanted to be an engineer in elementary school. I remember watching one of the last Apollo launches with my parents around the dinner table, and I said to them that I wanted to become an astronaut. My parents said once you really decide what you want to do, you can be anything you want to be.

*2. Who and what influenced you to become an engineer?*

Growing up with two parents as teachers, I always had role models. However, STEM became truly real for me when I was accepted into the Lincoln University Aerospace Engineering Recruitment Program for college. I had an assignment in the physics lab, and I couldn't imagine being a physicist. I thought of them as being nerdy, really smart with no fashion sense or people skills. I couldn't imagine how this would be fun. I went into the lab and there is this guy in the laser lab with a big white coat with big sleeves. He leans over the lab table and starts to adjust the mirrors when all of a sudden I hear Pop, Pop, Pop, Pop and see smoke!

I see holes in the sleeve of his coat and I say, YES! This is the field for me! You put holes in things; shoot lasers across the room; a little bit of danger but a whole lot of fun. This is where it started for me.

3. *What excites you about engineering?*
We are paid to think about solving exciting and impossible problems. Engineers are always looking to do the impossible. We are the ones who have to do the things people say cannot be done. I love it when people say this cannot be done. That's when I get excited because I'm happy to prove you wrong. That's what engineers and scientists do: we look for the solution to the impossible. That's where we get innovation.

4. *What is your job at NASA?*
I am currently the deputy center director for science and technology at the Goddard Space Flight Center in Greenbelt, Maryland. I manage the center's research and development portfolio, and I am responsible for formulating the center's future science and technology goals and leading an integrated program of investments aligned to meet those goals.

5. *How has engineering changed?*
I believe the risk posture has changed. Engineers are more reluctant to think outside of the box because the risk is too high.

6. *What do you see as the future of engineering?*
Engineering is a critical field for the future of our nation and the planet. We need to encourage more youths to become engineers, especially women and minorities. There are few women in the field for several reasons. For example, there are some counselors and teachers today who steer girls to softer fields. Therefore, it is very important that girls are exposed to women who are in the field. Plus, they need the opportunities to get hands-on experiences themselves that let them know that engineering is for everyone.

7. *What have been the keys to your success in school and on the job?*
The keys to my success have been first, believing that I can; having a strong conviction that you belong. Second, I have always had a strong support system. You can have your cake and eat it too, but you have to focus and make up your mind that this is what you want to do and do it.

8. *What are you most proud of as a NASA engineer?*
I could tell you it was some project that I led, but I am most proud of being able to make a difference in the lives of young people, especially young women. I was invited to speak at Oprah's school in South Africa

last year. One of the girls was not planning to go to college, so they asked me to speak with her. We talked, and I discovered that her parents had been discouraging her to go to college. They wanted her to work so she could help support the family. She did not realize that her family was looking very shortsightedly. I explained to her that she could help the family and the entire community if she focused now and committed to a degree in STEM. She needed to gather her inspiration from within and not from her family, who were not a positive influence. She had to look within for inspiration and on the outside to people who were doing this type of thing to be positive role models.

9. *How do you respond when someone asks, do you believe in "paying it forward?"*
Absolutely! There is no way we will be able to address our national issues without increasing the number of engineers and scientists we currently produce, especially among women and minorities. I love interacting with young girls and boys and talking to them about STEM. You must have that support system in order to achieve your goals.

**Leland D. Melvin.** Born: February 15, 1964, in Lynchburg, Virginia. Associate administrator for education (retired, 2014), NASA Headquarters, Washington, DC. NASA astronaut, August 1998 to January 2010. BS in chemistry from the University of Richmond and MS in materials science engineering from the University of Virginia in Charlottesville, Virginia.

INTERVIEW QUESTIONS

*1. How early did you know you wanted to be an engineer?*
I didn't know I wanted to be an engineer, but I began engineering in the sixth grade by building a skateboard and a bike. I really didn't know what engineering was, but I experienced it firsthand. However, my interest in STEM was stoked by a chemistry set I received in the sixth grade. I really got excited when one of my experiments resulted in a huge hole burned into my mom's carpet. Originally, my plans were to become a doctor, but I found that wasn't for me. I majored in chemistry at the University of Richmond, but before I graduated, I knew I wanted to consider a career in engineering. Once my NFL career didn't work out, I pursed my master's degree in materials science engineering.

*2. Who and what influenced you to become an engineer?*
My father indirectly influenced me to become an engineer by letting me know I could build things. He bought an old Merita bread truck and

made it into a family camper with an air conditioner on the top. I watched him design it, build it, and it was functional. I was also influenced by participating in the Minorities in Engineering Program at Virginia Tech.

*3. What excites you about engineering?*

To be able to move from an idea or vision to an actual product excites me about engineering. At Langley we made sensors, and in order to do that we planned everything from designing the building to making the optical fiber sensors. That was a big team effort that really excited me because that type of work had not been done before at NASA.

*4. What is your job at NASA?*

As the associate administrator for education, our vision is to use NASA's unique assets to mold, shape, produce, and inspire the next generation of explorers and teachers. I help to bring together the right people to leverage our assets to make dreams a reality.

*5. How has engineering changed?*

Engineering was a lot more insular back in the sixties and seventies. Also, when I did my master's work in the late eighties, we were more myopic. Today, the masses are used to solve engineering problems. There is a convergence of disciplines, plus engineering is more global. Additionally, engineering utilizes technology more now to solve problems.

*6. What do you see as the future of engineering?*

Breaking out of the traditional thought of engineering, how do we come together as a civilization to use cross disciplines to solve problems? It will take people from many disciplines to solve the world's problems. They will include engineers, artists, visionaries, etc.

*7. What have been the keys to your success in school and on the job?*

Success in school for me was ensuring I did both the even and odd problems. Always doing more problems than those assigned. Repetition and familiarity were part of the keys to my academic success. There was a joy of working hard and repetition. To be successful, one has to work hard and listen to people. Anyone may have the ability to give you the solution to a problem. Convergence of thought: taking off the filters of the world and using all available data to solve the problem.

*8. What are you most proud of as a NASA engineer?*

I am most proud of the early work I did at the Langley Research Center. As an aerospace research engineer I conducted research in the area of physical measurements for the development of advanced instrumentation for nondestructive evaluation. Responsibilities included using optical fiber sensors to measure strain, temperature, and chemical damage in both composite and metallic structures. What stood out about this project was the fact we designed the building from an alleyway between two buildings to create the sensor lab and conducted new and vital research for the agency. We had a vision, created a plan, and saw it through to completion.

9. *How do you respond when someone asks, do you believe in "paying it forward?"*
Not only do I believe in "paying it forward" but backwards, sideways, up, and down. We live in a three-dimensional world where we need to inspire and utilize all available talent. As such we can inspire not only students and teachers but colleagues as well. In other words, we need to utilize all of the available tools in our toolkit and all of the available space in our box. STEM is the key to solving our nation's and the world's problems, so we must pay it forward and in all directions.

**Dr. Woodrow Whitlow.** Born: December 13, 1952, in Inkster, Michigan. Associate administrator for mission support (retired, 2013) at NASA Headquarters in Washington, DC. Former center director at the NASA Glenn Research Center in Cleveland, Ohio. BS, MS, and PhD in aeronautics and astronautics from the Massachusetts Institute of Technology.

INTERVIEW QUESTIONS

1. *How early did you know you wanted to be an engineer?*
I wanted to be a chemist and would memorize the periodic table in my bedroom. However, the space missions in the 1960s captured my imagination. I was around 9 years old when I changed my career goals to become an astronaut.

2. *Who and what influenced you to become an engineer?*
The space program really influenced me to become an engineer. However, it was my African American high school counselor who encouraged me to attend the Massachusetts Institute of Technology (MIT). I received my BS, MS, and PhD in aeronautics and astronautics from MIT in 1974, 1975, and 1979, respectively.

3. *What excites you about engineering?*
I get excited about working on new stuff that most people have never done before.

## 4. What is your job at NASA?

I currently serve as the associate administrator for mission support. In this capacity, I am responsible for most NASA managerial functions, including human capital management, Headquarters operations, procurement, protective services, internal control, the NASA Shared Services Center, strategic infrastructure, cross-agency support, and construction and environmental compliance and restoration. More specifically, I enhance NASA's mission success by developing and implementing a strategy for an efficient, modern, streamlined infrastructure that is better aligned with NASA's programs and strategic direction.

## 5. How has engineering changed?

Engineering has changed significantly with the introduction of new technology such as information technology (IT) systems. Everything is automated. For example, when I was in school I used the slide rule to do manual calculations. However, IT automation takes you to a solution more efficiently and faster. In addition, the introduction of new technology is much faster and the rate in which things change occurs much faster.

## 6. What do you see as the future of engineering?

Midterm is to focus on people. When NASA went to the moon in 1957, only White males were included. We need to be more diverse and bring everyone into the fold to maintain US dominance in STEM disciplines and to maintain our standard of living. Additionally, IT, biotech, and nanotechnology focus on small things. We need to mimic birds and things in nature to improve aerospace flight design. For example, I believe the keys to getting to Mars include propulsion systems, materials, small/nano-materials, and energy systems.

## 7. What have been the keys to your success in school and on the job?

The keys to my success are good health and natural talent. My grandfather told me that if I show up to school I would learn something every day. I was in the fifth grade before I missed school because I was sick. In my 34-year tenure with NASA, I have only taken two sick leave days. The other key to my success is natural talent, and I have worked hard at taking advantage of it. Math and science have always been easy for me.

## 8. What are you most proud of as a NASA engineer?

I'm particularly proud of my research involving computational fluid dynamics (CFD). While at the Langley Research Center, I received the NASA Exceptional Service Medal for "The Development of Computational

Methods for Unsteady Aerodynamics and Aeroelastic Analysis" in May 1986. I am also proud to have been able to help people attain success in their careers.

9. *How do you respond when someone asks, do you believe in "paying it forward?"*
I believe in "paying it forward" and believe my actions have borne that out. I think I have helped a lot of people. As a manager, I really enjoy mentoring future engineers and scientists, and it's a great pride for me to see them advance in their careers.

The individuals highlighted above are a representative sample of the outstanding African American engineers at NASA. The following section highlights a representative sample of other African American engineers at NASA whose achievements were recently recognized by an external organization.

## Other African American Achievers at NASA

During the 2013 Black Engineer of the Year Awards (BEYA), African American engineers at NASA were prominently recognized. Ten distinguished NASA employees from three NASA centers received awards. The awards recognize the achievements of African Americans in STEM and encourage young African Americans to pursue careers in STEM fields. Of the 10 BEYA recipients from NASA, 8 were engineers. The objectives of the awards are to

1. create a network of role models who can serve as inspiration to others;
2. help senior leadership in companies identify exceptional talent;
3. promote better access of women and ethnic minority groups to STEM careers by showing them what it takes to get hired and to keep moving upward in their organizations;
4. help organizations promote and publicize the successes of women with the same intensity as male employees;
5. help organizations cultivate engaged employees; and
6. raise the profile of organizations as employers of choice among women and racial and ethnic minority groups.

The engineering awardees were:

*Senior Investigator Award*
JOHN W. HINES, former center chief technologist at NASA Ames Research Center, who designed, developed, tested, and evaluated space

systems and managed advanced technology development programs and projects during an exceptional career at NASA. Education: BS in electrical engineering, Tuskegee University; and MS in biomedical and electrical engineering, Stanford University.

*Special Recognition Award*

AISHA R. BOWE, aerospace engineer at Ames, whose work in Next Generation Air Transportation focuses on developing methods to maintain safe separation of air traffic and optimize fuel consumption within an automated system. Education: BSE in aerospace engineering and MEng in space engineering, University of Michigan.

*Trailblazer Awards*

KEN FREEMAN, a project engineer at Ames, has successfully implemented information technologies across NASA in varied IT fields for more than 20 years. Education: BS in electrical engineering and computer science, University of California, Berkeley; and MS in electrical engineering, San Jose State University.

KEVIN L. JONES, computer engineer at Ames, has been a pioneer in the field of networking at Ames, has managed several IT projects for various organizations, and is leading NASA's IPv6 implementation efforts as its transition manager. Education: BS in electrical engineering and computer science, MS in systems management, University of California, Berkeley; and MS in systems management, Notre Dame de Namur University.

*Modern-Day Technology Leaders Awards*

DR. OUSMANE N. DIALLO, an aerospace research engineer at Ames, focuses on the development of algorithms in support of advanced flight control and air traffic automation. Education: BS in mechanical engineering, City University of New York, College of Staten Island; and MS and PhD in aerospace, aeronautical, and astronautical engineering, Georgia Institute of Technology.

DR. MALCOLM K. STANFORD, a materials research engineer at the Glenn Research Center in Cleveland, works on complex problems in material engineering, especially when they involve extreme environments like high temperatures or the vacuum of space. Education: BS in engineering physics, Miami University; and MS and PhD in materials engineering, University of Dayton.

GILENA A. MONROE of Ames is the Log Aggregation Tool Lead and Agency Vulnerability Assessment and Remediation Project Manager for the NASA Security Operations Center, where she helps meet cyber security needs in the protection of NASA's technologies and information. Education: BS in computer science and MS in industrial and systems engineering, North Carolina A&T State University.

RAYMOND T. GILSTRAP is a network engineer at Ames who focuses on emerging networking technologies for a variety of NASA scientific research and mission support activities. As chief technology officer of IT for Ames, he has been responsible for identifying and evaluating key technologies and technology trends with the potential to enhance information technology services for NASA missions, projects, and users. Education: BS in electrical engineering, Florida A&M University; MS in electrical engineering, University of California, Berkeley; and pursuing PhD in electrical engineering, Stanford University.

BEYA recognizes STEM professionals and leaders who are committed to increasing the percentage of people from historically underrepresented communities in the technology workforce. The year 2013 marked the 27th anniversary of BEYA officially recognizing Black excellence in STEM fields (Schalkwyk 2013).

The following is a compilation of ideas by some of the survey participants for broadening participation of African Americans in engineering:

1. Get kids at an early age to realize what it means to build by using their hands to create things.
2. Meet kids where they are and use the things they enjoy doing, such as basketball, football, or music, and show them that engineering and science are intrinsically embedded in all of those things.
3. We need a coordinated pipeline approach that identifies bridge programs, determines what makes them successful, and identifies sustainable resources (both fiscal and human).
4. We have to institutionalize engineering within the Black community. There are Black legal institutions, religious institutions, economic institutions, and entertainment institutions, but scientific and technical institutions are lacking. It is not enough to merely enter and be successful in an engineering career. And while speaking to youths is a start, it also is not enough for true impact. We have to embed engineering activity in our culture. Imagine, for a moment, if basketball were not institutionalized within the Black community—take away the community basketball courts and community- and school-based teams, and effectively reduce its footprint to occasional television coverage of an NBA game and rare appearances by an

NBA player at a school to speak about his work. How many African Americans would then have any interest in basketball? This is what I [Robert Howard] am trying to do with organizations like the National Society of Black Engineers (NSBE)—leverage them to create engineering as an institution within the Black community.

5. It is imperative that we involve positive role models from the entertainment industry to stimulate interest in STEM among youths in African American and other underserved communities. NASA has been successful in getting artists like will.i.am, Pharrell Williams, and Mary J. Blige to participate in outreach endeavors that targeted mostly elementary and middle school kids. However, activities like those, though successful, need to be repeated more often. Going to an event and presenting NASA to kids is good, but it can't be done on a one-time-and-we're-out basis. We have to be a constant presence within certain communities in order to be effective. Once we leave a school without going back on a repetitive basis, we will fade from a child's memory, and the message that we presented will soon be lost. That's why I [Mark Branch] propose a more formal program that involves creating a pipeline from elementary to middle school to high school and then college. A 36-month pilot program can be created that starts in a particular city. If successful, the model can be repeated and expanded in other cities across the country.

6. A major overhaul in our elementary and middle school curriculums, especially in minority communities. There is a need for educators who love STEM subjects and know methods to make learning these subjects fun. Today's students want and expect tangible and instant results. Increased hands-on projects rather than routine theory and instruction are needed.

7. Outreach programs with African American role models who interface with elementary and middle school students and their families on a routine basis. Significant change will have to involve the family and the environment where students live.

8. Television networks, like OWN and BET, should be encouraged to produce shows displaying African American STEM professionals. STEM professionals are competing with professional sports and music professionals, who have more influence on our students. We are challenged to help young African Americans understand and believe that STEM careers are more obtainable than being sports or entertainment stars.

9. The key to increasing the participation of African Americans in engineering is exposure/outreach. Kids need to see the big picture and the path to get there. In other words, we need to show a fifth-grader all the steps it takes to become an engineer/rocket scientist.

NASA, like other federal agencies and technical organizations, must continue to recruit African American engineers from all sources. Special efforts should be made to recruit African American engineers from HBCUs. According to *Diverse Magazine,* four of the top five producers of African American engineers with BS degrees are HBCUs ("Top 100 Degree Producers" 2013). However, the reality is that we must have nationwide leadership and coordination from the president, Congress, state and local officials, nongovernmental organizations, school systems, community groups, and parents to substantially impact the participation of African Americans in engineering.

One example of national leadership is President Obama's fourth White House Science Fair in 2014. It is a part of the administration's continued efforts to promote STEM education. The White House has also made several announcements concerning new and ongoing initiatives in this area. In his opening remarks, President Obama told participating students, "Our job is to make sure you have everything you need to continue on this path of discovery and experimentation and innovation. . . . That's why we decided to organize these science fairs."

The 2014 science fair featured a focus on girls and women excelling in STEM. The efforts President Obama announced at the May 27 fair include:

- A new $35 million Department of Education competition to support the president's goal to train 100,000 excellent STEM teachers over the next decade, through a new round of the Teacher Quality Partnership grant competition;
- An expansion of STEM AmeriCorps to provide 18,000 low-income students with summer STEM learning opportunities.
- A national STEM mentoring effort, including new steps by companies and nonprofits to increase students' connectedness to STEM education and to keep youths engaged in STEM learning outside the classroom. With regard to this last item, US2020, which strives to mobilize one million STEM mentors annually by 2020, just announced that Chevron and Discovery Communications have joined as its newest partners. Additionally, the New York Academy of Sciences and its partners are launching the Global STEM Alliance to connect students from around the world.

Reiterating a line he's used before, President Obama said that STEM education and innovation should be given the same amount of attention as any sports teams, which are often congratulated at White House visits. "What's happening here is more important," he said, when referring to the science fair. "As a society, we have to celebrate outstanding work by young people in science at least as much as we do Superbowl winners. . . . They're what's going to transform our society" (Morones 2014).

## References

Allen, Lenell. 2013. "Tribute to Janez Yvonne Lawson Bordeaux." *Living the Dream: A Tribute to Dr. Martin Luther King Jr.* (Presentation), January 17.

American Mathematical Society. 1957. *American Mathematical Society Annual Meeting Conference Proceedings*, April 19–20, 257.

"Delta Mothers Honored Guests at Reception Sunday Afternoon." 1950. *California Eagle*, May 11, 11.

"Film Series Begins Sunday; Mutiny on the Bounty." 1954. *California Tech* 56 (2): 1.

The HistoryMakers. 2002. "Ed Dwight." www.thehistorymakers.com/biography /ed-dwight-39.

The HistoryMakers. 2004. "Harriett G. Jenkins." www.thehistorymakers.com /biography/harriett-g-jenkins-38.

The HistoryMakers. 2012. "Katherine G. Johnson." www.thehistorymakers.com /biography/katherine-g-johnson-42.

Hromas, Leslie A. 2005. "The Legacy of TRW and Space Park: A Summary with Key Dates and Milestones." http://tra-spacepark.org/docs/TRW_History.pdf.

Lab-Oratory. 1954. "Smiles of Happiness." *Jet Propulsion Laboratory (JPL) Lab-Oratory* 4 (2): 9.

Morones, Alyssa. 2014. "At Fourth White House Science Fair, Obama Unveils STEM Plans." *Education Week*, May 27. http://blogs.edweek.org/edweek /curriculum/2014/05/white_house_hosts_fourth_annua.html.

NASA (National Aeronautics and Space Administration). 1993. Biographical Data. Lyndon B. Johnson Space Center, National Aeronautics and Space Administration.

NASA (National Aeronautics and Space Administration). 2008. "NASA's Innovators and Unsung Heroes." *50th Anniversary Magazine.* www.nasa.gov/50th/50th _magazine/unsungHeroes.html.

NASA (National Aeronautics and Space Administration). 2009. "Charles F. Bolden, Jr., NASA Administrator." www.nasa.gov/about/highlights/bolden_bio .html.

NASA (National Aeronautics and Space Administration). 2011. "NASA 2011 Performance Plan." www.nasa.gov/pdf/533365main_NASAFY11_Performance _Plan-508.pdf.

Portree, David S. F. 1998. "NASA's Origins and the Dawn of the Space Age: Monographs in Aerospace History #10." www.hq.nasa.gov/office/pao/History/monograph10/.

Schalkwyk, James. 2013. "NASA Announces Black Engineer of the Year Awards." www.nasa.gov/topics/people/features/2013_beya.html.

Sylvia, Anna. 2008. "Biography for Guion S. Bluford." https://secureapps.libraries.psu.edu/PACFTB/bios/biography.cfm?AuthorID=7112.

"Top 100 Degree Producers." 2013. *Diverse Magazine* 30 (11). www.diverseeducation.com.

Tyson, Neil deGrasse. 2008. "Happy Birthday, NASA." www.nasa.gov/50th/50th_magazine/tysonLetter.html.

# CHAPTER EIGHT

# African American Engineers in Business and Industry

RODNEY ADKINS *and* GARLAND L. THOMPSON

IT WAS IN THE 1950S and 1960s that the "Space Race" inspired the world and captivated students everywhere, as the then Soviet Union and the United States competed in a race to successfully launch and safely return the first manned space mission to the moon (Siddiqi 2000).

As we know, the United States accomplished that feat on July 20, 1969, when Apollo 11 landed on the moon and returned to earth, splashing down in the Pacific Ocean on July 24, 1969 (Cadbury 2007). What was quite remarkable was not just the accomplishment of this monumental feat but how the technology that was developed by the engineers of Apollo 11 was then used and applied in the business and industrial worlds. As noted in *NASA Apollo 11 Manual*, the same technology that successfully launched the first manned space mission and inspired a generation also helped businesses and industries across the globe solve complex challenges and achieve success (Riley and Dolling 2011).

Today, like the "Space Race" of the midcentury, we find that a new revolution in technology has birthed a new generation of dreamers. These dreamers, engineers, scientists, mathematicians, and innovators are leading the world into unimaginable frontiers of discovery in today's awe-inspiring digital age.

We believe that today's new era in technology is not only inspiring engineers to understand the business of things but also igniting a new wave of businesses, global corporations, and many venture capitalists to identify and quickly secure their arsenal of bright, energetic engineers who can bring creativity and innovation to drive business growth and profits. As students reading this chapter, we believe that you are the torchbearers that we reference, who will lead this new and exciting digital age.

As noted in *Unheralded but Unbowed: Black Scientists and Engineers Who Changed the World*, whether it is the 1950s or the 2010s, engineers have had a tremendous impact on the business world, and Black engineers were prominent throughout the long stream of historical innovation (Thompson 2009). Americans, at every stage of development of this country, have lived through eras of accelerating technological change ignited by the imaginations of engineers. These changes have prompted, correspondingly, rapid shifts in the workings of American business and industry and in the societal response to the opportunities they present. In short order, Americans experienced stunningly rapid shifts in a number of areas owing to the contributions of Black inventors and engineers, including advancements in personal mobility in 1889, when L. R. Johnson developed the first bicycle frame; breakthroughs in safe transportation, with the invention of the traffic signal by Garrett Morgan in 1922 (Biography.com 2014a); the rapid spread of information via the new telegraphic communications and phone transmitter developed by Granville Woods (Biography.com 2014b); and advancements in refrigeration by inventors John Standard (About.com 2014) and Garrett Morgan (Fouché 2005).

Fundamentally, the radical shifts in the industry and commerce of the young United States also caused even more dramatic shifts in social relations. We find that the presence of African Americans among the corps of intrepid world changers, expanding cultural awareness while working to improve existing technologies and their application, or even to invent new ones, is often overlooked or, if noticed at all, dramatically undervalued.

Thus, this chapter explores the different achievements of Black engineers whose names regularly surface during Black History Month, but whose real importance may be underestimated, while also examining the economic impact of Black engineers—the benefits of the cultural and diversity influence, as well as the global competitive advantages gained by having Black engineers in business and industry. This chapter will also help cast a new light on the need for a new wave of such Black world changers in today's age of digital revolution. First, let us examine the economical impact that Black engineers have had in the business world.

## Changing Demographics and the Economic Impact of Black Engineers in Business

> On Wall Street, no one knew what the Internet was in 1993 and '94. . . . We couldn't raise the money.
>
> EMMIT MCHENRY, cofounder of Network Solutions Inc., one of the early leading Internet domain services providers (HistoryMakers 2014).

The journey to succeed as an engineer can be daunting. Luckily for engineers, we are innately designed to relentlessly pursue solutions to solve problems, no matter how large they may be. This determination has led many Black engineers in the business world to fight beyond any challenges that may present themselves.

We examine Norbert Rillieux, born in 1806 into a Creole family in New Orleans, Louisiana (Haber 1970). Although Rillieux, an African American, was born free, he emerged into society deep in the grips of race-based slavery. Rillieux's mother had been emancipated from slavery, and his father was a French-born engineer and inventor who sent the young Norbert off to school in Paris, where he studied engineering at the famous Ecole Centrale.

Steam engines were key high-tech products, and Rillieux stayed on at l'Ecole postgraduation to become an instructor and to publish papers on steam power and its uses. Growing up on a Louisiana sugar beet plantation, the young Rillieux watched slaves walking the "Jamaica Train" to produce refined sugar (Rillieux 1848). That is, they carried heavy iron ladles of scalding sugar beet juice from boiling cauldron to boiling cauldron. While in France, Rillieux researched ways to improve the refining process, and in 1833 Edmund Forstall, a business associate of Rillieux's relatives, invited Rillieux to return to Louisiana to serve as head engineer at a new sugar refinery. In 1843, Rillieux patented his "evaporative pan" process for the production of sugar (Low and Clift 1981). The multiple-effect evaporation system Rillieux developed could end the disastrous spills and danger of the "Jamaica Train," and it also offered better temperature control, ending the overcooking of the sugar beet juice.

Instead of multiple boiling cauldrons, Rillieux substituted a vacuum chamber that lowered the boiling point of the liquids, which moved between multiple evaporating pans, each heated by the pan below. Thus, Rillieux, the engineer, became a business leader whose exploits reshaped the Louisiana plantation economy. He convinced 13 sugar refineries to use his new system, and the word spread far and wide. A former US Agriculture

Department chemist, Charles Brown, once called Rillieux's evaporating pan technology "the greatest invention in the history of American chemical engineering" (Sammons 1990).

Granville Woods, born in 1856, went to grammar school in Columbus, Ohio, but left at age 10 to become a machine-shop apprentice, learning to be a blacksmith and machinist. In 1872, Woods became a fireman on the Danville and Southern Railroad, working his way up to locomotive engineer. But Woods was no ordinary dropout. In 1876, the year Alexander Graham Bell's telephone astonished the Brazilian emperor Dom Pedro II at the Centennial Exposition in Philadelphia—"It talks," the emperor exclaimed, generating headlines all over the world—Woods enrolled in college and studied electrical and mechanical engineering, taking classes for two years while also working in a steel rolling mill.

The year 1878 saw Woods working in the engineering department of the British steamship *Ironsides*, and two years later he was chief engineer. Back on land, Woods joined the Dayton and Southwestern Railroad. Then in 1880, Woods moved to Cincinnati, Ohio, and set up shop as an electrical engineer and inventor. Woods won a patent for a multiplex telegraph device, and he renamed his company the Woods Electric Company (Sammons 1990).

In 1885, Woods patented his telegraphony device, which permitted a telegraph operator to send voice as well as code-key messages over a single wire. Woods sold reproduction rights to the American Bell Telephone Company and moved on, having effectively presaged the twentieth century's development of many-channeled, multiplex telephony systems. Woods's deep investigation of the principles of electromagnetic induction led to the "Synchronous Multiplex Railway Telegraph," developing a viable product based on the pioneering 1884 discoveries of inventor Lucius Phelps (Fouché 2005).

In 1892, as the century came to its tumultuous close, with America building an all-steel navy and emerging as a world power with its defeat of Spain's wooden ships at Manila Bay in the Philippines and in the Caribbean, Woods moved his research and development operation to New York and partnered with his invention-minded brother, Lyates Woods.

Well-known inventor Thomas Edison filed a patent claim for Woods's invention and went to court to uphold it. The courts ruled, after two courtroom battles, that Granville T. Woods, not Thomas Edison, was the true inventor and rightful patent holder. Undaunted but definitely hampered in the commercial exploitation of his discoveries, Woods continued his business drive. Learning of an electric railway system pioneered by fellow inventor Charles Van Depoele, Woods built on Depoele's work to construct a similar system of catenary towers and overhead wires to power the newly developed electric railway cars being installed in cities such as New York, Philadelphia, Chicago, and St. Louis. And ever the steam engineer, Woods filed a patent for an improved steam boiler furnace in 1889.

In all, Granville Woods, sometimes called the "Black Edison," won more than 50 patents for devices such as an automatic brake and an egg incubator and improvements to other inventions such as electrical safety circuits, telegraphy and telephony devices, and a phonograph (Fouché 2005). When Woods died in 1910, he left a legacy as one of the major contributors to the mass transit industry. Hampered in getting financing to produce the devices he invented—and hampered as well in an environment of steadfast resistance to Black leadership in business—Woods had sold many of his patents to companies such as General Electric, Westinghouse, and American Engineering (Fouché 2005). But his role in the business growth of American industry cannot today be ignored.

As the son of a minister, Emmit McHenry grew up in Tulsa, Oklahoma. He would attend schools with names that would perhaps contribute to his astonishing legacy: Stewart Elementary School, Carver Middle School, and Booker T. Washington High School (HistoryMakers 2014). He continued his studies at the University of Denver, where he received his BS in communications. He then served as a lieutenant in the US Marine Corp and later returned to the academic domain to obtain an MS in communications from Northwestern University.

In 1979, Emmit McHenry and associates launched Network Solutions Inc., an Internet service provider. For 16 years, McHenry and his partners worked tirelessly to realize a unique and unprecedented vision: a complex computer code whereby ordinary people, not just computer scientists or engineers, could "surf" the web or have e-mails. McHenry created what we know today simply as ".com." On December 31, 1992, the National Science Foundation selected Network Solutions as manager of domain name registration services for the Internet.

In 1995, McHenry founded the award-winning telecommunications, engineering, consulting, and technical services company NetCom Solutions International. At the time of its founding, the evolution of the Internet was taking shape. In fact, in 1997 there were approximately 71,000 domain names in the entire world; by 2005, the number of domain names had grown to more than 46 million (Beyster and Daniels 2013; Zook 2014). It is unimaginable what the world would look like today without the Internet, and it is even more difficult to comprehend what the Internet would look like today without Emmit McHenry. His legacy is one that should never be forgotten.

In the late 1980s, Nigerian-born engineer and computer scientist/geologist Phillip Emeagwali successfully implemented the first petroleum reservoir model on a massively parallel computer (Thompson 2009). The technology aids the oil and gas industry in locating and recovering additional oil and gas. Emeagwali programmed a computer with more than 65,000 processors. The crux of the discovery was that Emeagwali had programmed each of the microprocessors to talk to six neighboring microprocessors at the same time, which was a tremendous achievement at the time.

These pioneers created concurrent tectonic shifts in American economy and industry and had partners among Africa's diasporic progeny, even while the bulk of the African American populace at the time was experiencing tremendous cultural and societal challenges.

## The Potential Economic Impact of Tomorrow's Black Engineers

As we move into today's decade of transformation, with US Census Bureau reports illustrating that White Americans' birth rate will decrease significantly and the American minority population will increase (US Department of Commerce, US Census Bureau 2012b), the challenge of motivating and preparing minority students, particularly African American students, to pursue careers in science, technology, engineering, and mathematics becomes vital. President Barack Obama's Council on Jobs and Competitiveness, responding to reports that nations such as India and China now out-produce the United States in graduating young engineers, launched the "Stay With It" campaign, calling for American higher education to boost the number of engineers graduated by 10,000 a year (White House Blog 2012).

Mortimer B. Zuckerman, writing in the September 27, 2011, edition of *US News and World Report*, published a commentary that supports the national call. The column, headlined "Why Math and Science Education Means More Jobs," argues that "the men and women who will make America's tomorrow are in school and college today. They are the human capital at the core of any productive economy. As the nation shifts into a new, nonindustrial economy, we will need a well-trained, technically competent workforce to manage and staff the science and technology businesses that create the high-paying jobs" (Zuckerman 2011).

Tomorrow's Black engineers, raised in an environment of open access to education which could not have been imagined by those Black pioneers of yesteryear, have an opportunity that has never before existed in American history to drive unprecedented economic development across the global landscape.

# Cultural Impact of Black Engineers in Business and Industry

> My engineering education gave me an approach to problem solving that has been applicable to every business problem I have ever faced.
>
> ANTHONY HARRIS, cofounder of National Society of
> Black Engineers (Purdue University School of Mechanical
> Engineering 2014).

It can be said that not only does a degree in engineering provide a unique engineering education, but it also provides a unique view of problems and opportunities that can lead to an organization's success. The power of diverse thinking and cultural backgrounds can, together, lead to unprecedented growth in business, technology, and industry. It was this thinking that led Anthony Harris in 1976 to launch the National Society of Black Engineers (NSBE). Then a sophomore at Purdue University, Harris, along with his five cofounders, established the mission of NSBE "to increase the number of culturally responsible black engineers who excel academically, succeed professionally, and positively impact the community" (NSBE 2014). Today, NSBE is one of the largest student-run organizations in the United States, with core activities focused on improving the recruitment and retention of Black and other minority engineers, in both academe and industry.

What Harris understood was that there is power in a student's unique perception of the world—how they view the world and their rich imagination of what this world could potentially be.

## SMALL PERCENTAGES, MAJOR DIFFERENCES

Today, Blacks working in architecture and engineering occupations make up 5.6% of the more than 2.8 million Americans in those fields. And lest that small percentage cause dismay—it's less than half the percentage of Blacks in the US population—in raw numbers it's still an impressive figure, at nearly 160,000 degreed Black professionals in the industrial workplace. Their presence has changed the social landscape in corporations all across the country, and their performance has reshaped product lines—and corporate bottom lines—all across the US economy.

Today, despite the challenges and barriers that still exist for many Black youths seeking access to the "meritocracy" so proudly discussed in presentations about America's open society, the US Census Bureau reports that 82% of the 42 million African Americans tallied in 2010 held high school diplomas, while 18% of Blacks aged 25 and older held bachelor's degrees or higher. Of these, 1.5 million held advanced degrees, and among the younger generation, 2.9 million Blacks were enrolled in college, an increase of 1.7 million over their enrollment in 1990 (US Department of Commerce, US Census Bureau 2012a).

That continuing drive toward educational growth, despite all obstacles, facilitates and supports today's necessary push for increased access by Blacks to careers in the technology enterprise. The first benefit of all this prodding, pushing, and driving for education is obvious, though many observers might not think so. Traditionally, the recruiting base for engineers and scientists in America has been young White males. But that group is a shrinking part of the American mosaic; US Census figures show that the production of children among non-Hispanic Whites has fallen below replacement of deaths, while Blacks continue to expand the size of their population (US Department of Commerce, US Census Bureau 2012b).

## THE NEXT FRONTIER FOR BLACK ENGINEERS IN BUSINESS AND INDUSTRY

US engineering schools produced just 579 African American engineers in 1972 and another 657 in 1973, according to Engineering Manpower Commission (EMC) data (NACME 2011). However, these figures may underestimate the actual number of degrees because not all institutions reported to the EMC. In 2009, the Integrated Postsecondary Education Data System (IPEDS) reported that African Americans earned 4.7% of all engineering bachelor's degrees—3,096 in total. In addition, African American representation among engineering bachelor's degree recipients grew to 5.6% in 2000, only to slightly decrease to less than 5% in 2009. Despite the decrease from 2000 to 2009, both years were an overall increase from the 3% reported in 1977 (NACME 2011).

The number as well as the percentage of engineering bachelor's degrees earned by African Americans has declined since 2005. The National Action Council for Minorities in Engineering (NACME) reported that in 2009 there were 128,042 Blacks working in the field of engineering (NACME 2011). That is greater than the number of African Americans enrolled in college in 1950, as higher education geared up to educate the millions of US war veterans coming back from their overseas assignments. In addition, between the years 2000 and 2010, African Americans—mostly through births—boosted the size of their population by 15.4%, even as the nation's

Hispanic population was making headlines with even bigger growth. The current projection is that in less than 40 years, Blacks will increase their numbers to 65.7 million (NACME 2011).

As the African American population has grown, along with the percentage of engineering bachelor's degrees, the technical and industrial contributions of African Americans have also developed. Corporations throughout the nation have utilized this increase in African American innovators to drive growth and development.

IBM, a major driver of the technology revolution that changed work styles, working relationships, and the speed of doing business all over the world, opened its doors to African American technology professionals. Recognizing earlier than many others that it had to become a global leader in diversity as much as in technological innovation, the company developed an entirely new talent pool to add to its existing roster of technology stars.

When IBM decided to jump into the desktop computer revolution, shaking business to its foundations during the late 1970s, it put a team to work in Boca Raton, Florida, far from the corporate headquarters in Armonk, New York, outside the arena of traditional IBM work styles, purchasing decisions, and product development programs (IBM Corporation 2011).

Among that team was a young University of Tennessee graduate named Mark Dean, who completed his university studies a scant 11 years after the assassination of Dr. Martin Luther King in Memphis, Tennessee. Dean quickly won attention as a student with a bright future. Dean's role in Boca Raton was to design a computer graphics adapter that could stand up to the competition of the Apple IIe, whose graphic capabilities were legendary among early adopters of desktop computers. VisiCalc, the first microcomputer spreadsheet, had opened up whole new vistas in "what-if" business planning, and IBM needed to go one better. During his early years at IBM, Dean was an integral team player in the development of the IBM Personal Computer.

Dr. Dean's expertise flowered in the architectural design of the PC Advanced Technology, producing a template that became the de facto standard for desktop computers all over the world and drove a new era of computing for the global economy. But his success was not only in advancing the technology. He rose into the prestigious ranks of the IBM Fellows, eventually being chosen to direct an IBM Research Lab in Almaden, California. Dean, now retired from IBM, holds three of IBM's nine patents for the original Personal Computer, the computer that in 1981 gained the unprecedented title of *Time* magazine's "Person of the Year" (IBM Corporation 2011).

Lloyd Trotter is a name that is probably a lot less well known than Mark Dean, but his work at General Electric (GE) took him right to the top of Corporate America, eventually landing him in a post at which he had complete charge of one of GE's signature business units. Trotter, a native of Cleveland, Ohio, came to public attention as one of the "50 Most Important African Americans in Technology," a list circulated and written about in such influential publications as the *Wall Street Journal* and *Fortune* magazine (Cobbs and Turnock 2003). But he did not begin at the top. In fact, after graduating from high school, at the height of the 1950s and 1960s Civil Rights movement, Trotter entered the workforce as one of the first few minority workers at Cleveland Twist Drill, a manufacturer of machining tools.

After completing a unique "journeyman" program with the organization, Trotter decided to pursue his college education and enrolled at Cleveland State University. During his studies, Trotter continued to work, and in 1970 he joined the GE team as a field service engineer at GE Lighting. In 1972, Trotter earned his bachelor's degree in business administration (Biography .com 2014c).

Throughout his decades-long tenure with the organization, Trotter rose to a top-level position as president and CEO of GE Industrial, a major division of GE. In this capacity, Trotter influenced the culture of the corporation, as a minority and, more importantly, as a leader. Through his leadership by example, he opened doors for other minorities and women. Despite his complex schedule, Trotter remained dedicated to the minority community, establishing GE's African American Forum, a collective where minorities within the organization could mentor one another and share best practices for leadership, development, and growth (Biography.com 2014c). Trotter served as a board member of the National Association of Manufacturers, NACME, and the GE Foundation. Trotter also oversaw GE's contributions to America's Promise Alliance, an organization devoted to helping create conditions for success for all young people and established by Colin Powell—US statesman, retired four-star general, the first African American appointed as the US Secretary of State, and the first, and so far the only, African American to serve on the Joint Chiefs of Staff, a body he chaired from 1989 to 1993 (America's Promise Alliance 2014).

Trotter left GE with a handful of engineering colleagues to establish GenNext360 Capital Partners, a New York financial powerhouse that buys technology companies, invests enough to ramp up their core capabilities, and launches them to new heights in industrial competition (GenNx360

Capital Partners 2014). Through his work at GE and continued dedication to the minority community, Trotter—an unsung hero—transformed the culture of a corporation, opened numerous doors for minorities, and established a platform for minorities to share and support one another.

## REENGINEERING THE CORPORATION

Linda R. Gooden is one who has stood above the rest in the technology arena as well. Gooden joined one of the world's leaders in defense technologies, Lockheed Martin, and helped the organization maintain its title during her 32-year career with the corporation. The Youngstown, Ohio, native joined the organization about the time its top executives were deciding to change the focus of its business, expanding from its traditional base of developing military hardware into the information technology (IT) world, morphing the company into a major provider of IT services to government agencies far from the fields of military contention. Gooden was recruited to Lockheed Martin in the mid-1990s as a computer science and business administration major (Lockheed Martin Corporation 2013).

Using her unique cultural and educational backgrounds, Gooden played an integral part in the organization's transition. Beginning as chief of a one-customer, one-contract organization, Lockheed Martin Information Technology, Gooden spent 20 years growing the division, by mergers and acquisitions, into a 14,000-person powerhouse. In 2006, she was promoted to executive vice president of Lockheed's Information Technology Services Business Area, with managerial responsibility for an estimated one-quarter of the defense giant's $32 billion annual revenues. Under her leadership, the company's Information Systems & Global Services division has remained the top provider of IT services to the US Government for the past 18 years and has expanded government IT business and operations into international markets. The organization has also reported that Gooden was instrumental in launching its commercial cyber security business (Lockheed Martin Corporation 2014).

Now retired from Lockheed Martin, Gooden was identified in 2009 as one of *Black Enterprise* magazine's 100 Most Powerful Executives in Corporate America and in 2010, 2011, and 2012 as one of *Fortune* magazine's Most Powerful Women in Business. Although retired, Gooden remains a servant of the community, volunteering for such entities as President Obama's National Security Telecommunications Advisory Committee.

These are just a few examples of what can be achieved when organizations and industries alike, throughout the world, tap into diverse thinkers and innovators to drive greater growth and success. It is only when an organization does so that the potential for unprecedented innovation is unleashed.

## The Future of Diversity in Today's Digital Age

> Sometimes when people do things that are courageous, it doesn't really mean that they're that courageous—it simply means that they believe it's important to do it.
>
> DR. FREEMAN HRABOWSKI, president of University of Maryland, Baltimore County (TED Talks 2013).

Today, Silicon Valley is buzzing with new ideas, technology, business ventures, and waves of youths in pursuit of actualizing their dreams. The uniqueness of Silicon Valley is that it is a conglomerate of start-ups, venture capitalists, and established corporations seeking to uncover the next pioneering technology that will change the way that we live, work, and thrive.

We have seen the positive impact that the inclusion of Black engineers has had in the business and industrial world. However, the diversity among technology users isn't reflected among the founders of Silicon Valley companies backed by venture capitalists. In a study conducted in 2010 by CB Insights, it was reported that 83% of founding teams were all-White, while only 1% of venture-capital-backed founders were Black (Butler 2013). From 2009 to 2011, per capita income rose by 4% for White Silicon Valley residents and fell by 18% for Black residents.

As the digital revolution evolves, with Silicon Valley at the forefront, it is important that we continue to focus on inclusion and overall diversification within the industry. Tristan Walker is answering this need as founder of Code2040, an internship program designed to bring Latino and Black engineering undergraduates to Silicon Valley. Walker is also working to resolve the technology industry's dearth of diversity.

The influence of diversity in Silicon Valley does not stop with Walker. Today, African American John W. Thompson serves as CEO of the privately held Virtual Instruments and chairman of the board of directors for Microsoft Corporation, replacing Bill Gates. Thompson has a long history as a Silicon Valley executive, serving in leadership positions for such institutions as IBM and Symantec, where in the span of a decade he increased the company's revenue from $632 million to $6.2 billion and labor force from 2,300 employees to more than 17,500. We can anticipate continued growth and unimaginable innovation sparked by his leadership.

Thompson, a product of a historically Black college or university (HBCU), has already shown a dedication to the minority community, as he returned to his alma mater, Florida A&M University, months after being named to Microsoft's board to deliver the fall 2014 commencement speech. As we anxiously await the great work from John Thompson, we can also anticipate new and exciting historical moments from African American engineers in business and industry in the future. With nearly six million Blacks employed today in management, professions, and related occupations, this recital could go on for a long space. But it is clear from Thompson and the exemplars discussed in this chapter that today's Black engineers, consummate professionals in technology, also occupy critical spaces in the constellation of business enterprises powering America's economy. Each is a success story in his or her own right, but each also is both a role model and a mentor for African Americans moving up the educational and corporate ladders. The effect on America's culture—not to speak of the corporate cultures where they work and now Silicon Valley and beyond—is clear: the face of American innovation is no longer a monochrome presentation. Indeed, the whole corporation, from the lowest entry-level office staff to the highest officers in the best and strongest industrial organizations, can see that excellence and high performance come from people of many colors and many ethnicities.

## Black Engineers and the Frontiers of Global Competitiveness

With today's digital revolution and leadership from Silicon Valley, the global marketplace has become more accessible. It has never been more important than today to remain globally competitive. It is on the minds of every corporate board member, venture capitalist, and entrepreneur. Today, global relevance is a key topic that is at the forefront for engineers as well. Engineers, as they have in the past, will remain key visionaries who will bring the ideas of global impact into fruition.

The opening of doors to the considerably larger numbers of Black engineers has produced major improvements in products and major global business opportunities for many companies, working in multiple industries throughout the world. In just one example, Silicon Valley, the launching pad of the desktop computer revolution that swamped old ways of doing business all over the world, might be a very different place if not for Roy Clay, a Black engineer who left Hewlett-Packard to form his own consulting company, Rod-L Consulting. Had Clay not been there to provide sharp technical analysis and insightful business projections on what new product ideas and companies to support by the Valley's seminal "venture capital firm"

Kleiner, Perkins, Caufield & Byers, companies such as Compaq Computer Corp., developer of the "PC-compatible" computer market, might not have been funded.

Thus, it is clear that the era of greatest participation by African Americans in engineering is yet to come. Creativity comes in all colors; talent knows no color line. The roles of early pioneers such as Lewis Latimer, Granville T. Woods, and Frederick M. Jones are well understood, if not so widely acknowledged among the general public.

Lewis Latimer helped bring Alexander Graham Bell's dream of instantaneous voice communication across the continent to reality, changing the way people do business all over the world. Latimer then joined Thomas Edison and played a key role in lighting the homes, offices, and factories of the world, radically altering the worlds of study, work, and play everywhere. Frederick Jones, applying his knowledge of small-engine capabilities, built on the earlier development of mechanical refrigeration and miniaturized the refrigeration device to make it portable on trucks, railcars, and ships, changing not only the world's eating habits but also the agricultural industry's ability to market its goods to distant consumers.

With these examples, it is clear that Black engineers have escaped the pigeonholes of days gone by and that forward-thinking corporate leaders have recognized the true value of diversity: that talented people, from very different backgrounds, cultural outlooks, and personal styles, add more to a creative team than the consciousness of differences can subtract. Not only the for-profit industry but also nonprofits like Black Girl Code—an organization devoted to teaching young girls of color computer coding and programming languages—can impact the future of Black engineers in the global marketplace. .

The sky is just the baseline for where the next generation of engineers, many of whom will be Black, will take the world. However, the face of the global marketplace must be represented by a multicolored mosaic of thinkers and innovators to surpass the innovations of the past. As a student or young professional, you may face challenges throughout your career, but remember that you have a unique background and story that will aid you in surpassing any mountain you may face. Whether it is obtaining funding sources, like Emmit McHenry, or fighting for your voice, like Granville Woods, your ideas matter, and you have a unique place to drive unprecedented innovations.

Today's Black student engineers, the people who figure out the meaning of new discoveries in science and then apply the principles revealed to solve problems and make it possible for greater numbers of their fellow citizens to pursue active, productive lifestyles once reserved for only the wealthiest individuals, are the ones who will play a key role in making it all happen.

## References

About.com. 2014. "John Standard." http://inventors.about.com/od/blackinventors/p/JohnStandard.htm.

America's Promise Alliance. 2014. "About." www.americaspromise.org/about.

Beyster, J. Robert, and Michael Daniels. 2013. *Names, Numbers and Network Solutions: The Monetization of the Internet.* North Charleston, SC: CreateSpace Independent Publishing Platform.

Biography.com. 2014a. "Garrett Augustus Morgan Sr." www.biography.com/people/garrett-morgan-9414691.

Biography.com. 2014b. "Granville T. Woods." www.biography.com/people/granville-t-woods-9536481.

Biography.com. 2014c. "Lloyd G. Trotter." http://biography.jrank.org/pages/2994/Trotter-Lloyd-G.html.

Butler, Will. 2013. "Addressing Silicon Valley's Race Problem." *New Yorker.* www.newyorker.com/online/blogs/currency/2013/12/addressing-silicon-valleys-race-problem.html.

Cadbury, Deborah. 2007. *Space Race: The Epic Battle between America and the Soviet Union for Dominion of Space.* New York: HarperCollins.

Cobbs, Price, and Judith L. Turnock. 2003. *Cracking the Corporate Code: The Revealing Success Stories of 32 African-American Executives.* New York: American Management Association.

Fouché, Rayvon. 2005. *Black Inventors in the Age of Segregation: Granville T. Woods, Lewis H. Latimer, and Shelby J. Davidson.* Baltimore: Johns Hopkins University Press.

GenNx360 Capital Partners. 2014. "Lloyd Trotter." http://gennx360.com/team/operations/Lloyd_Trotter.

Haber, Louis. 1970. *Black Pioneers of Science and Invention.* New York: Harcourt, Brace and World.

HistoryMakers. 2014. "Interview with Emmit McHenry." www.thehistorymakers.com/biography/emmit-j-mchenry-38.

IBM Corporation. 2011. "IBM100 The PC: Personal Computing Comes to Age." www-03.ibm.com/ibm/history/ibm100/us/en/icons/personalcomputer.

Lockheed Martin Corporation. 2013. "Lockheed Martin Names Rick Ambrose Executive Vice President, Space Systems and Sondra Barbour Executive Vice President, IS&GS." www.lockheedmartin.com/us/news/press-releases/2013/january/0124hq-leadership-changes.html.

Low, Augustus, and Virgil A. Clift. 1981. *Encyclopedia of Black America.* New York: McGraw-Hill.

NACME (National Action Council for Minorities in Engineering). 2011. "African Americans in Engineering." *NACME Research and Policy Brief* 1 (4). www .nacme.org/publications/research_briefs/2011AfricanAmericansinEngineering .pdf.

NSBE (National Society of Black Engineers). 2014. "NSBE Founder: Anthony Harris." www.nsbe.org/about-us/nsbe-founders.aspx#.VK8RlCvF_9Y.

Purdue University School of Mechanical Engineering. 2014. "Anthony Harris: 2008 Mechanical Engineering Distinguished Engineering Alumni." https:// engineering.purdue.edu/ME/People/DEAs/harris.html.

Riley, Christopher, and Philip Dolling. 2010. *NASA Apollo 11 Manual: An Insight into the Hardware from the First Manned Mission to Land on the Moon*. Sparkford, UK: Haynes.

Rillieux, Norbert. 1848. "Sugar Making in Louisiana." *DeBow's Review* 5:285– 288.

Sammons, Vivian O. 1990. *Blacks in Science and Medicine*. New York: Hemisphere.

Siddiqi, Asif. 2000. *Challenge to Apollo: The Soviet Union and the Space Race, 1945–1974*. Saffron Walden, UK: Books Express.

TED Talks. 2013. "Dr. Freeman Hrabowski: 4 Pillars of College Success." www .ted.com/speakers/freeman_hrabowski.

Thompson, Garland. 2009. *Unheralded but Unbowed: Black Scientists & Engineers Who Changed the World*. Garland Thompson.

US Department of Commerce, US Census Bureau. 2012a. "Profile America Facts for Features: Black (African-American) History Month." www.census.gov /newsroom/releases/archives/facts_for_features_special_editions/cb12-ff01 .html.

US Department of Commerce, US Census Bureau. 2012b. "2012 National Population Projections." www.census.gov/population/projections/data/national /2012.html.

The White House Blog. 2012. "NASA Joins Campaign to Encourage the Next Generation of American Engineers and Innovators." www.whitehouse.gov /blog/2012/03/19/nasa-joins-campaign-encourage-next-generation-american -engineers-and-innovators.

Zook, Matthew. 2014. "History of gTLD Domain Name Growth." www.zooknic .com/Domains/counts.html.

Zuckerman, Mortimer. 2011. "Why Math and Science Education Means More Jobs." *US News and World Report*, September 27. www.usnews.com/opinion /articles/2011/09/27/why-math-and-science-education-means-more-jobs.

# CHAPTER NINE

# Socializing African American Female Engineers into Academic Careers

## The Case of the Cross-Disciplinary Initiative for Minority Women Faculty

CHERYL B. LEGGON *and* GILDA A. BARABINO

O VER THE PAST 20 YEARS, concerns have increased about the United States' capacity for innovation and global competitiveness (Pearson and Fechter 1994; BEST 2004; NAS 2007). Responses to these concerns frequently focus on increasing the number of US citizens who become engineers; consequently, earned degrees become a proxy for increasing the engineering workforce. Degrees are indicators of the successful completion of the formal academic curriculum—that is, substantive disciplinary knowledge. Equally important is the successful mastery of the informal or hidden curriculum—that is, the unwritten "rules of the game" which govern the engineering profession and define career pathways. Sociologists refer to the process through which the values, norms, and expectations for professional behavior are transmitted as *professional socialization.*

This chapter presents a case study of a three-year initiative—the Cross-Disciplinary Initiative for Minority Women Engineering Faculty (XD for short), funded by the National Science Foundation (NSF grant no. 07502300). Although the initiative included women from other ethnic groups, this case study focuses on African American women—who comprised the majority of participants in the initiative. XD was conceptualized as an initiative rather than an intervention because of the significant distinctions between the two. An intervention refers to a series of actions or activities designed to alter a course of action leading to undesirable or negative consequences. An initiative refers to a series of actions or activities designed to enhance the effects of a course of action leading to desirable or positive consequences. Therefore, XD is an initiative insofar as the participants are already on a path to success as indicated by the fact that they are in tenure-track positions on engineering faculties.

This initiative is interdisciplinary in terms of the research literature on which it is based. One innovative feature of this research-driven initiative is that it is informed by the integration of literature from multiple disciplines: sociology, social studies of science, science technology studies, engineering, and engineering education.

## Reframing the Issues

Research on groups underrepresented among engineering faculties tends to focus on race, ethnicity, or gender. The XD reframes the issue of diversity in academic engineering in terms of the intersection of race, ethnicity, and gender because the results of the intersections of race and gender are not additive but synergistic: race/ethnicity influences how one experiences gender, and, conversely, gender influences how one experiences race/ethnicity (Leggon 2006, 2010a, 2010b). Although African American women share some of the same experiences as other women, the confluence of race and gender often intensifies the impact of these experiences.

Increasing the numbers is a necessary first step to enhancing the participation of African American women on engineering faculties (Jackson 2004). An African American woman may be the only one of her race, gender, or both on an engineering faculty. Research indicates that feelings of inclusion and "belonging" are critical factors in professional success (Lim 2009). Being the "only one" is an indicator of a peripheral status, and it engenders feelings of isolation (Zuckerman, Cole, and Bruer 1991). Professional socialization is critical to enhancing the quality of one's participation in the engineering enterprise.

The XD initiative is based on a holistic view of the professoriate—not in isolation, but rather as part of a complex system that includes students, administrators, and the engineering enterprise overall. XD focused on faculty, because they function as gatekeepers insofar as they influence how and what engineers are taught and socialized.

## Diversity among the Professoriate

Engineering faculty exert a great deal of influence on both the quantity and quality of the engineering workforce by choosing which students they will mentor and advise and which they will not. This is especially significant because "identification as a 'chosen' student targeting a faculty position" has a major impact on the early socialization of new faculty (Eddy and Gaston-Gayles 2008). Being chosen to be a research assistant provides opportunities to publish as a student; predoctoral publications are one of the strongest predictors of academic career success.

Increased diversity among the professoriate enhances the educational experiences not only of students from groups that are underrepresented in engineering but of all students. Moreover, increasing the participation of African American females on engineering faculties enhances innovation and competitiveness (Malcom, Chubin, and Jesse 2004; Bassett-Jones 2005; CPST 2006; Leggon 2010a). Gains in the percentages of engineering doctorates earned by African Americans are not reflected in their representation among engineering faculties (Leggon 2010b). This is particularly pronounced for African American women.

The XD initiative uses an innovative approach to study, enhance understanding of, and address the unique challenges facing African American women by drawing on the interdisciplinary knowledge base of the social sciences, physical sciences, engineering, and engineering education. The primary goal of the XD initiative is to enhance academic engineering career success through a series of research-based activities. Moreover, the initiative facilitates among participants the understanding of two critical components of an academic career: engineering in the context of the academy, and an academic engineering career in various institutional contexts—in sum, understanding the rules governing academic engineering in general and the "rules of the game" in specific institutional settings.

The primary objectives of XD are twofold: professional socialization and enhanced development of professional skills. The professional socialization component consists of career planning—devising and implementing a research agenda in conjunction with a career plan—and strategies for networking with senior faculty, representatives from funding agencies, publishers, and policy stakeholders. The professional skills component consists of laboratory management, grant management, time management, and proposal writing. XD objectives are presented in figure 9.1.

## Background Data

An initial list of participants was developed using snowball sampling— i.e., a nonprobability sampling technique in which one person names others from among their colleagues. Snowball sampling is often used in populations that are difficult for researchers to identify and/or access. Those on the original list of 32 women of color in tenure-track positions on college/ university engineering faculties were asked for a copy of their most recent curriculum vitae (CV). The authors explained the purpose of the initiative

| Professional Socialization | • Career Planning<br>• Research Agenda<br>• Networking |
| Skills Development | • Proposal Writing<br>• Grants Management<br>• Laboratory Management<br>• Time Management |

FIGURE 9.1 XD objectives

and asked for their commitment to participate for all three years of the initiative. This participation consists of three components: attending the annual conference, submitting CVs annually, and responding to minimal requests for information. This chapter focuses on nine African American women, who comprised 47% of all initiative participants and 75% of African American women participating in the initiative. It is important to note that the very fact that the number of African American women is so small illustrates the scope of the problem.

Of the nine African American XD participants, 56% are on the engineering faculties of public institutions. The geographic distribution of the colleges/universities in which all of the African American XD participants are employed is evenly divided among the Northeast, Southeast, and Midwest—33% each. The field distribution of African American XD participants is as follows: 38% in mechanical engineering, 37% in chemical engineering, 13% in electrical/computer engineering, and 12% in bioengineering.

## XD Initiative

Results of the XD initiative are based on both qualitative and quantitative data collected each year from surveys, focus groups, personal interviews, and analyses of CVs. The initiative was structured so that the authors' role would be greatest in the first year and then would decrease each subsequent year. In other words, the original plan was to structure the initiative so as to facilitate a gradual transition from the authors setting the initiative agenda to the participants setting that agenda; however, throughout the initiative, the authors were responsible for organizing and implementing both the

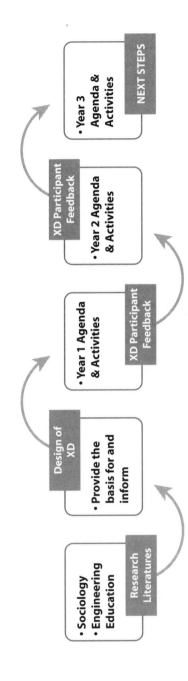

FIGURE 9.2  XD initiative conceptualization schema

agenda and activities. Indeed, the entire XD initiative can be conceptualized as an iterative process informed by research and practice, as shown in figure 9.2. This strategy is designed to enhance participants' sense of empowerment and ownership in the initiative.

## Inaugural Year

During the first year, the authors created and implemented the XD activity agenda. This agenda was informed by research literature on women in academic engineering careers. The first year of XD culminated with a two-and-a-half-day national conference, "Succeeding and Thriving in Academic Engineering." Held at the Georgia Institute of Technology, the conference brought XD participants together with engineering deans and faculty from institutions across the United States. Based on relevant research literature, the authors crafted the conference agenda. Figure 9.3 shows the conference

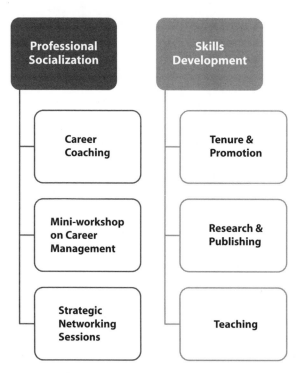

FIGURE 9.3  Year 1 conference agenda

elements in the context of XD's two major objectives: professional socialization and skills development.

## RESEARCH BASE

To set the tone for the initiative in general and the conference in particular, the conference began with a session entitled "Women of Color in STEM: The Research Base."* The main purpose of this session was to provide XD participants with an introduction to and overview of the relevant literature informing the XD initiative, including sociology, engineering, and engineering education. The authors felt strongly that it was important to let the XD participants know that their issues and career experiences are important topics for scholarly research and to tell them about the best sources for this research literature.

## CAREER COACHING

The career coaching/speed mentoring activity was adapted from a program instituted under the Georgia Tech ADVANCE institutional transformation program. During the first day of the conference, there was a session providing an overview of career coaching—in terms of purpose, structure, and process. During the second day of the conference, the career coaching/speed mentoring activity lasted three and a half hours. Based on reputation and experience as faculty mentors, the authors selected eight engineering faculty members to serve as career coaches. Before the conference, the authors provided an overview of the initiative and detailed information about the career coaching activity. Each coach met individually with each XD participant to discuss their CV in terms of both content and structure. Each individual session lasted approximately 20 minutes and ended with formulating the outline for a career plan. Each XD participant rotated to the next table until she had consulted with every career coach. The purpose of the activity was to provide feedback from a variety of faculty to enhance both the CV and development of a career plan. Results from the conference survey are an indicator of how useful and effective the career coaching activity was: all XD participants "agreed/agreed strongly" with the statement "I plan to keep in touch with the career coaches."

---

*The presenters were one of the coauthors, Cheryl Leggon (Georgia Institute of Technology), and Anne MacLachlan (University of California, Berkeley).

FOCUS GROUPS

At the end of the first day of the conference, one hour was allocated for focus groups with XD participants. Consistent with rules governing the protection of human subjects, each participant was told that her participation was strictly voluntary and that at any time she could stop participating without penalty or pressure to continue. In addition, the authors asked each participant for permission to tape the focus group session, assuring them that only the authors would have access to the focus group data in both electronic and hard-copy forms, and that the data would be reported in such a way as to preclude identification of any individual. Each focus group participant was given an informed consent form to sign, indicating their consent to tape the session; all signed.

One interactive session, entitled "Nuts and Bolts," focused on critical components for success in academic engineering careers: research, publishing, teaching, tenure, and promotion. This session, conducted by tenured professors from the Massachusetts Institute of Technology and Georgia Institute of Technology, included discussions of developing a research agenda, developing professional strategies, and building a portfolio for promotion and tenure.

Networking opportunities were built into each day of the conference. At the end of the second day, a two-hour networking reception was held to provide the opportunity for XD participants to meet with peers, senior faculty, administrators, and minority women graduate students and postdocs.

CONFERENCE SURVEY

At the end of the conference, XD participants completed an electronic survey assessing conference activities, organization, presentations, and presenters. The survey also contained open-ended questions about their perceptions of the conference in terms of both interest and utility of the topics—including what should have been presented that was not presented. In addition, the survey contained questions to ascertain XD participants' perceptions of the initiative. Respondents were asked to indicate the extent to which they agreed with a series of statements about the conference using a five-point scale: Strongly Disagree, Somewhat Disagree, Neither Agree nor Disagree, Somewhat Agree, and Strongly Agree. It is noteworthy that for each of the items in table 9.1, 100% of respondents reported that they "strongly agree" or "somewhat agree."

TABLE 9.1    Selected survey results from XD inaugural conference 2008
(N=17)

| Response | Somewhat Agree (%) | Strongly Agree (%) |
|---|---|---|
| Adequate networking opportunities during conference | 18 | 83 |
| Learned things I did not know before | 24 | 76 |
| Plan to stay in touch with career coaches | 29 | 71 |
| Plan to stay in touch with other junior faculty | 12 | 88 |

NOTE: Percentages may not total 100% because of rounding.

Also noteworthy is the response to the question, "Was there something included in the conference that you think should not have been?"—93% of respondents reported that there was nothing included that should have been excluded. The following quotes are representative of respondents' assessments of the conference overall:

- "This is a great conference. I look forward to using this information in my action plan."
- "This conference has been inspirational and motivating. I think the most important part of this conference is developing the tools and strategies necessary to successfully manage a career."
- "This is a wonderful opportunity to interact with others; exciting projects may come out of this."

Participants also expressed satisfaction with the speed mentoring activity, which provided multiple opportunities to receive feedback on career plans and CVs from senior faculty and administrators across disciplines. Receiving feedback from individuals both inside and outside an individual's own discipline was deemed helpful. We also learned that participants valued having sufficient time for networking and the ability to coalesce as both a community of scholars and a community of support. Overall, the conference was very well received, and participants reported that their time was well spent. From our inaugural conference we have found that junior faculty XD participants were primarily concerned with three issues:

- time management;
- laboratory management; and
- scholarly collaboration.

Findings from year 1 of the initiative underscore the need to specifically target issues that are not unique to but heightened for African American female engineering faculty:

- networking;
- professional socialization and support in early career development; and
- community building.

The intensity of these issues is heightened by feelings of isolation resulting from being the only—or one of a few—African American and/or female engineering faculty member.

A goal that participants set for themselves as a community is to produce something tangible that is uniquely identified with the group and would not have been possible otherwise; what this would be was not yet determined by the end of year 1. This both reflects and reinforces the importance that XD participants attribute to community building. Feedback from participants was used to shape subsequent meetings and intervening activities.

Each year, XD participants are surveyed about strategic planning for successful academic engineering careers. The majority of survey respondents identified issues concerning collaboration on grants and publications—especially developing guidelines for successful national and international collaborations. Of critical importance is that survey respondents said that they hope that the XD initiative "continues to provide a forum for us to know that we are not alone." This provided opportunities for XD participants to expand their professional network—especially in terms of meeting professional engineers in industry as well as in academe.

Data from both the surveys and focus groups were used to make adjustments to components of the initiative agenda. After the first year, the participants drove initiative activities and set the agenda for the annual conference based on their interests—such as managing a lab and funding research. XD participants set the agenda, and the authors implemented it.

## Year 2

In early 2009, the XD authors accepted an invitation from the Anita Borg Institute* (ABI) to hold our second meeting in conjunction with the ABI annual conference and Grace Hopper Celebration of Women in Computing.

---

* According to its website (www.anitaborg.org), the Anita Borg Institute for Women and Technology was founded in 1994 by renowned computer scientist Anita Borg, PhD.

There is a synergy between the XD initiative and ABI's mission in general, which is to increase the impact of women on all aspects of technology and increase the positive impact of technology on the world's women (ABI website). Moreover, there is additional synergy between the XD initiative and one component of ABI—that is, Underrepresented Women in Computing, which is a community created for and by underrepresented women in computing that seeks to promote the representation of underrepresented women in computer science and other fields. The authors conducted sessions for XD participants only, during which participants presented updates on professional activities and progress—as well as pitfalls—in following their career plans. In addition, participants discussed "next steps" for the XD initiative.

This "conference-in-a-conference" structure provided ample time for XD participants to attend XD-only sessions, as well as ABI conference sessions. Attending ABI conference sessions not only exposed XD participants to sessions focused on women in industry and IT but also provided opportunities to network with women in nonacademic engineering careers. This arrangement yielded major benefits for XD faculty, including:

- opportunities to expand professional networks to include women engineers in industry;
- exposure to a variety of issues; and
- opportunity to initiate ongoing collaborations between and among participants.

## Year 3

During the third year, there were two major XD activities:

1. a policy forum, consisting of meetings with science and engineering (S&E) policy actors in Washington, DC; and
2. the third XD conference: "Succeeding and Thriving in Academic Engineering."

### POLICY FORUM

At the request of XD participants, the authors arranged a meeting in Washington, DC, with engineering (and science) policy consultants. In addition to two independent consultants, the following organizations were represented: the American Association for the Advancement of Science (AAAS), the American Society for Engineering Education (ASEE), and NSF. The purpose of the meeting was to serve as a catalyst to expand XD participants' professional socialization in terms of science/technology policy and policy makers. The meeting lasted one and a half days, during which time the consultants briefed XD participants on policy and the availability of funds. Participants completed a follow-up survey to assess their perceptions and impressions of the Washington, DC, policy forum. However, because these survey data were not disaggregated by ethnicity of respondent, they cannot be presented in this chapter.

XD CONFERENCE: SUCCEEDING AND THRIVING IN
ACADEMIC ENGINEERING II

Part II of the XD conference—succeeding and thriving in academic engineering—was held in May 2011 at the Georgia Institute of Technology. This one-day conference had one presentation on thriving in the academy from the executive director of the National Center for Faculty Development and Diversity. The rest of the conference consisted of three interactive sessions:

- group updates: participants' achievements, activities, and plans;
- funding opportunities; and
- brainstorming: future of the cross-disciplinary initiative.

## Update on XD Participants

Given the significance of the issues addressed in this initiative, the researchers continue to stay in contact with XD participants. No one among the nine African American XD participants had tenure at the start of the XD initiative. It is noteworthy that the most recent CVs of African American XD participants indicate that six of the nine have been tenured and promoted. Of these six tenured faculty, two are in material sciences, two are in chemical engineering, one is in materials science and engineering, and one is in electrical and computer engineering. Of the remaining three, two will come up for tenure, and one was denied tenure.

## Lessons Learned

Over the course of the XD project, not only the XD participants but also the authors learned a great deal. This section summarizes the lessons learned.

- *Importance of managerial skills.* Initiative participants wanted to learn how to start, staff, and manage a lab. Managerial skills also included grant management.
- *Importance of mentors.* Over the course of the initiative, issues concerning mentors were discussed. These included:
  - *One size does not fit all*: simultaneous and sequential mentors.
    - *Simultaneous or multiple mentors*: because one person cannot mentor another in terms of every aspect of personal and professional life, it is wise to have multiple mentors at a given point in time.
    - *Sequential or serial mentors*: often, it is necessary to have different mentors at different points/stages in one's career.
  - *Sponsors and champions.* Mentors are necessary but not sufficient. In addition to mentors, junior faculty need sponsors and champions—that is, senior, experienced individuals who not only know the rules of the game but also have the clout and the contacts that can be used to advance the career of a protégé.
- *Communalism.* A recurring issue over the course of the initiative concerned the extent to which one should/could share one's ideas with one's colleagues. This is what the sociology of science literature calls *communalism*—that is, the free exchange of ideas with members of the scientific community (Merton 1973). At least two of the participants had very negative experiences with sharing with colleagues, in which those colleagues took their ideas, wrote a grant proposal, and did not include them on the proposal.
- *Community and community building.* As the only African American and/or woman faculty member in one's department, building and maintaining a sense of community is especially important. The community can serve as a "sounding board" insofar as it provides a "reality check"—for example, participants can share their experiences and request feedback from the community. At first, XD participants shared individually with the authors their professional experiences and asked for another assessment. As the initiative progressed, a sense of community developed not only between the participants and the authors but also among the participants themselves.
- *Unanticipated spillover effects.* XD participants took it upon themselves to mentor others not in XD.
- *Longitudinal data.* Periodic long-term follow-up with participants is critical in at least two ways: (1) to assess the impact(s) of having

participated in XD at various points in the career, and (2) to test the hypothesis that participating in an initiative like XD increases the likelihood of participating in other initiatives.

► *Context for evaluation and assessment.* Funders—both public and private—are coming under increasing scrutiny to document/demonstrate that the projects they fund actually achieve their stated goals and objectives. Appropriate evaluation for such initiatives requires more than metrics. Lessons learned from the XD initiative indicate the criticality of understanding and placing both the initiative and its target group(s) in appropriate context—specifically, understanding both the issues and ramifications of underrepresentation in the engineering workforces in general and especially in the contexts of groups disaggregated by race, ethnicity, and gender. Although some strategies may work for more than one group, the better our understanding of the confluence of race, ethnicity, and gender, the more precisely strategies can be tailored to specific groups.

► *Iterative approach.* The iterative structure of XD proved to be effective. At the outset, the XD initiative activities were planned by the authors based on relevant research literature: sociology, sociology of science, science and technology studies, engineering, and engineering education. As the initiative progressed, the authors' role in terms of planning activities decreased (although the authors continued to implement activities), while XD participants' role increased. This iterative approach gradually transferred ownership to the participants and thereby empowered them.

In sum, this chapter is a case study. The lessons learned from XD have proven to be beneficial to both the participants and the authors. XD initiative participants learned skills including time management and laboratory management. In addition, they learned some of the "unwritten rules" of successfully pursuing an academic career in engineering. The authors learned a great deal as well—especially in terms of how to structure and assess initiatives designed to socialize African American women into academic careers in engineering. The authors hope that the lessons learned from XD can enhance the knowledge base and inform the creation, implementation, and evaluation of future initiatives.

# References

Bassett-Jones, Nigel. 2005. "The Paradox of Diversity Management, Creativity, and Innovation." *Creativity and Innovation Management* 14:169–175.

BEST (Building Engineering and Science Talent). 2004. *A Bridge for All: Higher Education Design Principles to Broaden Participation in Science, Technology, Engineering and Mathematics*. San Diego, CA: BEST.

CPST (Commission on Professionals in Science and Technology). 2006. *Comments (June 2006)*.

Eddy, Pamela L., and Joy L. Gaston-Gayles. 2008. "Faculty on the Block: Issues of Stress and Support." *Journal of Human Behavior in the Social Environment* 17:89–106.

Jackson, Judy J. 2004. "The Story Is Not in the Numbers: Academic Socialization and Diversifying the Faculty." *NWSA Journal* 16 (1): 172–185.

Leggon, Cheryl B. 2006. "Women in Science: Racial and Ethnic Differences and the Differences They Make." *Journal of Technology Transfer* 31:325–333.

Leggon, Cheryl B. 2010a. "Diversifying Science and Engineering Faculties: Intersections of Race, Ethnicity, and Gender." *American Behavioral Scientist* 53 (7): 1013–1028.

Leggon, Cheryl B. 2010b. "Women of Color in Science, Technology, Engineering and Mathematics (STEM): Refining the Concepts, Reframing the Issues." *International Encyclopedia of Education* 1:686–690.

Lim, Victoria. 2009. "A Feeling of Belonging and Effectiveness Key to Women's Success." *Diverse: Issues in Higher Education* 26 (2): 17.

Malcom, Shirley M., Daryl E. Chubin, and Jolene K. Jesse. 2004. *Standing Our Ground: A Guidebook for STEM Educators in the Post-Michigan Age*. Washington, DC: American Association for the Advancement of Science.

Merton, Robert K. 1973. "The Normative Structure of Science." In *The Sociology of Science: Theoretical and Empirical Investigations*, edited by Robert K. Merton, 267–278. Chicago: University of Chicago Press.

NAS (National Academy of Sciences). 2007. *Rising above the Gathering Storm: Energizing America for a Brighter Economic Future*. Washington, DC: National Academies Press.

Pearson, Willie, Jr., and Alan Fechter, eds. 1994. *Who Will Do Science? Educating the Next Generation*. Baltimore: Johns Hopkins University Press.

Zuckerman, Harriet, Jonathan R. Cole, and John T. Bruer, eds. 1991. *The Outer Circle: Women in the Scientific Community*. New York: W. W. Norton.

# CHAPTER TEN

# Race for the Gold

## African Americans—Honorific Awards and Recognition

TYRONE D. TABORN *and* LANGO DEEN

Allow me to say a word about the theme of tonight's celebration: reflections. The premise on which this theme is based is that telling the stories of yesterday's experiences galvanizes our strength for tomorrow's encounters. I don't know about you, but after hearing these marvelous stories, and reflecting upon Mr. Anderson's role in my life, I am made more ready for tomorrow.

JOSEPH N. BALLARD (2006)

IT WASN'T UNTIL 1791 that the first African American inventor was believed to receive a patent. Thomas Jennings was 30 years old when he was granted a patent for a dry cleaning process. Because slaves could not apply for patents, many of their inventions and innovations were erased from history and appropriated by others. The first African American woman to receive a patent was Sarah Goode, in 1855, for her invention of the folding cabinet bed. Between 1863 and 1913, Black inventors patented approximately 1,200 inventions. Many more were unidentified because they hid their race to avoid discrimination. Some sold their inventions to White men ("Faces of Science" 2014). Black innovation has not stopped or slowed; many have stood on the shoulders of those forgotten creators and have contributed greatly to American progress. They include Elijah McCoy's oil-drip cup; Lewis Latimer's work on electric filament manufacturing techniques; Granville Woods's devices sold to such companies as Westinghouse, General Electric, and American Engineering; Otis Boykin's wire precision resistor and a control unit for the pacemaker; and, more recently, Herbert C. Smitherman, whose products that he helped create and develop include, among others, Crest toothpaste, Safeguard soap, Bounce fabric softeners, Biz, Folgers Coffee, and Crush soda flavors ("Dr. Herbert C. Smitherman Sr. Broke Barriers at P&G" 2010).

The drive for individual achievement and distinction has long been recognized with honors and titles. For American institutions and organizations, as well as local, state, and federal government, awards have proved useful in inspiring innovation. Equally, recognition of success in science, technology, engineering, and mathematics (STEM) has been critical in encouraging historically underrepresented students to pursue STEM careers. However, the lack of visible role models (as in the case of Lt. Gen. Ballard's Mr. Anderson) has presented an enormous problem for changing the face of engineering—in part because good role models help lead aspiring students into scientific and technical fields. For White youths, the odds are on their side for opportunities to meet role models who reflect their values and common culture. The situation is not quite the same for African Americans. The small pool of African American technology professionals makes it unlikely that many African American youths will have meaningful, life-changing interactions or encounters with people who look like them in STEM fields. In a *US Black Engineer & Information Technology* "Professional Engineer" survey conducted in 2008, 23% of respondents cited an advisor's role as a key factor in influencing their career choice, 20% mention the encouragement of a family member, and 22% said having a role model working in the technology field was a big factor. For minority youths, many of whom lack access to professional role models, contemporary media images play a part in shaping not only their perceptions in the digital culture that defines much of their everyday lives but also their relationship with technology. Underlying this problem is the fact that significant achievements of African Americans in scientific and technical fields have been obscured by history (see the introduction and chaps. 7 and 8 in this volume).

Even more sinister is collective memory loss of the embrace and adaptation of scientific advancement and a record of the role of African Americans in producing some of the greatest technological innovations. Therefore, having a dearth of identifiable role models and limited or in some cases nonexistent historical records to consult, young African Americans see technology and the world of science as places where few of their kind are reflected, and consequently spaces where none of them can enter and flourish.

The underrepresentation of African Americans in scientific and technical careers in America has historical roots. According to Eugene DeLoatch, dean and professor of engineering in the School of Engineering at Morgan State University, "During the period of early growth of engineering education, Black American involvement was practically non-existent. The Morrill Act, which authorized the use of public lands and money to subsidize and support the scientific and technical education of U.S. citizens, did not favorably impact Blacks." As a result, African Americans had few examples in these fields to emulate. Engineering education was not a viable study path until the establishment of an engineering school at Howard University in 1910 (see chap. 1 in this volume).

Six years earlier, in 1904, in a speech given to mark Abraham Lincoln's birthday and presentation of Hampton Institute (former name of Hampton University), Booker T. Washington challenged politicians of the day who were of the view that "it does not pay, from any point of view, to educate the Negro; and that all attempts at his education have so far failed to accomplish any good results" (Washington 1904). In a passionate, 4,000-plus-word address, Washington used "testimony of the best Southern white men" (Washington 1904)—such as the state superintendent of education in Florida, a representative for one of the largest accident and casualty companies in New York, and *Atlanta Constitution* journalist Joel Chandler Harris, author of *Uncle Remus*, a series of folk stories from slave plantations—to push back the stereotypes. Harris observed that the Negro should not be judged by a minority of negative individuals, but by the best of the race: "I am bound to conclude from what I see all about me, and from what I know of the race elsewhere, that the Negro, notwithstanding the late start he has made in civilization and enlightenment, is capable of making himself a useful member in the communities in which he lives and moves, and that he is becoming more and more desirous of conforming to all the laws that have been enacted for the protection of society" (Washington 1904).

Against pervasive negativity about Black contributions to American society, individual honors and awards in the African American community have shored up a communal narrative of achievement and actuated various forms of recognition. An early example was the Spingarn Medal instituted in 1914 by Joel Elias Spingarn, chairman of the board of the National Association for the Advancement of Colored People (NAACP). The purpose of the NAACP Spingarn Medal was "first to call the attention of the American people to the existence of distinguished merit and achievement among American Negroes and secondly to serve as a reward for such achievement, and as a stimulus to the ambition of colored youth." The first medal was awarded to Ernest E. Just in 1915 for research in biology. At the time, Dr. Just was head of physiology at Howard University's Medical School (Epsilon Mu Mu 2014).

In the fall of 1925, the movement to highlight African American achievement was further advanced with the incorporation of the National Technical Association (NTA). Founded by Charles S. Duke, the first African American to receive an engineering degree from Harvard University, NTA leadership and its members have had a remarkable history of achievement in the areas of education, research, engineering, science, and technology (National Technical Association 2014).

Five decades later, the National Society of Black Engineers (NSBE) launched with the following mission: "To increase the number of culturally responsible Black engineers who excel academically, succeed professionally and positively impact the community." The idea of NSBE's Golden Torch Awards came much later, in 1998, under the vision of Robert Ingram, former *US Black Engineer & Information Technology* magazine sales executive, and NSBE executive director Charles Walker. Walker believed that since the student-run organization is a major force in increasing the number of African Americans in STEM fields, NSBE should also highlight the best and brightest in the industry. The Golden Torch Awards link the accomplishments of distinguished African American engineers and technical professionals with college-bound dreams of K–12 students (NSBE 2014).

## National Recognition

African American organizations were not alone in honoring African American accomplishments. Notable is the American Physics Association Edward A. Bouchet Award, given to an African American, Hispanic American, or Native American physicist who has made remarkable contributions to physics. Edward Bouchet graduated as valedictorian of the Hopkins Grammar School class of 1870, and in 1874 he became the first African American to receive a PhD in physics at Yale University (Bio of A&E Television Networks LLC 2014). Although organizations like the NTA had done much to create a medium for channeling growing concern over the underrepresentation of African Americans in scientific and technical fields, by 1987 Career Communications Group Inc. recognized an imperative to found the Black Engineer of the Year Awards (BEYA; "Welcome to the First Annual Black Engineer of the Year Awards Conference" 1987). The employee recognition program, now in its 28th year, came on the heels of a number of demographic projections of a shortage of scientists and engineers as supply of homegrown talent in the educational pipeline declined. BEYA is designed to honor African Americans with significant contributions in STEM fields and promote better access of historically underrepresented groups in STEM careers. Over the past 28 years, the BEYA event (held each February during Black History Month) and Career Communications Group's *US Black Engineer & Information Technology* magazine have highlighted the work of hundreds of Black Engineer Award–winning engineers, innovators, and inventors.

BEYA winners are found in almost every professional engineering society and association across the United States. Currently, a few of them are members of the National Academy of Engineering (NAE), which has more than 2,200 US members and 200 foreign associates. Although the NAE does not track such information, the number of members who could be counted as African American since 1965 is estimated to be 35. They include John B. Slaughter (1987 Black Engineer of the Year), Arnold F. Stancell (1992 Black Engineer of the Year), James W. Mitchell (1993 Black Engineer of the Year), Mark E. Dean (2000 Black Engineer of the Year), Shirley Ann Jackson (2001 Black Engineer of the Year), Lester L. Lyles (2003 Black Engineer of the Year Lifetime Achievement Award winner), Robert Whyms (2006 Lifetime Achievement), Martin Proctor (2011 Lifetime Achievement), and Kelvin R. Mason (2012 Lifetime Achievement). These BEYA winners, as well as hundreds of others, continue to be a source of inspiration for a new STEM generation.

## Dreamers and Dream Makers

Awards and recognition of success have been critical in encouraging historically underrepresented students to pursue careers in STEM fields. Over the three decades, BEYA winners and Women of Color STEM winners (Women of Color magazine is a sister publication to US Black Engineer & Information Technology magazine) have strived to be recognized as role models for others and continue to make themselves "visible" long after they have walked across the stage. They have also led policy coalitions, community-based efforts, and national STEM movements. However, there is a need for more professionals of color as visible role models, a need for more achievers who have successfully navigated career paths in engineering-related fields to serve as examples for those who might follow. It will take many more industrial-, academic-, and government-based engineers like those below to help close the gap.

William Redmond, 27, is a poster child for the impact of honorific awards and recognition as a catalyst for changing the face of engineering. The 2005 BEYA Conference sparked Redmond's dream of becoming an engineer. After his first BEYA experience and exposure to a raft of honorees drawn from diverse fields in the STEM workforce, Redmond went on to obtain a bachelor's degree in electrical engineering from Morgan State University and a master's degree in business administration from the University of Massachusetts, Amherst. Today he is an engineer with Naval Air Systems Command (NAVAIR) Core Avionics Division. Redmond is responsible for providing engineering support and technical recommendations to program management for the communications systems on the US Marine Corps H-1 helicopter (McCoy 2013).

**Lt. Cdr. Donnie Cochran**, 1989 Black Engineer of the Year, had always dreamed of flying. Growing up on a farm near Pelham, in Georgia's southwestern section, young Donnie, the fifth of 12 children, would pause while working in the fields to watch Navy jets fly over. Those dreams took him from the Georgia farm to becoming the first Black Navy Blue Angel (Wells 1989). Cochran majored in civil engineering technology and might well have continued the tradition begun by pioneers such as Archibald Alexander, an internationally known African American architect and bridge builder responsible for the reflecting pool at the Jefferson Memorial, Washington's Whitehurst Freeway, and the Alabama airfield where the Tuskegee Airmen got their wings.

"I started thinking about flying when I was about 12 years old," he told *US Black Engineer & Information Technology* magazine in 1987 (Wells 1989). Out working in the fields, Cochran had the opportunity to see Navy jets flying over, in what he figured out was low-level navigation training. "And as I'm out there, working in the heat of the day, I see those jets screaming by, and I wondered, 'Now which is better, me down here or those jets up there? Which is more exciting?' It didn't take me long to realize that if I had an opportunity to pursue flying, that was what I was going to do" (Wells 1989).

Just after Cochran finished high school in the summer of 1972, his oldest brother came home wearing Naval Reserve Officers' Training Corps (NROTC) whites. Three brothers had gone on to historically Black Savannah State University and majored in engineering. When young Donnie saw his oldest brother wearing this real sharp white uniform and with the opportunity to travel outside the state of Georgia, it convinced him to go to college. Cochran also enrolled in Savannah State's NROTC and joined the FLIP program, which put him into the cockpit of a Cessna 152, following a similar path as that of Tuskegee flight instructor "Coach" Roscoe Draper and the Air Force's Chappie James. By the time Cochran graduated, he had even made a "hop" in a TA-4 tactical jet trainer, as well as completing 30 hours of flying Cessnas. Using his degree and NROTC credits, Cochran skipped the rigorous Aviation Officer Candidate School and graduated into a commission.

After that came flight school at Pensacola, Florida, the closest Naval Air Station from which the planes Cochran admired as a youth could have come. After jet training at Kingsville, Texas, Cochran went to sea, flying RF-8 Crusader photographic reconnaissance aircraft off the decks of the nuclear-powered carrier *Nimitz* in the Mediterranean Sea. Cochran shifted to the F-14 Tomcat, the Navy's premier air superiority fighter, for 38 months of sea duty aboard the USS *Enterprise*, based just north of San Diego, California, at Miramar Naval Air Station, memorialized in Top Gun as "Fighter Town, U.S.A." The conditions under which the Blue Angels fly are more hazardous. The Angels—the world's most famous precision aerobatics team— took the art of close-quarters formation flying to unprecedented levels.

Lt. Cdr. Cochran became the first African American to fly in the Blue Angels formation in the team's 40-year existence. He was 31 years old. On July 4, 1986, Cochran and his teammates flew Sky Hawks in a ceremonial celebration to salute the restoration of the Statue of Liberty. The young officer, finding himself doubly the center of attraction because of his historic posting, told reporters, "What I am doing is not just a job, it's an opportunity. I would like to show young people the roads that are open to them in America. Nobody said, 'here, Donnie, apply for the team, and they will give it to you.' You have to earn it" (Reef 2010). Capt. Cochran spent more than 4,350 hours in seven different types of aircraft, completed 570 carrier landings, and was awarded the Meritorious Service Medal, the Air Medal, and the Navy Commendation Medal, among other awards.

In 1987, **Dr. John Brooks Slaughter** became the first person to win the Black Engineer of the Year Award for his contributions to diversity in STEM education. Writing in *US Black Engineer* magazine in 2000, Slaughter said he felt privileged to be the first Black Engineer of the Year. "I recall the evening of the award presentation vividly," he wrote. "On that occasion, none of us present could have envisioned how significant and widely recognized this annual program would become" (Slaughter 2006).

Dr. Slaughter had big dreams and created dreams for others. He would go on to be the first African American to lead the National Science Foundation and, later, the first African American to ever lead the University of Maryland, College Park. Beginning as an electronics engineer at General Dynamics, after earning a bachelor's degree in electrical engineering at Kansas State University in 1956, John Slaughter moved up the ranks, crossing the lines from industry, to government, to academia, and back again. Bonnie Winston reported in a 1987 *US Black Engineer & Information Technology* story that when he joined the San Diego–based Naval Ocean Systems Center in 1960, he was refused a supervisory position. Before leaving to head the Applied Physics Laboratory (APL) of the University of Washington in Seattle, Slaughter was running a department with 250 engineers and researchers. His team developed some of the early theories for computer control systems in naval weapons. During his time in San Diego, he completed studies for his master's degree in engineering from University of California, Los Angeles (UCLA), in 1961 and for his PhD in engineering physics from University of California, San Diego, in 1971 (Thompson 2006). At APL, a Slaughter-led initiative delved into underwater acoustics research and applications that had critical implications in the decades-long confrontation between US forces and submarines from the Communist-led Soviet Union. In 1977, Dr. Slaughter, then one of only three African Americans elected to the prestigious NAE, moved east to Washington, DC, as assistant director of astronomic, atmospheric, earth, and ocean sciences at the National Science Foundation. In 1980, President Carter named him director. When he stepped up to the leadership of the agency driving American science—like his older contemporary, Lincoln Hawkins, the first Black member to be elected to the NAE—Dr. Slaughter became a mentor and driving force for promotion of African Americans and other minorities in the research establishment. As fellow of the American Association for the Advancement of Science (AAAS), he served on the Committee on Minorities in Engineering for the NAE and, later, as cochair of the NAE Action Forum on Workforce Diversity. In 1973, he became the founding editor of the *International Journal of Computers and Electrical Engineering*. He's also a fellow of the American Academy of Arts and Sciences and the Institute of Electrical and Electronics Engineers, and in 1993 the American Society for Engineering Education inducted him into its Hall of Fame.

Slaughter served as chancellor/president of the University of Maryland, College Park, from 1982 to 1988. Later, after a long, distinguished tenure as president of Occidental College in Los Angeles, he moved back east in 2000 to take over as president and CEO of the National Action Council for Minorities in Engineering (NACME), an organization founded in 1974 out of an NAE inquiry on the low status and small numbers of Blacks, Latinos, and Native Americans in the engineering profession. Since 2010, Slaughter has been professor of education and engineering at the University of Southern California (Slaughter 2014).

A retired executive vice president of Lockheed Martin Information Systems & Global Services and officer of the Lockheed Martin Corporation, **Linda Gooden** won the Black Engineer of the Year Award in 2006, exactly 20 years after John Slaughter, the first Black Engineer of the Year. Like Dr. Slaughter, Gooden had long participated in the national effort to increase the numbers of underrepresented students on the pathways to an engineering education and career, as she noted in a *US Black Engineer & Information Technology* article titled "A Pause for Reflection: Black Engineer of the Year," published in February/March 2006. The digital divide is real, Gooden said, and as the country becomes more diverse, the problem is not just for African Americans. It's a problem for the nation, as far as where the next generation of computer engineers and scientists is coming from. During her career, Gooden worked with various institutions, including Morgan State and Hampton Universities, Prince George's Community College Foundation, the Maryland Business Roundtable for Education, and the Boy Scouts of America. Appointed in 2009 by the Maryland governor, she served a five-year term on the 17-member Board of Regents. "She is, first and foremost, one of the most driven executives I have ever encountered in my professional life," noted Michael Camardo, then Lockheed executive vice president for Information and Technology Services ("Pause for Reflection" 2006). Part of that drive comes from growing up on the south side of Youngstown, Ohio, in the midst of America's steel manufacturing heartland during a period of transition. "It was a place," Gooden recalled in the same article, "where a person with a sixth grade education could get a good job with a decent wage and raise a middle class family. But when I was in high school in the 1970s, the steel mills started to shut down, and a lot of people lost their way of life" ("Pause for Reflection" 2006). By the time Gooden graduated, half the steel mills were gone, and she said that a lot of people didn't have the opportunity to change their lives, and they lost their homes. "It pointed out to me that people really needed an education so they would have a choice in life." She made her choice when she entered Youngstown State University and saw the installation of a new IBM 360 computer. "When I saw that," Gooden told *USBE&IT* magazine, "I decided that was what I wanted to do in life. I was mesmerized by the size of the machine. It took up six normal-sized rooms, though the computing power you have on your wrist today is probably greater than they had in those early machines. But it was clear to me that that was the direction the nation was moving in, and that's where I wanted to be. The heavy industries were going down, and this computer was offering a bright new future" ("Pause for Reflection" 2006). Gooden received her bachelor's degree in computer technology from Youngstown State University in 1977 and went to work writing software for General Dynamics. Three years later, Martin Marietta (in 1995, it merged with Lockheed Corporation to form Lockheed Martin) offered her a job developing the software for the Peacekeeper Missile. Then she was offered a chance to switch from military to business software and develop and install a corporate-wide payment and personnel system. That experience set the stage for the rest of her corporate life. In 1988, Gooden won a contract to modernize the Social Security Administration. She knew from her research that the federal government intended to automate its massive infrastructure (Deen 2009). A company with the right approach to automating federal office functions could grab a lot of future business. Her initial request was small; she needed $200,000 in seed money and 57 employees to pursue outsourcing contracts with a target return of $11 million at the end of the first year. The company staked her group to $600,000, and the new venture brought in $24 million in 12 months. Gooden's team modernized the Social Security system, ensuring that the nation's elderly receive their benefits; digitized the FBI's fingerprint database, so millions of prints can be searched in minutes; automated the Navy's payroll system, so all personnel get paid, wherever they are in the world; developed the communications infrastructure for 25,000 employees at the Pentagon; and eliminated mountains of paperwork by making the patent application process electronic.

Gooden realized that there was a potential market for Lockheed's information technology (IT) services throughout the federal government, and she held the planning session that launched Lockheed's information service. Three years later, Gooden sought and received permission to consolidate all IT into her business area and create the extensive information services company that dominates government today. On September 11, 2001, when terrorist attacks on the United States hit strategic sites in New York City and Washington, DC, Gooden recalled (as she told *USBE&IT* in 2006), "We had 400 people working in the Pentagon when the airplane hit the building. It took until about nine that evening to account for all of the employees. We had a team of people who walked from the Pentagon to our offices with the customers and worked all night so the Pentagon could be live the next morning. Those guys went into the Pentagon while it was burning and brought up those computers. You can't buy that kind of commitment." Gooden retired in 2013 as executive vice president of Lockheed Martin's Information Systems & Global Solutions. In 2010, President Barack Obama appointed her to the National Security Telecommunications Advisory Committee. Gooden also serves on the boards of the Eisenhower Fellowships Board of Trustees, the Armed Forces Communications and Electronics Association International, Automatic Data Processing Inc., and TechAmerica, which represents 1,200 companies within the public and commercial sectors to provide a "grassroots to global" advocacy program in all 50 US state capitals, Washington, DC, and international locations.

**Erroll Davis**, 1988 Black Engineer of the Year, grew up in a working-class neighborhood in Pittsburgh, Pennsylvania. His grandfather and strongest role model in his youth, John Boykin, was a Georgia farmer who had migrated to Pittsburgh and worked as a chauffeur. It was that influence that pushed him to graduate from high school at age 16 and to become the first in his family to receive a college degree from Carnegie Mellon University. That humble beginning also created a fire in him that would drive him to the highest ranks of the utility industry (Black Entrepreneur Profiles 2014). The military opened opportunities for Davis. Before launching his career in industry, Davis, an ROTC cadet at Carnegie Mellon, served in the Army's Tank Automotive Command at a maintenance facility in Warren, Michigan, from 1967 to 1969. The Vietnam conflict was raging, and many in Davis's generation were heading off to fight in Southeast Asia, but Davis jokes that the Army sent him to Michigan because of his poor eyesight. He didn't go far from his Michigan posting when he mustered out.

After fulfilling ROTC active duty obligations, Davis joined the Ford Motor Company, an outfit whose reputation for opening career opportunities for minorities goes all the way back to Henry Ford's leadership during the 1930s. Davis spent his time learning how to manage a multimillion-dollar technology organization and digging into what it took to keep his hardware running at optimal levels (Williams 1996). A bachelor of science degree holder in electrical engineering from Carnegie Mellon University in his native Pittsburgh, he won an IBM Fellowship to complete his MBA studies at the University of Chicago in 1967 before launching a career that made him the highest-ranking Black American executive in the utility business. By 1987, when Davis rose from executive vice president to president of Wisconsin Power & Light—the nation's only African American leader of a publicly held company—he had reached the rank of chairman of the United Way of Dane County, Wisconsin. He was a director of the Wisconsin Manufacturers and Commerce organization, a member of the Madison Police and Fire Commission, a director of the Madison Capital Corp. and the Madison Development Corp., and past president of the Madison Urban League Board of Directors (Thompson 2006).

Davis was now the highest-ranking African American in any business field. Asked about his new status by a Madison news reporter, Davis said he was proud to be Black, but he didn't dwell on race. Davis, who was now running the utility's operations, stepped into the top job at a time when then-CEO James R. Underkofler and the board of directors were pushing restructuring of the corporation. Their plan, now complete, created a holding company for utility operations and subsidiary corporations, to pursue developing opportunities in such businesses as landfill development and communications. They had to move fast. WPL Holdings, as it became known, faced growing competition from other gas and electric suppliers as energy deregulation opened doors to nonutility generators and the selling of excess power from private corporate systems into the public net. Large customers were beginning to buy natural gas directly from its producers, using utilities such as WPL only to transport the gas through their pipelines. In other words, WPL, a company employing 2,600 workers supplying power and natural gas to customers spread over a 16,000-square-mile section of southern Wisconsin, faced large challenges. Revenues had slipped from $170.2 million in the first quarter of 1986 to $157.4 million, only partially because of an unusually warm winter. Davis, a key player in the transition to a holding company and then in the merger with several other utilities to become the main electric power and gas supplier for southern Wisconsin and parts of Iowa, Illinois, and Minnesota, was ready for his move up to the last rung of the corporate ladder. Davis was promoted to chairman and chief executive of Alliant Energy, a Fortune 1000 company with 8,500 employees across the country and abroad. Alliant, traded on the New York Stock Exchange under the ticker symbol LNT, had operating revenues of $3.1 billion in 2003 and total assets of more than $7.7 billion. The company served more than 1.4 million customers over a 54,000-square-mile territory that contained nearly 10,000 miles of electric transmission lines and 8,000 miles of natural gas mains. Alliant operated fossil fuel, nuclear, and renewable generating facilities across the upper Midwest, generating more than 31 million megawatt hours of electric power a year.

In 2003, Davis was named by *US Black Engineer & Information Technology* and *Blackmoney.com* as one of the "50 Most Important African Americans in Technology," and in 2004 as one of the "50 Top Blacks in Technology," again by *USBE&IT.* He was a man at the top of his game. From 2006 to 2011, Davis served as chancellor of the University System of Georgia. He was appointed superintendent of Atlanta Public Schools in July of 2011 and retired in July 2014 (Bloom 2014).

Col. Guion Bluford, 1991 Black Engineer of the Year, attended Philadelphia's inner-city Overbrook High School. In 1991, Bluford told *US Black Engineer* that, while growing up, he developed a curiosity about airplanes as a kid, read a lot of books about airplanes, and built a lot of models; by junior high school he decided he wanted to be an engineer ("1991 Black Engineer of the Year Awards Finalist Black Engineer of the Year" 1991). Along the way, he joined the Air Force and became a pilot and eventually an astronaut. According to the mission biographies of the National Aeronautics and Space Administration (NASA), Bluford, who logged 4,000 hours in jet planes, became a NASA astronaut in 1979, working with engineering systems for the new space shuttle, which had not yet flown. These included the remote manipulator system, Spacelab-3 experiments, the Shuttle Avionics Integration Laboratory, and the Flight Systems Laboratory. Bluford wrote and presented scientific papers in computational fluid dynamics, critical to rocketry and to improvements in powered air flight at the Air Force Research Lab, while continuing his studies to complete a PhD in aerospace engineering, with a minor in laser physics. That moved him up to the front ranks in space exploration. Four years later, the name and face of "Guy" Bluford became famous around the world when he flew on the space shuttle *Challenger* for three days in 1983, deploying the Indian communications satellite INSAT-1B on NASA Mission STS-8. The ability of America's broadcast news media to put color TV pictures in homes, schools, and offices in countries on every continent made all the difference in the world. The Soviets were never able to capitalize on their success in beating the United States to yet another space milestone. In October 1985, Bluford flew into orbit again on Mission STS-61A, a seven-day flight that deployed the German D-1 Spacelab mission and launched the Global Low Orbiting Message Relay Satellite. That was the last time the shuttle *Challenger* went into space before its destruction in a fireball early in 1986.

Delivering a keynote speech at the 40th International Science and Engineering Fair held in 1989, Bluford told 746 high school student contestants that being an astronaut was the "best job in the world." But he also sounded an alarm: "There's going to be a shortfall of scientists and engineers in this country in the years to come," he told a press conference during the competition. "If we're going to be competitive, the leaders in developing new ideas and new products, we're going to need more scientists and engineers" (Thompson 2006). Now in civilian clothes, Bluford has had a career that spanned nearly the entire arc of the superpower competition. He is still deeply involved with space science.

**Dr. Shirley Ann Jackson,** the 2001 Black Engineer of the Year, is a theoretical physicist and currently president of Rensselaer Polytechnic Institute, the oldest technological research university in the United States (Thompson 2001). Earlier in her career, from 1980 to 1995, Dr. Jackson promoted the advancement of women in science as a member of the National Research Council's Committee on the Education and Employment of Women in Science and Engineering and its Committee on Women in Science & Engineering. She has also served as a member, board member, or officer of other scientific organizations. Jackson, the first woman to win the Black Engineer of the Year Award, spoke of her early challenges in STEM in a speech at the 2001 Black Engineer of the Year Awards: "Sometimes a veil of discrimination comes in the form of discouraging words. While I was still deciding on a major at MIT, a professor offered me a bit of career advice. 'Colored girls,' he said, 'should learn a trade.' I pondered this advice and again thought of the choices available to me. Chance had made me colored, I was by chance a girl, and I was here to learn. As for the trade? Well, I chose physics. And I have been trading very well in that domain for these many years. Discouraging words can deflate career ambitions, but only when those ambitions are not anchored in rock-solid principles. Mine were—thanks to my family—and I was able to overcome" (Jackson 2006).

Arthur Schurr, a *Diversity/Careers* contributing editor, tells us that in 1976, Shirley Malcom, Paula Hall, and Janet Brown conducted a landmark study for the AAAS (Schurr 2014). In their report, "The Double Bind: The Price of Being a Minority Woman in Science," Schurr said that they focused attention on the challenges that women of color faced in science careers, as well as in the years leading up to their careers. Now head of the directorate for education and human resources programs at the AAAS, Dr. Malcom observes that the double bind is no less problematic today: "The numbers are not good for women of color in engineering and IT. We were making good progress for a while, but then the numbers started going south again. And the biggest problem with low numbers is the low numbers. A critical mass is necessary for any kind of profound change. Many people credit mentors. And most successful women have several mentors along the way. Mentors are important, but I think the role of community may be more important. Building community that can support study through mentoring is the way to go."

Jackson, "a theoretical physicist with a long string of 'first' behind her name" (Thompson 2001), overcame her early challenges to become the first African American woman to earn a PhD at the Massachusetts Institute of Technology (MIT) in 1973. Her theoretical work took her to AT&T Bell Laboratories in Murray Hill, New Jersey, in 1976. During her 15 years there, the practical effects of her explorations in solid-state physics led to rapid improvements in the signal-handling capabilities of semiconductor devices. That helped keep Bell Laboratories in the forefront of the advancing field of electronic communications (Thompson 2001). In 2001, Jackson was elected to the NAE for her contributions to the formation of the International Nuclear Regulators Association and to industry and her breakthrough research in electronic and opto-electronic materials used in semiconductor materials that are now in many devices (Rensselaer Polytechnic Institute 2014). As the first African American on the Nuclear Regulatory Commission, and first to head the agency, Dr. Jackson crisscrossed the nation, reviewing troubled nuclear facilities and unsnarling knotty environmental safety issues, while also bringing down the temperature of critical debates (Rensselaer Polytechnic Institute 2014). During the mid-1970s, she also served as a lecturer at the Advanced Study Institute run by the North Atlantic

Treaty Organization (NATO) at Antwerp, Belgium. She lectured at the Stanford Linear Accelerator, in California, and served as a visiting scientist at the Aspen Center for Physics in Aspen, Colorado (Rensselaer Polytechnic Institute 2014). A member of MIT's board since 1975, she was selected to be on the board of the nation's oldest historically Black college, Lincoln University, in 1980. Rutgers University's Board of Trustees also welcomed her membership in 1986, and she moved up, in 1990, to the Board of Governors, serving on the Educational Planning and Policy Committee. In 1993, she joined the board of Associated Universities Inc., operator of the Brookhaven National Laboratory and the National Radio Astronomy Observatory (Rensselaer Polytechnic Institute 2014). She joined the board of the Brookings Institution in June 2000. Dr. Jackson, who also has continued active teaching duties during her career, was inducted into the National Women's Hall of Fame in 1998 and won the NSBE's Golden Torch Award for Lifetime Achievement in Academia in March 2000. She also is on the board of trustees of her alma mater, MIT.

Dr. Mark E. Dean was an IBM Fellow and IBM's director of Advanced Systems Development when he won the 2000 Black Engineer of the Year Award (Dean 2000). Dean's contributions to the fields of computer science and IT are exemplary. While you may not have heard of Mark Dean, almost everything in your life has been affected by his work (Taborn 2001). Dr. Dean holds three of the original nine patents that all personal computers are based on. He has more than 30 patents pending. Given all of the pressure mass media are under about negative portrayals of African Americans on television and in print, you would think that it would be a slam dunk to highlight someone like Dr. Dean (Taborn 2001). According to Dean, after integration, one White friend in sixth grade asked if he was really Black. Dean said that his friend had concluded that he was too smart to be Black. "That was the problem—the assumption about what blacks could do was tilted," Dean said. History is cruel when it comes to telling the stories of African Americans. Dr. Dean isn't the first Black inventor to be overlooked. Consider John Stanard, inventor of the refrigerator; George Sampson, creator of the clothes dryer; Alexander Miles and his elevator; and Alice H. Parker and her heating furnace. All of these inventors share two things: (1) they changed the landscape of our society, and (2) society relegated them to the footnotes of history. Hopefully, Dr. Dean won't go away as quietly as they did. He certainly shouldn't. Dr. Dean helped start a digital revolution that created people like Microsoft's Bill Gates and Dell Computer's Michael Dell. Millions of jobs in IT can be traced back directly to Dr. Dean. More importantly, stories like Dean's should serve as inspiration for African American children. Already victims of the "digital divide" and failing school systems, young Black kids might embrace technology with more enthusiasm if they knew that someone like Dr. Dean already was leading the way. Although technically Dr. Dean can't be credited with creating the computer—that is left to Alan Turing, a pioneering twentieth-century English mathematician widely considered to be the father of modern computer science—Dr. Dean rightly deserves to take a bow for the machine we use today. The computer really wasn't practical for home or small business use until he came along, leading a team that developed the interior architecture (ISA systems bus) that enables multiple devices, such as modems and printers, to be connected to personal computers. In other words, because of Dr. Dean, the PC became a part of our daily lives. Frustrated by the bulkiness of newspapers, Dean came up with the idea for a rugged, magazine-sized device that could download any electronic text, from newspapers to books (Crawford 2000). Crawford wrote that the device would be a DVD player, radio, and wireless telephone and provide access to the Internet. It would recognize handwriting (written directly on the screen), be voice activated, and even talk back. But while it could accomplish all of those things, Dean thought that the tablet could be produced cheaply enough so that every student could get one in lieu of books, and publications could give one to every person who buys a subscription. As far-out as the tablet may have seemed, Dean said that it could be available soon. "We are almost there. The only technology left to conquer is the display, we have the other pieces," Dean said. "We will see it pretty soon—easily within 10 years." With technology, "if you can talk about it, that means it's possible," he said (Crawford 2000). Dr. Dean recently made history again by leading the design team responsible for creating the first 1 GHz processor chip. It's just another huge step in making computers faster and smaller, Tyrone Taborn once noted. "As the world congratulates itself for the new Digital Age brought on by the personal computer, we need to guarantee that the African-American story is part of the hoopla surrounding the most stunning technological advance the world has ever seen" (Taborn 2001). Dean, 57, joined University of Tennessee's College of Engineering in the fall of 2013. He is the John Fisher Distinguished Professor in the Department of Electrical Engineering and Computer Science. He last served as chief technology officer for IBM Middle East and Africa, based in Dubai.

Dr. Lydia Thomas, 2003 Black Engineer of the Year, built an extraordinary career based on one simple concept: curiosity about the world we live in. Thomas had an unquenchable thirst for knowledge about the earth and its infinite variety of inhabitants. This is what led her to earn her bachelor's degree in zoology at Howard University in 1965 and a master's degree in microbiology from American University in 1971. She returned to Howard in 1973 to earn her PhD in cytology, which is the study of cells.

As she said in her acceptance speech at the 2003 Black Engineer of the Year Awards, "I have been very fortunate in my life. The people who wanted me to soar have always far outnumbered those who would clip my wings. It was they who taught me to always be aware that there is no force conceivable that can long endure what you put your heart and mind into achieving. So while my gratitude is overflowing, to thank them all would take all night" (Thomas 2006).

Her determination and sense of purpose are what led her to join the MITRE Corporation, where she specialized in the people sciences and environmental issues. As but one example of her many accomplishments in this arena, she led the effort to select the Superfund sites, areas of our country targeted for a much-needed cleanup of toxic waste. Her talent and intellect were immediately recognized at MITRE, and after a rapid rise through the management ranks, Dr. Thomas was tapped to lead a spin-off company called Mitretek Systems (now Noblis Inc.) in 1996. She retired from the position as president and CEO of Noblis, a public interest scientific research, technology, and strategy company, in 2007, and she has served as director since 2008. Thomas has served as a director of Cabot Corporation, a performance materials company, since 1994 (Cabot Corporation 2014), and she serves as a trustee of Washington Mutual Investors Fund, a mutual fund.

**Arnold Stancell**, 1992 Black Engineer of the Year, majored in chemical engineering at the City College of New York. Like his contemporary John Slaughter, Stancell grew up without knowing much about Black scientists and engineers who came before (Kastre 1992). Stancell graduated with a bachelor's in chemical engineering and moved on to MIT in 1958. Four years later, Stancell began his own career of "firsts." He was MIT's first Black doctorate in chemical engineering. During a 10-year stint at Mobil Research, Dr. Stancell developed nine patented processes for making plastics. Still in his research mode, Dr. Stancell opened up a new field of investigation into plasma reactions. He rose to manager of chemical process development and then took a short break to return to his alma mater. On leave from Mobil, Dr. Stancell served as associate professor of chemical engineering at MIT for the academic year 1970–1971. He enjoyed early success mentoring and inspiring a doctoral student, David Lam, who later founded Lam Research, leading maker of plasma etchers for the making of computer chips. Impressed, MIT offered him a tenured post, but Dr. Stancell was headed for bigger things as a manager back at Mobil. Working as the general manager of Mobil's plastics business, Dr. Stancell revolutionized the packaging industry with a clear plastic that totally replaced the ubiquitous cellophane. He also developed the plastic base for PVC pipes, a vastly cheaper competitor to copper for indoor plumbing. Dr. Stancell also played a key role in the communications revolution. His research in thin-film deposition greatly affected the computer industry, and one of his plastic products became the "cladding," or outer coating, for optical fibers. With cladding that has the proper refractory characteristics, most of the energy in the light beam stays within the fiber. This enables the tiny laser beams used in optical fiber communications to travel thousands of miles, still carrying intelligible signals (Thompson 2006). Retiring from Mobil Oil after a 31-year career, Dr. Stancell joined the faculty of Georgia Institute of Technology as professor of chemical engineering, focusing on polymer and petrochemical processes and plasma reactions in microelectronics processing. He was appointed to the National Science Board in 2011 and is currently Emeritus Professor and Emeritus Turner Servant Leadership Chair in Georgia Tech's School of Chemical and Biomolecular Engineering.

**Dr. Wanda M. Austin,** 2009 Black Engineer of the Year, learned to fight for her dreams early. There was a time when she wanted to excel in literature. But after a passionate discussion with a teacher over varying interpretations of a novel, Austin wasn't so sure literature was for her. She became drawn to a discipline where, she says, "someone could take issue with me if the logic was wrong or a mistake had been made—but it had to be more than just his or her opinion" (Witherspoon 2009). That discipline was math. In 1965, when Austin entered the Bronx High School of Science, there weren't many other females or Blacks. But that didn't bother a teenage girl who grew to love mathematics because, as she recalled, "It wasn't subjective like other fields."

Austin earned a bachelor's degree in mathematics from Franklin & Marshall College in 1975 and, two years later, two master's degrees in systems engineering and mathematics from the University of Pittsburgh. She earned a doctorate in systems engineering from the University of Southern California in 1988 (Witherspoon 2009). Since joining the Aerospace Corp. in 1979, Austin has accomplished much while rising through its ranks. She also benefited from the leadership of visionary Aerospace CEO Dr. Eberhardt Rechtin, founding father of systems architecture engineering (Witherspoon 2009). He insisted that jobs were publicly posted and all candidates interviewed, diminishing the influence of the old boy network. He encouraged Austin to earn her doctorate in industrial and systems engineering. Once, she recalls, she worked on a team that launched a satellite from Cape Canaveral.

Austin's ascent was steady. From managing development of communications systems, to general manager of the military satellite program, to senior vice president of the engineering and technology group overseeing 1,000 engineers, Dr. Austin excelled as a technical leader. In January 2008, she was named Aerospace's president and CEO. She now presides over development of the nation's military space projects, as head of a firm with 4,000 engineers and scientists and annual revenues of more than $850 million. "Who would have thought that a young girl growing up in New York City would come this far? But, I was lucky. . . . I grew up with supportive parents that were very encouraging, and told me there were no boundaries to what I could accomplish, and not going to college was never an option for me," reflected Austin. "I didn't set out to be the best African-American female engineer, just to be the best engineer I could be. I have never operated out of the context that I am African-American, except to the extent that I can be a role model" (Austin 2009).

**Olabisi Boyle**, 2012 winner of the Women of Color STEM Award for Managerial Leadership in Industry, is director of engineering planning at Chrysler Group LLC. Boyle received the Women of Color STEM Award in recognition of her "accomplishments in leading and managing a significant part of a technology enterprise and whose career choices serve as an example to women working to move beyond what are considered traditional roles for women" (Chrysler Group LLC 2012). She worked for the Ford Motor Company from 1995 through 2004 in a variety of engineering and manufacturing positions ("Olabisi Boyle 2012 Women of Color Award Winners" 2012). Since joining Chrysler, she has served as senior manager in strategy and product engineering and director of powertrain product management and vehicle engineering planning and technical cost reduction. Prior to that, she was chief engineer for 2011 Chrysler Town & Country and Dodge Grand Caravan minivans. The Chrysler Town & Country won the JD Power "Best Initial Quality" award in the minivan segment. As a senior manager for Dodge SUVs in Truck Engineering at DaimlerChrysler Corporation, she was responsible for leading the product team's business objectives for strategy, budget, timing, and quality. Boyle was cochair of Chrysler's Engineering Talent Management Committee, which focuses on career development for entry and midlevel employees. She also cochaired the Chrysler African American Network, which supports recruitment, retention, mentoring, and community service. Since 2006, she has been a board member on the Detroit Area Pre-College Engineering Program, which provides STEM support in metropolitan Detroit.

But her life hasn't always been easy. When Boyle was 14, growing up in New York City, she found herself in the wrong place at the wrong time: she was caught up in a store robbery, an innocent victim of gunshot wounds in one shoulder and both hands. The incident left her physically and emotionally scarred, initially fearful of going about her normal life as she had before the attack. With the help and loving support of her mother, Boyle overcame these fears and went on to lead successful academic and business careers (Philips 2006). Boyle didn't let tragedy stop her from earning a BS in physics from Fordham University in 1987, a BS in industrial engineering from Columbia University in 1988, and an MS in mechanical engineering from Columbia University in 1991. Boyle told *USBE&IT*, "I have learned that you must work hard not to let events steal your peace, and remember that with some distance, the events do not break you. Ultimately, the events will make you." She approaches her life mindfully, not carelessly, and urges others to do the same: "Happiness is a side product of living your life around good values. Everything we do each day—from the mundane to the grand—is a decision we make. . . . Over the long term, we are summed up positively or sadly based on all of those little decisions we make each day. That said, I suggest we make our daily decisions in such a way that what we value is impacted in a positive way."

She advises students and young managers to develop a "brand" as a leader. "Be the one they know. Show up every day. . . . Be the one they can count on. Be the one who is sought out for leadership, influence, knowledge, vision, and energy." Boyle believes that success comes more easily to those who have a pleasing personality. As she told *USBE&IT* in 2006, "Don't fight the negativity, it's wasteful. Instead, fight to hold onto your peace. Love life. Inhale, breathe deeply, and live" (Philips 2006).

She also advises new leaders to review their own experiences, training, and education to see "where you would like to end up in the future and determine what gaps you have. . . . Always have a current plan in mind. Then relax and enjoy them." What is her advice on how best to do this? "Understand that this too shall pass. . . . Remember, you become a person of character not just when everything goes well, but when you pick up your head after things go poorly."

**Stephanie C. Hill,** 2014 Black Engineer of the Year, like Linda Gooden, works to inspire young people to pursue their dreams and careers in STEM fields. As Lockheed Martin's Information System and Global Solutions–Civil business president, she established a program within her business unit to encourage volunteering and mentorship with local K–12 students, and she sits as a board member of the Maryland Business Roundtable for Education. Most recently, Hill championed the partnership of Lockheed with the creators of a cyber and STEM mentoring program to help students become more interested in STEM-related professions. "My family didn't know any engineers, so I didn't have a concept of what an engineer did and how much you could contribute in that kind of role. It wasn't a natural option as it is for some young people today, but when I think of the things I've had the opportunity to work on, things that have benefited the military and the civilian world, it's amazing! I never dreamed I'd have the kind of opportunities that Lockheed Martin has provided me" ("2014 Black Engineer of the Year Stephanie C. Hill" 2014).

Hill, a 1986 double-major graduate (computer engineering and economics) of the University of Maryland, Baltimore County, led a team that developed the software that directed cruise missiles to targets with such exactitude that amazed TV news viewers could watch one veer around a hotel full of journalists to reach its intended Baghdad landing place ("2014 Black Engineer of the Year Stephanie C. Hill" 2014). Hill won recognition as "Most Promising Engineer" at the 1993 Black Engineer of the Year Awards, and from there she continued her steady climb, taking on ever-bigger technology challenges, in ever-bigger career postings. In 1994 Hill, then a systems engineer, dug deeply into the Navy's Sea Sparrow missile program, working to maintain its multiple warfare capabilities while making sure it would not turn and target "friendly" naval units. Hill also performed trade studies to identify hardware platforms to upgrade missile launch control unit equipment and developed the software and interface requirements for the vertical launch system extant on Aegis missile cruisers.

By 2008, Hill had reached vice presidential levels, first leading Lockheed Martin's corporate internal audit unit, then, in 2012, its Information Systems and Global Solutions–Civil business unit, reporting to another Black Engineer of the Year, Linda Gooden, who described Hill as "a true leader throughout our corporation, industry and community." Today, Hill leads an organization with some 10,000 employees in the United States and eight other countries. Now working in the "civil" side of the world's biggest defense contractor, Hill oversees teams that work to provide solutions to the FBI, including advanced biometric scanning systems that can identify suspects by their palm prints, and one of her team's strongest contributions came in their success winning authorization for Lockheed's SolaS Cloud Solution, a software set that allows government contractors to operate within a secure "cloud computing" environment under the Federal Risk and Authorization Management Program. Hill's teams also support the Federal

Aviation Administration (FAA) to guarantee air safety for global travelers; support NASA's programs to explore space and perform scientific research; help manage claims processing and disability examinations for millions of US military veterans; and address energy challenges by implementing energy efficiency programs. "There were some very pivotal things in the first 14 years of my career that kind of me put me on the career path I am on today," she noted. "I spent years doing hardcore engineering—writing code, testing software, writing requirements for systems. I spent time in the trenches, getting my hands dirty; understanding how to do engineering and how to solve problems with technology. I think that is so important, because I hear sometimes people entering the workforce wanting to move quickly and dramatically into leadership. That time I spent just doing the work of engineering enabled me to be a leader of engineering."

In a letter to the BEYA selection panel, Lockheed Martin's president and CEO noted, "The mark of a great leader is the ability to develop other leaders, and this is an area where Stephanie excels. She genuinely invests her time in mentoring high-potential employees, and she is passionate about promoting STEM education. Stephanie Hill's contributions to her company, her community and her nation [make] her a worthy recipient of this honor. It's a privilege to present her recommendation."

**Professor James Mitchell,** 1993 Black Engineer of the Year, was appointed dean of Howard University's College of Engineering, Architecture and Computer Sciences in 2010. A member of the faculty since he joined the Department of Chemical Engineering as the David and Lucille Packard Professor of Materials Science in 2002, he has also served as director of the CREST Center for Nanomaterials in the School of Engineering since 2003. Before coming to Howard, Mitchell had a 31-year career at Lucent Technologies, Bell Laboratories, and AT&T Bell Laboratories. Mitchell grew up the only boy in a family with four sisters in the home of tobacco factory workers and attended segregated schools. "In the era that I grew up in the South, parents did not have to be well educated to prepare their children to be well educated. All the parents had to do was send a disciplined child to school and support them when they came home. Once I earned a doctorate degree, my decision was that I wanted to be known in the circle of the best people on earth who work in my field." The standards and efforts would make Mitchell a world-class researcher and a member of the NAE ("Kastre, Michael F." 1993).

Mitchell joined Bell Labs as a member of the technical staff in 1970 after receiving his bachelor's degree in chemistry from the Agricultural and Technical State University of North Carolina at Greensboro and a doctorate in analytical chemistry from Iowa State University at Ames. Later, he accepted the position of supervisor of the Inorganic Analysis Group, and in 1975 he was promoted to head of the department. He retired in 2001 as vice president of the communications materials research lab before heading back to academia.

According to Robert A. Laudise, director of the Materials and Processing Research laboratory, "he [was] responsible for a number of important accomplishments that contribute to the increased global competitiveness of the U.S. electronics industry. Using the concept he invented, 'on-demand' reagent generation, which integrates turn-key chemical synthesis with real-time purification and online analysis, dangerously toxic arsine has been produced at precisely determined part per billion levels in order to produce the highest quality silicon wafers for device manufacture" (Thompson 2006).

A major supplier of electronic reagents is negotiating terms for the commercialization of this analytical process system. Mitchell explained the impact on the industry: "In telecommunications, one needs materials that are extraordinarily pure. Silicon was the first example of a broadly used material where it couldn't have trace impurities. The modern equivalent of that is an optical wave-guide material that must be even purer than the specifications we have for silicon. If the optical wave guide materials are not pure, the light pulses that you attempt to transmit through the glass fibers to carry the telecommunications will be absorbed by the impurities" (*US Black Engineer* 1993).

He added, "I am really pursuing a line of materials engineering where I am looking at *in situ* generation chemistry and fabrication of thin films with extraordinarily high purity. When one takes excruciating care to prepare materials in states of purity that exceed those that existed previously, you find that they have extraordinary properties. Materials that were opaque will now be very transmissive. Materials that were easily decomposed by high voltage are now very stable. Or materials that normally were non-conducive, now become conducive by eliminating impurities." Mitchell's personal research is one of the cornerstones of modern trace analysis, and his work, together with that of his group, has been absolutely vital to understanding and manipulating the chemistry of electronic and optical communication materials. Without his remarkable analytical chemistry research, neither optical fibers nor high-purity semiconductors would have advanced to their present stage.

"Substantial quality and bottom-line economic improvements of optical fiber technology are accruing from remedies resulting from Mitchell's process analytical team's diagnostics," says Laudise. "This practical work is exceptionally important in view of the need to greatly enhance the global competitiveness of the U.S. optical fiber industry." Mitchell has also directed a diamond material program. Comprehensive characterization protocols, including new thermal conductivity methods and infrared luminescence techniques, permit polycrystalline diamond materials to be quality assured for use as heat sinks for high-power laser devices. He has been granted a key patent for the selective patterning and nucleation of diamond and has initiated new research in carbon vapor transport methods to further improve the properties of diamond. As a result of his global objectives, he has established "firsts" and received various corporate and professional awards. He is one of only a handful of African Americans to be inducted into the NAE and the first African American to be made a Bell Labs fellow. In addition, he has received the Pharmacia Industrial Analytical Chemistry award, the Percy L. Julian Research award, and two IR-100 awards.

Mitchell has also lectured internationally. In addition, he coauthored a book called *Contamination Control in Trace Analysis*, published more than 60 scientific papers, and invented instruments and processes. Mitchell also served as a member of the editorial advisory boards of three international analytical chemistry journals. A native of Durham, North Carolina, Mitchell has a love of chemistry that goes back to his high school days. While participating in a college program for high school students, he says that a chemistry professor with a doctorate "showed me a mathematical way of balancing equations." This was something his high school chemistry teacher had limited knowledge about. "I was just fascinated by the fact that there was order to chemistry. You didn't have to memorize things. You could figure it out based on established rules. That demonstration showed me that chemistry was also math. So I wanted to learn as much about chemistry as I could" (Thompson 2006). The rest, of course, is history. Mitchell excelled on the leading edge of technology at one of the world's most prestigious research organizations.

**Walt Braithwaite**'s work was critical to the Boeing 777, the world's largest twinjet (Mitchell 1995). Braithwaite, 1995 Black Engineer of the Year, helped perfect Boeing's use of computer technology in the design and manufacturing process. Through the use of computer-aided design (CAD), engineers now are able to assemble entire planes, down to the smallest bolts, and trap any flaws before the actual model is built. These innovations have saved millions of dollars in man-hours by cutting the research and development phase in half.

In his capacity as vice president, Information Systems, Braithwaite was responsible for information systems activities for Boeing's Commercial

Airplane Group. This meant that the commercial airplanes that Boeing produced wouldn't make the cut unless he got the design, computer technology, and resources in place on the front end. Also, he had to ensure that these component parts stayed in place during the entire production process—no small task. In 1985, Braithwaite received an award in recognition of outstanding effort and guidance in the creation of the Initial Graphics Exchange Specification. In 1987, he received the Joseph Marie Jacquard Memorial Award presented by the American Institute of Manufacturing Technology for outstanding technical contributions to the science of computer-integrated technology.

"We came up with the standard to allow different computer-aided design systems to communicate," said Braithwaite. "We did it at Boeing first and I took it to the Academy of Sciences in Washington, D.C., where we discussed it with other people from industry. As a result, we decided to use it as the foundation from which to develop a standard" (Thompson 2006).

"This guy is amazing," wrote Garland Thompson, then editor of *Black Issues in Higher Education*, and a member of the conference's awards selection committee. "What stood out in his package was the fact that he helped to set industry standards for CAD, which today is a commonplace tool in technical industries" (Thompson 2006).

## Changing the Face of Engineering

While there has been much progress in growing African American scientists, engineers, and technologists since Howard University School of Engineering opened in 1910, African Americans remain one of the most underrepresented minority groups in engineering-related fields, according to NACME. African Americans comprise only 5 percent of all engineering bachelor's degrees achieved, with the same percentage of career holders in the engineering workforce. Awards and recognition of success have been critical in encouraging historically underrepresented students to pursue careers in STEM fields. Over three decades, BEYA and Women of Color STEM winners have strived to be recognized as role models for others and continue to make themselves "visible" long after they have walked across the stage. They have also led policy coalitions, community-based efforts, and national STEM movements. However, as indicated at the start, there is a need for more professionals of color as visible role models; a need for more achievers who have successfully navigated career paths in engineering-related fields to serve as examples for those who might follow. It will take many more industrial-, academic-, and government-based engineers to help close the gap.

Here are some things you can do:

- Keep informed about national initiatives focused on STEM education, such as Change the Equation, a CEO-led effort to improve education in STEM, as part of President Obama's "Educate to Innovate" campaign. Change the Equation is a nonprofit organization dedicated to mobilizing the business community to improve the quality of STEM education in the United States.
- Volunteer to teach STEM in your local school or community college, or become an adjunct professor.
- Sponsor a STEM student.
- Join the hundreds of BEYA or Women of Color STEM Award winners who have joined forces to change the face of STEM. In addition to their many other affiliations with STEM organizations around the country, BEYA alumni serve as professional mentors and coaches at sessions for high school students during annual BEYA STEM conferences held each February during Black History Month. They also network with college students from historically Black colleges and universities and other minority institutions looking for internships, co-ops, and other job opportunities with STEM employers.

## References

Austin, Wanda M. 2009. Acceptance Speech, Black Engineer of the Year Award, February.

Ballard, Joe N. 2006. "Lt. General Joe. N. Ballard (Ret.) 1998 Black Engineer of the Year." In *20 Years at the Top: A Generation of Black Engineers of the Year*, edited by Tyrone Taborn and Garland Thompson, 62. Lulu.com.

Bio of A&E Television Networks LLC. 2014. "Edward Alexander Bouchet Biography." www.biography.com/people/edward-alexander-bouchet-21317497# synopsis.

Black Entrepreneur Profiles. 2014. "Erroll Davis." www.blackentrepreneurprofile .com/profile-full/article/erroll-davis/.

Bloom, Molly. 2014. "See the Good-bye Gift the APS School Board Gave Erroll Davis." *Atlanta Journal Constitution*, June 3. www.ajc.com/news/news/local /see-the-good-bye-gift-the-aps-school-board-gave-er/ngC48/.

Cabot Corporation. 2014. "Board of Directors: Lydia W. Thomas." www.cabot -corp.com/About-Cabot/Governance/GN200808251502PM5524.

Chrysler Group LLC. 2012. "Chrysler Group 'Women of Color' Gain Major Recognition at Annual Event." http://media.chrysler.com/newsrelease.do;jsessionid =EC965FEDCEDB3521C255C31D037BE099?&id=13301&mid=1.

Crawford, Amanda J. 2000. "Top Black Engineer Is Computer Whiz." *SUN*, February 19. http://articles.baltimoresun.com/2000-02-19/business/0002190100_1_mark-dean-ibm-black-engineer.

Dean, Mark. 2000. "First Byte Divided Revolution." *USBE Information Technology*, March/April, 10.

Deen, Lango. 2009. "Making America a Better Place to Live." *US Black Engineer & Information Technology*, Fall/Winter, 13.

"Dr. Herbert C. Smitherman Sr. Broke Barriers at P&G, Was Innovative Educator." 2010. *Cincinnati Herald*, October 16. http://cincinnatiherald.our-hometown.com/news/2010-10-16/Front_Page/Dr_Herbert_C_Smitherman_Sr_broke_barriers_at_PG_wa.html.

Epsilon Mu Mu Military Chapter of Omega Psi Phi, Fraternity Inc. 2014. "Omega Psi Phi Founders." www.emm-opp.org/content.aspx?page_id=22&club_id=458224&module_id=152640.

"Faces of Science: African Americans in Science." 2014. https://webfiles.uci.edu/mcbrown/display/Women_inventors.html.

Jackson, Shirley Ann. 2006. "Dr. Shirley Ann Jackson: 2001 Black Engineer of the Year." In *20 Years at the Top: A Generation of Black Engineers of the Year*, edited by Tyrone Taborn and Garland Thompson, 79. Lulu.com.

Kastre, Michael F. 1992. *US Black Engineer*, February, 10.

"Kastre, Michael F.: James W. Mitchell 1993 Black Engineer of the Year." 1993. *US Black Engineer*, 1993 Black Engineer of the Year, 26.

McCoy, Frank. 2013. "Recruited by NAVAIR." *US Black Engineer & Information Technology*, Conference Edition 2013, 99.

Mitchell, Travis E. 1995. "Walt Braithwaite: The 1995 Black Engineer of the Year." *US Black Engineer*, Conference Issue, 37–38.

National Technical Association. 2014. "History: The Minority Technical Voice . . . since 1925." http://ntaonline.org/history/.

"1991 Black Engineer of the Year Awards Finalist Black Engineer of the Year." 1991. *US Black Engineer*, Conference Issue, 51, 52.

NSBE (National Society of Black Engineers). 2014. "Golden Torch Awards." www.nsbe.org/Programs/NSBE-Programs/Golden-Torch-Awards.

"Olabisi Boyle 2012 Women of Color Award Winners." 2012. *Women of Color*, Fall, 36.

"A Pause for Reflection: Black Engineer of the Year." 2006. *US Black Engineer & Information Technology*, February/March, 33–34.

Philips, Bruce E. 2006. "Olabisi Boyle Stays Positive." *USBE & Information Technology*, February/March, 18.

Reef, Catherine. 2010. "Cochran, Donnie." *African Americans in the Military, Revised Edition, A to Z of African Americans*. New York: Facts on File Inc. www.fofweb.com/activelink2.asp?ItemID=WE01&iPin=AAM0036&SingleRecord=True.

Rensselaer Polytechnic Institute. 2014. "Profile of Shirley Ann Jackson, Ph.D." www.rpi.edu/president/profile.html.

Schurr, Arthur. 2014. "Women of Color in Engineering: Progress, but Still a Long Way to Go."

Slaughter, John B. 2006. "John Brooks Slaughter 1987 Black Engineer of the Year." In *20 Years at the Top: A Generation of Black Engineers of the Year*, edited by Tyrone Taborn and Garland Thompson, 9–12. Lulu.com.

Slaughter, John Brooks. 2014. "A Task of Preparation." *Black Engineer*. www.blackengineer.com/commentary/slaughter.shtml.

Taborn, Tyrone. 2001. "MicroCHIPs on My Shoulder: America's 'Masked Inventor.' " *USBE & Information Technology*, November, 10.

Thomas, Lydia W. 2006. "Lydia W. Thomas: 2003 Black Engineer of the Year." In *20 Years at the Top: A Generation of Black Engineers of the Year*, edited by Tyrone Taborn and Garland Thompson, 88. Lulu.com.

Thompson, Garland. 2001. "Engineering's First Lady." *US Black Engineer & Information Technology*, March/April, 28–29.

Thompson, Garland. 2006. "Erroll Davis Jr.: 1988 Black Engineer of the Year." In *20 Years at the Top: A Generation of Black Engineers of the Year*, edited by Tyrone Taborn and Garland Thompson, 15. Lulu.com.

"2014 Black Engineer of the Year Stephanie C. Hill." 2014. *USBE&IT*, Conference Edition, 18–20.

*US Black Engineer*. 1993. Conference Issue, 36.

Washington, Booker T. 1904. "Negro Education Not a Failure." www.advances.umd.edu/LincolnBirthday/btwash1904.xml.

"Welcome to the First Annual Black Engineer of the Year Awards Conference from Morgan State University." 1987. *US Black Engineer*, Conference Issue, 10.

Wells, Grady. 1989. "Black Engineers of the Year Lieutenant Commander Donnie Cochran of the Blue Angels." *US Black Engineer*, Conference Issue, 41–45.

Williams, Donne C. 1996. "Following the Merger Road to the Top: Errol Davis, 1988 Black Engineer of the Year to Chair Mega-utility." *US Black Engineer*, Conference Issue, 16–17.

Witherspoon, Roger. 2009. "Meet the Face of the National Security Space Programs." *USBE&IT*, Winter, 16–17.

# Policies and Programs to Broaden Participation

The four chapters in part IV focus on selected strategies, policies, and programs to increase the representation of African Americans in engineering. In chapter 11, Darryl N. Williams and Angelicque Tucker-Blackmon identify and highlight examples of three highly successful out-of-school engineering programs and initiatives that have historically focused on African Americans. Using a common rubric or assessment metric, the authors examine the programs' strategies for increasing African American participation in engineering. The authors emphasize the need to better understand how to create cohesive learning experiences across formal and informal education. This is both critical and timely given the recent adoption and implementation of the Next Generation Science Standards and the Common Core State Standards.

In chapter 12, Irving Pressley McPhail points to community colleges as a potentially rich pool from which to increase the degree production of African Americans in engineering. He builds on a 2005 National Academy of Engineering report to frame a discussion of promising and best practices for transitioning African American students from the community college to completion of the bachelor's degree in engineering.

In chapter 13, Carmen K. Sidbury, Jennifer S. Johnson, and Retina Q. Burton present a case study of the origins of Spelman College's dual-degree engineering program. The initial program, a partnership with the Georgia Institute of Technology, was designed to increase the number of African American women in engineering. Today, Spelman College has 13 partner engineering institutions. The chapter also discusses other pathway programs at Spelman. The case study provides insights into promising practices that are effective in enhancing the participation of African American women in engineering.

In chapter 14, Kenneth I. Maton, Karen M. Watkins-Lewis, Tiffany Beason, and Freeman A. Hrabowski III examine the educational outcomes of African American engineering students enrolled in the Meyerhoff Scholarship Program at the University of Maryland, Baltimore County

(UMBC). The authors point out that the Meyerhoff Program is an effective intervention for enhancing the participation of African Americans in engineering, and that UMBC is one of the top baccalaureate-origin institutions of African American engineering PhD recipients. The authors conclude by discussing key features of the program which account for its success.

# CHAPTER ELEVEN

# College Me, Career Me

## Building K–12 Student Identities for Success in Engineering

DARRYL N. WILLIAMS
*and* ANGELICQUE TUCKER-BLACKMON

THERE ARE NUMEROUS SCIENCE, technology, engineering, and mathematics (STEM) focused precollege programs with empirical evidence showing that student access to informal learning experiences, in spaces equipped with STEM curricula, equipment, and materials, is an effective approach to increasing students' awareness of the practices of STEM professionals. Further, there is substantial literature demonstrating that precollege programs that allow students to interact with STEM professionals serving in roles of mentors or program facilitators improve students' chances for future STEM college and career success. Similar to the literature on precollege programs, the literature on out-of-school time (OST) programs reports that such programs are an essential component in any strategy to improve the life chances (i.e., academic and social outcomes, career preparation, and social and emotional development) for African American and other underrepresented youths (Russell et al. 2007; Bodilly et al. 2010).

Participation in high-quality OST programming is associated with higher self-concept, self-esteem, resilience, and academic success of youths (e.g., American Youth Policy Forum 2006; TASC 2007). There is mounting evidence demonstrating the impact of OST on students' engagement in STEM (NRC 2009, 2012; Krishnamurthi, Ottinger, and Topol 2013). According to the National Research Council (NRC), the strengths of OST and informal science education programs include increasing students' engagement in science and helping students develop identities as science learners (NRC 2009). As such, after-school and summer learning programs can be very effective in improving access to STEM fields and careers among populations that are currently greatly underrepresented in STEM (i.e., African Americans), especially since these students are most likely to attend OST programs

(Krishnamurthi, Ballard, and Noam 2014). Also, research shows that students who have expanded opportunities, beyond school, to participate in STEM activities are more likely to follow STEM career pathways and excel in them (Wai et al. 2010). Key characteristics of high-quality STEM-based programs include low-risk opportunities for students to engage in content-building activities where they can be wrong without penalty or fear of negative assessments and opportunities to explore beyond the confines of the schools' science and mathematics curricula. This means that students have opportunities to generate questions and ideas on their own in a context that is supportive of their tangential explorations.

There are several underexplored areas with regard to STEM precollege and OST education. First, although there is increased focus on STEM education inside schools, most teachers and administrators engage solely in STEM education activities. There are limited opportunities for students to participate in engineering-type activities inside schools. Therefore, in response to this omission, precollege and OST programs have increasingly provided students with engineering design and innovation opportunities. All of these activities provide students with opportunities to better understand engineering as a discipline and career, while providing them with opportunities to learn engineering design and processing skills. However, few programs focus on building students' efficacy toward engineering. In fact, the engineering education literature does not speak in depth to programmatic practices that lead toward increasing the engineering efficacy of African American students. Second, few studies examine changes in African American students' engineering identity development over time. In contrast, there are studies that explore the development of African American and other minority students' science identity. As with engineering efficacy, an extrapolation of evidence-based practices from science education may prove fruitful for engineering education.

## Empirical Evidence

In light of the limitations in engineering education and precollege programming, educators can look to the literature on OST and precollege programs designed to increase students' knowledge, skills, attitudes, motivation, and efficacy toward science, mathematics, and technology to understand the essential programmatic elements needed in designing precollege engineering programs that increase engineering efficacy. This approach is feasible given that science and mathematics knowledge, skills, attitudes, and efficacy serve as antecedents to students' future engagement in engineering. Although there are a number of empirical studies on precollege and OST programs, designed to assess activities that lead to increases in STEM knowledge and skills, efficacy is not as robust as desired. Findings from evaluation studies of projects funded by the National Science Foundation (NSF) are revealing interesting patterns and opportunities to improve students' STEM motivation and efficacy (Blackmon 2012). Empirical data from underrepresented high school students participating in science and technology OST programs (after-school and summer) reveal statistically significant differences across pre-/posttests that assess students' content knowledge. The studies also show that high school students' knowledge of careers in STEM increases as a result of their participation in these programs (Blackmon and Crawford 2014). These findings were statistically significant across pre-/posttest means. However, data from some of the studies reveal a reverse pattern with regard to African American students' interest and motivation toward science and technology.

In a study of African American and other underrepresented high school girls in a multiyear information technology program, 30 students took both the pre– and post–Motivated Strategies for Learning Questionnaire (MSLQ) test, answering all questions (Morris, Austin, and Davis 2013). Seventeen students were in the treatment group, and 13 students were in the control group. Gain scores on the MSLQ indicated a loss in motivation for the treatment group ($M=-10.82$, $SD=31.78$, $n=17$), while the control group reported a slight gain in motivation ($M=3.54$, $SD=19.25$, $n=13$). Although there was a decrease in scores for the MSLQ treatment group, the decrease was not statistically significant ($p \geq .05$), $t(28)=-1.44$, $p=.92$ (two-tailed). The difference in decline between the two groups was approximately one-half of one standard deviation (Cohen's $d=-.54$). In this same program, 63 students took both the pre– and post–Computer Attitude Questionnaire (CAQ), with all questions answered; 39 students were in the treatment group, and 24 students were in the control group. Gains scores on the CAQ indicated decreases in computer interest for the treatment group ($M=-5.26$, $SD=9.70$, $n=39$), while the control group reported a slight gain in interest ($M=2.54$, $SD=10.24$, $n=24$). Although there was a drop in scores on the CAQ treatment group, the difference in decline between the two groups was not statistically significant ($p \geq .05$), $t(361)=-3.03$, $p=1.00$ (two-tailed). The difference in decline between the two groups was approximately three-fourths of one standard deviation (Cohen's $d=-.78$).

These independent studies reveal a pattern that is rarely discussed in STEM education. In essence, the OST and precollege programs designed, funded, and executed for African American student groups increase their content knowledge and awareness of careers in the STEM disciplines; however, the experiences appear to have a significant negative impact on students' motivation and interest in these areas. It could be concluded that STEM-based programs designed by educators and engineers in the absence of experts in youth development, social cognitive theorists, and/or educational psychologists are not attending to the nuanced experiences necessary to increase motivation and interest, nor are the program designers ensuring that students have positive vicarious interactions with scientists and mentors which would enhance students' efficacy. Further, in the absence of social science experts who know the tenets that drive increases in efficacy, program planners are less likely to intentionally include positive verbal persuasions as a critical element of STEM precollege programming. In essence, efforts to increase interest, motivation, and efficacy become a hit-or-miss practice or are executed purely on a trial-and-error basis.

The aforementioned has been the practice even with most STEM precollege programs situated on resource-rich university campuses where the human, financial, and STEM intellectual capital is readily available; yet, the intellectual capital of individuals with the sociocultural expertise (e.g., anthropologists, social psychologists) to devise strategies that lead to self-efficacy among precollege students consistently goes untapped. Because precollege programs and youth development do not constitute the core business of higher education, the chances of students in such programs engaging in year-round activities necessary to build their STEM efficacy, and subsequent identity, are limited.

There is ever-growing research and policy interest geared at understanding how and why fewer African American youths are cultivating and sustaining interest in STEM-related subjects leading toward increased efficacy and identity as STEM-competent individuals (Gottfried and Williams 2013). The end goal is to determine strategies that will substantially prepare and increase the number of new, diverse people in the STEM workforce. The aforementioned new next steps are juxtaposed to the decades of research findings and reports disseminated by academia, government agencies, and industry bellowing the alarming conclusion that there is a serious deficit in our education system. The question "Can we, as a nation, provide African American students with the necessary opportunities to prepare them for the rigors of STEM?" can be answered with a resounding "Yes," but the response must move beyond blame and deficit depictions of students. The deficits may be attributable to infrastructure and practice, hence the need for practice and perspective interventions, particularly in K–12, such that the US economy can sustain global competitiveness in the future (Committee 2007).

If we take this a step further by focusing exclusively on engineering, findings show that the trends mentioned above have implications for whether African American students elect to pursue engineering college majors and, subsequently, persist toward engineering careers. Williams and Gottfried (2010) point out in their investigation that nonminority male students overwhelmingly receive sustained preparation, exposure, and external motivation in mathematics and science early and often throughout K–12 and beyond, and as a result they are more likely to pursue engineering courses of study than their counterparts. Preparation, early exposure, and external motivation and support are the types of experiences that all students pursuing engineering need, yet African American students, for a multitude of reasons, are not receiving them as regularly and as consistently as nonminority male students. The fact that nonminority male students receive sustained preparation and support is further substantiated in the findings from a study conducted on engineering degree matriculation, where it was shown that only 22% of the total number of students matriculating in engineering were females, and that this encompasses every ethnicity (Lord et al. 2009). Thus, the challenges that come to bear deal with the complexity of early, consistent formal academic preparation and the degree to which various informal support structures in the form and quality of faculty/staff-student interactions are designed, implemented, and reinforced (or not). An analysis of the support and external motivation for some students and not others should be addressed in order to create holistic environments that cultivate and nurture students' efficacy and identity leading toward successful engineering career trajectories for African Americans and other traditionally underrepresented youths.

It seems that pressures continue to mount as researchers try to identify the keys to successful matriculation into any postsecondary discipline or domain, pointing to sufficient levels of preparation coupled with a host of other internal and external factors (i.e., motivation, persistence). Data sources from federal agencies, such as the US Department of Education, and other organizations, such as the National Action Council for Minorities in Engineering (NACME), outline the challenges related to the preparation that many African American students face in K–12 leading up to college entrance (NACME 2011). These indicators build deeper awareness and help explain why greater attention should be given to precollege experiences that include deliberate and intentional strategies to enhance African American students' interest, motivation, and efficacy in engineering.

From a theoretical perspective, there have been countless studies on both self-efficacy and motivation to explain how students come to develop and strengthen their cognitive abilities and persist toward (and in) college and beyond (Dweck 1986; Dweck and Leggett 1988; Elliot and Dweck 1988). Self-efficacy, as defined by Bandura (1977), is a belief regarding personal capabilities to conceptualize, organize, and execute the actions required to accomplish a particular task(s). He outlined the four primary tenets of efficacy: vicarious experiences, verbal persuasion, content mastery, and physiological state. Bandura (1993) goes further to relate the tenets of self-efficacy to academic performance by describing how self-efficacy shapes the choices students make and the level of tenacity they exude in their pursuits. Social cognitive career theory (SCCT) builds on Bandura's model by investigating how an individual integrates self-efficacy, outcome expectations, and personal goals. These three components form the basis for one's career development and identity (Brown and Lent 1996). But, in general, the level of individual competencies based on experiences is a key barometer for determining interest levels in a given domain and has an impact on whether one chooses further exploration (Blustein and Flum 1999). Current educational research points to evidence that suggests that individuals display agency and tenacity leading to deep pursuit when they display noncognitive characteristics, such as perseverance (Duckworth et al. 2007).

However, what is still not clear in the literature is how opportunities to experience and persist in a discipline and the beliefs about one's ability to succeed in that discipline relate to each other in a longitudinal fashion, particularly for African American students (Simpkins, Davis-Kean, and Eccles 2006). Other factors, such as prejudice and negative social stereotypes, may create extra pressures that interfere with performance and persistence (Spencer, Steele, and Quinn 1999). Currently, there is limited understanding of the language used to verbally persuade or convince African American students of their competencies in STEM. Little is known about specific vicarious experiences that lead toward increased STEM efficacy or identity. Lastly, we do not know the physiological state of students as they encounter and engage with scientists or engineers. The literature does not speak to how encounters with disciplinary-based mentors or a disciplinary practice may or may not induce physiological states of anxiety, frustration, and fear that may drive students' efficacy in the opposite direction or increase their efficacy and motivation.

Applying evidence-based frameworks of success with research-based practices to explore unanswered questions can be used to clarify the challenges and experiences of many African American students in the current K–12 academic environment. It might also help to explain how and why many African American students miss out on opportunities to become fully engaged in STEM (and engineering in particular) beyond K–12, owing to a lack of rigorous preparation, internal and external motivation, and exposure to engineering in both formal and informal learning environments (Williams and Gottfried 2010). These functional and behavioral components influence the development of student identity, which, according to Capobianco, French, and Diefes-Dux (2012), is not fixed but linked to lived learning experiences involving engineering-related tasks. Their study further concludes that identification with engineering is related to the level and quality of exposure, which must be contextual, contingent, and conditional.

Another important factor that should be emphasized is how African American students come to identify and commit to a particular academic trajectory and subsequently navigate the hurdles that present themselves along the way to college entrance. To this point, Hagedorn and Fogel (2002) state that less attention has been paid to the impact self-efficacy (in terms of beliefs and abilities) may have on a student's ability to maneuver particular academic environments, and that a key feature of many precollege programs is their focus on increasing students' self-confidence and academic tenacity. Although much of the engineering education literature describes support and external motivation as factors toward nonminority male students' success, there were three organizations launched in the early 1970s to specifically respond to the needs of African American students in STEM. These organizations were launched in response to the push for policy makers, educators, and administrators to devise and implement programs targeting African American precollege students as a means of increasing their interest and motivation to pursue STEM-related careers. Other programs and initiatives across the country sprouted up to engage African American and other underrepresented students in STEM. Many programs initiated partnerships with stakeholders from academia, industry, and the federal government. Of these programs, some have placed emphasis on attracting and engaging African American students in activities involving engineering to facilitate how they learn and apply core mathematics and science concepts. These programs have also found innovative ways not only to engage students but also to include participation from the community by reaching out to parents and caregivers. As a result, they have developed a holistic approach to preparing and guiding students through a trajectory that can lead to the successful pursuit of engineering and other STEM-related career pathways.

## Three Models of Successful Precollege Engineering

This section identifies and highlights examples of three successful precollege programs and initiatives that have historically targeted African American youths by exploring their strategies for broadening participation through engineering activities, college preparation, and career awareness. A methodology for assessing these programs employs a modified framework outlined by Tierney and Hagedorn (2002). Each organization was analyzed based on the following variables: (a) date of inception and purpose, (b) goals and/or mission of program, (c) model framework, (d) target grade level(s), (e) demographics, (f) instructional processes, and (g) sources of funding. This approach was applied to identify exemplars that have been developed and sustained through various combinations of support from the federal government, academia, and/or industry.

## MESA

Founded in 1970, Mathematics Engineering Science Achievement* (MESA) uses a pathway model to increase the number of African American middle and high school students prepared to pursue degrees leading to STEM or medical careers. The intervention framework incorporates individual student academic plans monitored by counselors, PSAT/SAT preparation, study skills training, local and regional math and science competitions, career and college exploration, parent leadership development, in-school academic support and participation in MESA-sponsored activities, and teacher training opportunities. Funding sources for MESA include federal, state, school district, corporate, and private contributions.

## SECME

Like MESA, the mission of the Southeastern Conference for Minorities in Engineering† (SECME) was to increase the pool of historically underrepresented (i.e., African Americans) and underserved students who would become prepared to enter and complete postsecondary studies in STEM. Similar to MESA, SECME relies on competition and teacher professional development as a framework for promoting preparation for STEM degrees and the workforce. While MESA engages middle and high school students, SECME engages students earlier, starting in third grade, and allows them to remain engaged through high school. The SECME model is teacher-centric, placing heavy emphasis on K–12 teacher professional development as a means of supporting deeper content knowledge for all students. Teacher

* http://mesausa.org.
† http://secmeinc.wix.com/secme.

professional development takes place in the form of an eight-day residential Summer Institute, which rotates among member universities and is supported by industry and government entities. Some participants receive financial support directly from their respective schools or districts, including Title I, Title II, and other funding. Similar to MESA, SECME funding sources include federal, corporate, and private foundation contributions. SECME also receives funding from its K–12 school and university partners.

DAPCEP

The Detroit Area Pre-College Engineering Program* (DAPCEP) was founded six years after MESA and just one year after SECME. Like both MESA and SECME, DAPCEP was launched to address the low number of underrepresented students (i.e., African Americans) receiving engineering degrees. DAPCEP's mission is nearly identical to those of MESA and SECME, that is, to increase the number of historically underrepresented students who are academically ready to pursue degrees in engineering. While MESA primarily works with students in the western region of the United States and SECME works with students across six southeastern states, students attending DAPCEP programs are predominantly from schools in the Detroit metropolitan area. Unlike MESA and SECME, DAPCEP is not a multicity or multistate STEM program provider. Another major difference is that DAPCEP starts even earlier, introducing students as young as pre-K to engineering and engineering design principles. The DAPCEP intervention framework allows students to explore many aspects of engineering and technology, such as chemical engineering, nanotechnology, robotics, and computer programming. They focus on enhancing students' quantitative and qualitative skills. Like the other two organizations, DAPCEP provides students with sessions to prepare for college and offers workshops for parents and caregivers. In addition, DAPCEP funding sources include school districts, university partners, corporate and private foundations, and federal dollars.

## Pinpointing Success

It is clear from the model analysis of the aforementioned programs that there are a series of key variables that have worked in intricate detail to enable program sustainability for more than 30 years. These programs began with solid commitment from corporate donors and private foundations that

* www.dapcep.org.

expressed concern about the underrepresentation of African Americans in college engineering programs and subsequent underrepresentation in the engineering profession. It was also deemed critical to establish strong relationships between universities that provided access to accredited engineering programs and K–12 schools (or school districts) that identified a need to provide supplemental enrichment for their students and teachers. Participating universities, through their institutional leadership, have made it clear in their strategic plans to support the academic enhancement of K–12 students and communities and view support of these organizations as a means for developing a cohort of students to facilitate the diversification of their future student populations.

The model frameworks for these programs vary to some degree but ultimately focus on precollege activities that supplement formal science and mathematics learning inside the classroom. Most of the organizations' extended STEM programs are in urban school districts and deal with predominantly African Americans and students from low-income families. All of the programs offer summer components, some of which provide teacher professional development opportunities. The professional development activities often give teachers the chance to (a) work with STEM professionals (be they in academia or industry), (b) enhance their pedagogical skills, and (c) learn science and engineering concepts from real-world contexts. As stated by SECME, the premise is that a teacher- centric approach to implementing precollege STEM programs can translate into deeper learning opportunities for students. This framework now creates avenues for industry professionals to engage and collaborate with teachers and students to articulate details and nuances about what it means to practice STEM. Some of the programs report strong ties with former student participants (alumni) who have gone on to industry and now work as industry professionals, yet find time to serve as mentors to the newer student cohorts and as ambassadors for the programs at their respective companies and institutions. This feedback loop adds another dimension of sustainability and reinforces the partnerships that have been established.

Not only are teachers and industry professionals essential to the successful development and engagement of African American students in STEM, specifically engineering, but parents and primary caregivers are critical as well. The purpose for advocating their engagement is to provide a supportive infrastructure that enables the success of the students beyond the corridors of formal institutions, whether they are formal or informal STEM-based institutions. Key features of parental participation include equipping parents with the skills to work with and encourage their children, identifying the steps necessary for academic achievement and matriculation into college, and increasing their knowledge about college and the resources available to support college entrance and completion. By combining academic enrichment, mentorship, and parental involvement, the program models described have delivered and continue to deliver intervention opportunities in the form of STEM programming which are consistent and robust, thus leading to model replicability.

It is important to note that while the original mission of the three exemplary STEM-based organizations was to focus on precollege engineering, most have modified their scope to explore STEM more broadly to meet the needs of K–12, the needs of higher education, and the desires of funders. One particular program, DAPCEP, also includes the medical sciences and has adopted the acronym "STEMM" to reflect the additional foci. Further, although all three programs have 30-plus years of success, over the past two years the programs have been documenting success and the processes that led to success so that others can replicate the program models. For instance, evaluation data from DAPCEP show that they are uncovering specific individual and combined structural support components that influence the sustained engagement and science efficacy of African American girls. The program reports that, for example, programmatic structural support includes ensuring that each class session is interesting to participants and that classroom activities are conducted in short intervals and include a multimedia component. The program also reports that girls, in particular, are more inclined to work cooperatively in groups. Therefore, sessions should allow for cooperative grouping of students and that facilitators work to create an environment that is comfortable for girls, that maintains friendships, and that allows them to ask questions without fear of judgment.

While success indicators from DAPCEP are more qualitative in nature, data from SECME provide a quantitative measure of student success. Specifically, 59% of SECME 2012 juniors and seniors intend to major in engineering when they go to college. For African American students, the SECME SAT composite average was 1720, while the national African American student composite average was 1273. Although these findings are not exhaustive, this general overview reveals the level of historical and contemporary success of students in each organization. One primary reason that success data from the aforementioned organizations are not easily accessible is the dearth of research and evaluation studies that these programs provide in the STEM education literature.

## Policy Implications

Swail and Perna (2002) conducted a cross-comparison evaluation of pre-college programs by analyzing survey responses from the National Survey of Outreach Programs, which was originally designed and administered in 1999 by the College Board, the Education Resources Institute (TERI), and the Council for Opportunity in Education (COE). The National Survey of Outreach Programs, designed by both Swail and Perna, is comprehensive in scope and does not focus on any particular discipline or domain. Their investigation and analysis of responses provide some important takeaways that could help build stronger research and evaluation of STEM precollege programs and interventions and have implications for current and future programs. In their analysis, they concluded that programs should contain the following elements to truly create lasting and sustainable impacts: (1) provide a clear, focused mission and vision, (2) start well before high school, (3) motivate and set expectations, (4) support parental involvement, (5) foster a collaborative infrastructure, (6) design effective funding strategies, (7) support staff development, (8) identify and implement "best practices," (9) adopt standard processes, and (10) infuse technology throughout the program and administration. Mapping across MESA, SECME, and DAPCEP, it is clear that these programs meet, if not exceed, these components effectively. However, as programs begin to evolve from exposing and preparing underrepresented students for college and future engineering careers toward increasing their engineering efficacy and identity development, additional recommendations beyond the macrolevel ones presented are required. These midlevel recommendations include the infusion of strategies that will lead toward intentional building up of efficacy and identity as opposed to serendipitous movement toward these constructs.

Precollege programs have been supplementing formal teaching and learning of mathematics and the sciences for nearly four decades. These two domains have had clearly defined roles in K–12 education, while engineering has had much less of a presence in formal teaching and learning on a larger scale until the early 2000s (National Academy of Engineering 2009). At this time, a number of K–12 engineering projects and programs began to emerge owing to increased federal attention and support (namely, NSF), and a few states began to articulate standards in the required curriculum. However, a paradigm shift is unfolding, and soon new pressures will be placed on K–12 education as a result of the adoption and implementation of the Next Generation Science Standards (NGSS) and the Common Core State Standards (CCSS). NGSS emphasizes engineering and technology concepts and practices in a more substantive way than students have been previously taught. As a result, there will be significant demand for more robust teacher professional development, curricular materials, and assessments and new ways of engaging students in formal and informal learning around STEM. Precollege engineering programs are poised to take on the challenges presented through the introduction of common state standards and can support students in improving their competencies and performance in mathematics and science—all leading to their ability to perform well on current and newly created assessments that will measure their proficiency in STEM (again, with more emphasis on engineering and technology) more rigorously.

Recently, an investigation was conducted to explore the landscape of K–12 engineering education by looking across the United States to identify and understand the challenges and opportunities for engineering as a mechanism for formal teaching and learning (Carr, Bennett, and Strobel 2012). The authors compared various states by looking at metrics that pertain to how engineering is articulated in state standards across K–12, breaking up the analysis by elementary school (ES), middle school (MS), and high school (HS) grade bands. For the purposes of this chapter, results from the study were used to map the states and regions served by MESA, SECME, and DAPCEP to see whether or not engineering has a presence in the states' curriculum. The rationale for this assessment is to acknowledge those states with the infrastructure in place to deliver engineering content versus those that lack the resources. This state-by-state analysis has implications for the level of exposure students from these areas will have related to engineering and may further validate the need for informal program models, such as the three highlighted in this chapter, to supplement what might be missing. Table 11.1 is a modification of the data depicted in the study used for this analysis (Carr, Bennett, and Strobel 2012). It excludes those states not served by MESA, SECME, and DAPCEP. From analysis, many of the states served by these programs articulate explicit engineering standards, either as stand-alone components to complement mathematics and science standards, or through standards developed by the International Technology and Engineering Educators Association (ITEEA) or Project Lead the Way (PLTW). Some states have very weak representations of engineering in formal learning, while others demonstrate no engineering focus in their state curriculum at all. For the states lacking any representation of engineering in their curriculum, the absence of engineering curriculum standards further validates the need for targeted support through informal programming from long-standing nationally recognized organizations like MESA, SECME, and DAPCEP.

TABLE 11.1 Comparison of states providing MESA, SECME, and DAPCEP

| States | Engineering Standards | Program(s) Offered |
|---|---|---|
| California (HS), Georgia (HS), Mississippi (HS), New York (MS, HS), Oregon (K–12), Tennessee (K–12), Texas (HS) | States with explicit engineering standards | MESA, SECME Inc. |
| Alabama (HS), Colorado (HS), Hawaii (K–12), Illinois (K–12), Maryland (K–12), North Carolina (HS), Pennsylvania (HS) | States with explicit engineering/ITEEA | MESA, SECME Inc. |
| Florida (MS, HS), Iowa (MS, HS), Utah (MS, HS) | States with explicit engineering/PLTW | MESA, SECME Inc. |
| Washington (K–12) | States with engineering in the context of technology design | MESA |
| Arizona (K–12), South Carolina (HS) | States with mention of technology design components (large variance; often very weak) | MESA, SECME Inc. |
| Michigan (HS) | States with mention of engineering components (large variance; often very weak) | DAPCEP |
| Louisiana, New Mexico, Virginia | None | MESA, SECME Inc. |

Although this chapter discusses three major programs that cater to African Americans and other students of color, there have been a number of other programs across the country established to address similar issues. In some cases, professional societies, such as the National Society of Black Engineers (NSBE), have created mechanisms to engage K–12 students through summer programs, design competitions, and science fairs. The Harris Foundation, in partnership with ExxonMobil, hosts a two-week residential summer camp on college campuses across the country for select middle school students from underserved and underrepresented backgrounds. From a formal learning perspective, NACME has recently partnered with the National Academy Foundation and PLTW to establish Academies of Engineering at participating high schools to "create small learning communities" that encourage and cultivate interest in STEM careers. These are just a few additional examples of younger programs that seek to build student interest and motivation to pursue engineering-related careers.

## Challenges and Opportunities

The apparent challenges deal with developing a deeper understanding of how to create cohesive learning experiences across formal and informal STEM education which intentionally lead toward preparing students academically for STEM degrees and careers. Equally important, educators and policy advocates must attend to factors that lead directly toward enhanced efficacy and identity in engineering. The literature informs us that nonminority male students receive sustained engagement and support throughout their academic trajectory leading toward an engineering identity and future career in engineering. Although we have presented organizational models and macrolevel program components specifically for underrepresented students, the core issue is that these experiences are not sustained over multiple years during most underrepresented students' formal academic preparation. Additionally, most of the effort to increase their engagement focuses on academic preparation and enhancement during after-school hours, on Saturdays, or in the summer—periods of time when other students are involved in extracurricular sports or recreational activities.

Because of the demographics served, most programs rely on tremendous support from external dollars, which places limitations on how well they are able to accommodate all underrepresented students who are in need. Coupling informal learning experiences with effective mentorship and career guidance throughout K–12 may facilitate greater academic tenacity, efficacy, and STEM career pursuit for a larger cohort of African American students. However, infusing immediate activities that lead to efficacy, like verbal persuasion and vicarious experiences, is essential. For future models to have a true impact, program developers and facilitators have to consider efficacy- and identity-building nuances in order to get the outcomes they intend. Program models need to be developed with these nuances in mind, through the lens and advice of content experts, practitioners, and social scientists.

Engineering has shown that it can play a vital role in the manifestation of how African American students experience science and mathematics in real-world contexts. The opportunities provided by MESA, SECME, and DAPCEP are invaluable to students who otherwise would not have access to quality resources, role models, and supportive learning environments. It is evident that for these opportunities to continue to have meaningful and lasting impact, collaborations involving practicing engineers, social scientists, and educators are ideal to create robust and transformative learning experiences. NRC (2011) makes this point very clear and stresses the need for ensuring that all students have equal access to high-quality STEM learning opportunities, as well as authentic experiences that fully engage them in STEM practices, behaviors, and habits of thinking. Striking a balance between professional collaboration and high-quality STEM learning experiences is imperative. Designing sustainable and replicable programs that produce the maximum impact for students should be a top priority.

## References

American Youth Policy Forum. 2006. "*Helping Youth Succeed through Out-of-School Time Programs.*" Washington, DC: American Youth Policy Forum. www.aypf.org/publications/HelpingYouthOST2006.pdf.

Bandura, Albert. 1977. "Self-Efficacy: Toward a Unifying Theory of Behavioral Change." *Psychological Review* 84 (2): 191.

Bandura, Albert. 1993. "Perceived Self-Efficacy in Cognitive Development and Functioning." *Educational Psychologist* 28 (2): 117–148.

Blackmon, Angelicque. 2012. "Biotechnology and Bioinformatics—Institute of Food Safety Science Program." Evaluation Report produced by Innovative Learning Concepts LLC under the grant from the National Science Foundation Award no. 0903158.

Blackmon, Angelicque, and Brandi Hinnant-Crawford. 2014. "AP Biology Alternative: Influence of Neuroscience Research on Urban Students' Science Knowledge and Efficacy." Paper submitted to the 2015 American Educational Research Association Annual Meeting, Chicago, Illinois, April 16–20.

Blustein, David L., and Hanoch Flum. 1999. "A Self-Determination Perspective of Interests and Exploration in Career Development." In *Vocational Interests: Meaning, Measurement, and Counseling Use*, edited by Mark L. Savickas and Arnold R. Spokane, 345–368. Palo Alto, CA: Davies-Black Publishing.

Bodilly, Susan J., Jennifer Sloan McCombs, Nate Orr, Ethan Scherer, Louay Constant, and Daniel Gershwin. 2010. *Hours of Opportunity, Volume 1: Lessons from Five Cities on Building Systems to Improve After-School, Summer School, and Other Out-of-School-Time Programs*. Monograph. RAND Corporation.

Brown, Steven D., and Robert W. Lent. 1996. "A Social Cognitive Framework for Career Choice Counseling." *Career Development Quarterly* 44 (4): 354–366.

Capobianco, Brenda M., Brian F. French, and Heidi A. Diefes-Dux. 2012. "Engineering Identity Development among Pre-adolescent Learners." *Journal of Engineering Education* 101 (4): 698–716.

Carr, Ronald L., Lynch D. Bennett, and Johannes Strobel. 2012. "Engineering in the K–12 STEM Standards of the 50 US States: An Analysis of Presence and Extent." *Journal of Engineering Education* 101 (3): 539–564.

Committee on Prospering in the Global Economy of the 21st Century. 2007. *Rising Above the Gathering Storm: Energizing and Employing America for a Brighter Economic Future*. Washington, DC: National Academies Press.

Duckworth, Angela L., Christopher Peterson, Michael D. Matthews, and Dennis R. Kelly. 2007. "Grit: Perseverance and Passion for Long-Term Goals." *Journal of Personality and Social Psychology* 92 (6): 1087.

Dweck, Carol S. 1986. "Motivational Processes Affecting Learning." *American Psychologist* 41 (10): 1040.

Dweck, Carol S., and Ellen L. Leggett. 1988. "A Social-Cognitive Approach to Motivation and Personality." *Psychological Review* 95 (2): 256.

Elliott, Elaine S., and Carol S. Dweck. 1988. "Goals: An Approach to Motivation and Achievement." *Journal of Personality and Social Psychology* 54 (1): 5.

Gottfried, Michael A., and Darryl N. Williams. 2013. "STEM Club Participation and STEM Schooling Outcomes," *Education Policy Analysis Archives* 21:1–24.

Hagedorn, Linda Serra, and Shereen F. Fogel. 2002. "Making School to College Programs Work." In Tierney and Hagedorn 2002, 69–193.

Krishnamurthi, Anita, Melissa Ballard, and Gil Noam. 2014. "Examining the Impact of Afterschool STEM programs." Paper commissioned by the Noyce Foundation.

Krishnamurthi, Anita, Ron Ottinger, and Tessie Topol. 2013. "STEM Learning in Afterschool and Summer Programming: An Essential Strategy for STEM Education Reform." *Power of Afterschool and Summer Learning for Student Success* 31.

Lord, Susan M., Michelle Madsen Camacho, Richard A. Layton, Russell A. Long, Matthew W. Ohland, and Mara H. Wasburn. 2009. "Who's Persisting in Engineering? A Comparative Analysis of Female and Male Asian, Black, Hispanic, Native American, and White Students." *Journal of Women and Minorities in Science and Engineering* 15 (2).

Morris, LaDonna K., Linda J. Austin, and Amaya M. Davis. 2013. "Sparking Girls' Interest in Technology: The NSF Tri-IT Project." *National Social Science Journal* 39 (2): 60.

NACME (National Action Council for Minorities in Engineering). 2011. *2011 NACME Data Book.* www.nacme.org/user/docs/NACMEDatabookReprint Final2Post.pdf.

National Academy of Engineering and National Research Council of the National Academies. 2009. *Engineering in K–12 Education: Understanding the Status and Improving the Prospects.* Washington, DC: National Academies Press.

NRC (National Research Council). 2009. "Learning Science in Informal Environments: People, Places, and Pursuits." Washington, DC: National Academies Press.

NRC (National Research Council). 2011. "Successful K–12 STEM Education: Identifying Effective Approaches in Science, Technology, Engineering, and Mathematics." Washington, DC: National Academies Press.

NRC (National Research Council). 2012. "Assessment of Informal and After School Science Learning." Paper commissioned by the National Academies of Science. http://sites.nationalacademies.org/DBASSE/BOSE/DBASSE_071087# .ULkF72eJun4.

Russell, Christina A., Monica B. Mielke, Tiffany D. Miller, and Jennifer C. Johnson. 2007. "After-School Programs and High School Success: Analysis of Post-
program Educational Patterns of Former Middle-Grades TASC Participants."
*Policy Studies Associates.*

Simpkins, Sandra D., Pamela E. Davis-Kean, and Jacquelynne S. Eccles. 2006.
"Math and Science Motivation: A Longitudinal Examination of the Links
between Choices and Beliefs." *Developmental Psychology* 42 (1): 70.

Spencer, Steven J., Claude M. Steele, and Diane M. Quinn. 1999. "Stereotype
Threat and Women's Math Performance." *Journal of Experimental Social Psychology* 35 (1): 4–28.

Swail, Watson S., and Laura W. Perna. 2002. "Precollege Outreach and Early
Intervention Programs: A National Imperative." In Tierney and Hagedorn
2002, 15–34.

TASC (The After-School Corporation). 2007. "Meeting the High School Challenge:
Making After-School Work for Older Students." New York: The After-School
Corporation.

Tierney, William G., and Linda Serra Hagedorn, eds. 2002. *Increasing Access to
College: Extending Possibilities for All Students.* Albany: SUNY Press.

Wai, Jonathan, David Lubinski, Camilla P. Benbow, and James H. Steiger. 2010.
"Accomplishment in Science, Technology, Engineering, and Mathematics
(STEM) and Its Relation to STEM Educational Dose: A 25-Year Longitudinal
Study." *Journal of Educational Psychology* 102 (4): 860.

Williams, Darryl N., and Michael A. Gottfried. 2010. "Who Chooses the E in
STEM?" *Proceedings of Engineering Education* 34:1–20.

# CHAPTER TWELVE

# Enhancing the Community College Pathway to Engineering Careers for African American Students

## A Critical Review of Promising and Best Practices

IRVING PRESSLEY MCPHAIL

NEARLY HALF OF ALL US undergraduates enroll in community colleges. Community college students constitute 40% of first-time freshmen and 52% of American Indian, 45% of Asian/Pacific Islander, 43% of African American, and 52% of Latino undergraduates (American Association of Community Colleges 2012). For many of these students, a community college education is the gateway to a four-year college degree.

Although the collegiate function (transfer and liberal arts) of the community college has been well documented (Carnegie Commission on Higher Education 1970; Townsend 2001; Cohen and Brawer 2003), new data from the National Student Clearinghouse Research Center (2012) have shown the impact of the community college in providing an educational foundation for students who transfer successfully and earn a four-year degree. The study showed that nearly 75% of the students who earned an associate's degree and then moved to a four-year college graduated with a bachelor's degree within four years of transferring (National Student Clearinghouse Research Center 2012). The report also demonstrated the importance of tracking the outcomes of community college graduates over a longer period.

Less well acknowledged is the role of the community college in the education of engineers in the United States. Adelman (1998) revealed that 20% of engineering degree recipients began their academic careers at community colleges, earning a minimum of 10 credits from these institutions. As presented in table 12.1, data from the 2008 National Survey of Recent College Graduates (NSRCG) documented that 44.4% of recent graduates with bachelor's degrees and 25% with master's degrees in engineering attended community college (National Science Foundation, National Center for Science and Engineering Statistics 2012).

TABLE 12.1 Community college attendance and associate's degree receipt among recent graduates with bachelor's and master's degrees in science, engineering, or health, by major field of degree, October 2008

| Major Field | Recent Graduates with Bachelor's Degrees | | | Recent Graduates with Master's Degrees | | |
|---|---|---|---|---|---|---|
| | All Graduates | Community College (%) | Associate's Degree (%) | Graduates | Community College (%) | Associate's Degree (%) |
| All fields | 1,128,000 | 52.2 | 16.2 | 309,000 | 43 | 14.6 |
| Science | 808,000 | 49.1 | 14.5 | 146,000 | 39.6 | 12.3 |
| Biological, agricultural, and environmental life sciences | 173,000 | 50 | 9.7 | 22,000 | 39.8 | 7.5 |
| Agricultural/food sciences | 14,000 | 60.1 | 17.9 | 2,000 | 39.5 | D |
| Biological sciences | 147,000 | 49.7 | 8.6 | 18,000 | 37.4 | 6.4 |
| Environmental life sciences | 12,000 | 42.5 | 13.5 | 2,000 | 57.9 | D |
| Computer and information sciences | 85,000 | 51.2 | 28.4 | 30,000 | 41 | 13.6 |
| Mathematics and statistics | 33,000 | 43.5 | 14.4 | 9,000 | 30.6 | 9.5 |
| Physical and related sciences | 41,000 | 35.9 | 6.3 | 11,000 | 24 | 5.2 |
| Chemistry, except biochemistry | 22,000 | 33.8 | 5.2 | 4,000 | 17.4 | 9.5 |
| Earth, atmospheric, and ocean sciences[a] | 9,000 | 43.5 | 8.9 | 4,000 | 40.4 | S |
| Physics/astronomy | 10,000 | 33.4 | 6.5 | 3,000 | 14.3 | D |

| | | | | | | |
|---|---|---|---|---|---|---|
| Psychology | 184,000 | 52 | 15.5 | 41,000 | 48.6 | 18.4 |
| Social and related sciences | 292,000 | 48.8 | 13.7 | 34,000 | 34.8 | 9.8 |
| Economics | 48,000 | 39 | 9.1 | 5,000 | 18.8 | 5.4 |
| Political and related sciences | 98,000 | 46.8 | 9.4 | 12,000 | 33.1 | 4.9 |
| Sociology/anthropology | 87,000 | 52.8 | 15 | 6,000 | 40.8 | 12.5 |
| Other social sciences | 59,000 | 54 | 22.8 | 10,000 | 41.9 | 16.4 |
| Engineering | 126,000 | 44.4 | 8.1 | 58,000 | 25 | 6.2 |
| Chemical | 8,000 | 46.3 | 6.7 | 2,000 | 17.6 | 11.4 |
| Civil/architectural | 19,000 | 42.2 | 6.6 | 7,000 | 34.1 | 8.3 |
| Electrical/computer | 37,000 | 43.5 | 7.7 | 22,000 | 21.6 | 5.1 |
| Industrial | 6,000 | 44.7 | 10 | 4,000 | 17.6 | 6.1 |
| Mechanical | 30,000 | 46.5 | 12.4 | 8,000 | 20.9 | 8.3 |
| Other | 26,000 | 44 | 4.7 | 16,000 | 30.2 | 5.3 |
| Health | 194,000 | 70.3 | 28.8 | 105,000 | 57.6 | 22.6 |

SOURCE: National Science Foundation/National Center for Science and Engineering Statistics, National Survey of Recent College Graduates, 2008.

NOTES: Numbers are rounded to nearest 1,000. Percentages are based on unrounded numbers and are rounded to nearest 0.1%. Detail may not add to total because of rounding. Estimates are from sample survey of college graduates who received bachelor's or master's degrees in science, engineering, or health fields in 2006 and 2007 academic years; estimates may differ from degree counts published elsewhere. D=suppressed to avoid disclosure of confidential information. S=suppressed for reliability; coefficient of variation exceeds publication standards.

[a]Other physical sciences are included in earth, atmospheric, and ocean sciences.

TABLE 12.2   2004 and 2005 S&E bachelor's and master's degree recipients by attendance at community college and race, 2006

| Race | All Recipients | Attended Community College | |
|---|---|---|---|
| | | Number | Percent |
| All races | 1,940,900 | 934,800 | 48 |
| Asian only | 344,500 | 150,200 | 44 |
| American Indian/Alaska Native only | 14,000 | 9,000 | 64 |
| Black only | 134,600 | 67,900 | 50 |
| White only | 1,365,800 | 667,700 | 49 |
| Native Hawaiian/other Pacific Islander only | 19,000 | 9,800 | 52 |
| Multiple race | 63,000 | 30,200 | 48 |

SOURCE: National Science Foundation, Division of Science Resource Statistics, National Survey of Recent College Graduates, 2006.
NOTE: Details may not add to total because of rounding.

TABLE 12.3   2004 and 2005 S&E bachelor's and master's degree recipients by attendance at community college and ethnicity, 2006

| Race | All Recipients | Attended Community College | |
|---|---|---|---|
| | | Number | Percent |
| All ethnicities | 1,940,900 | 934,800 | 48 |
| Hispanic | 148,700 | 81,900 | 55 |
| Mexican American | 64,600 | 44,700 | 69 |
| Puerto Rican | 31,700 | 9,900 | 31 |
| Cuban | 8,900 | 4,500 | 51 |
| Other Hispanic | 43,600 | 22,800 | 52 |
| Other | 1,792,200 | 852,900 | 48 |

SOURCE: National Science Foundation, Division of Science Resource Statistics, National Survey of Recent College Graduates, 2006.
NOTE: Details may not add to total because of rounding.

Analysis of the 2006 NSRCG data by Tsapogas (2007), presented in tables 12.2 and 12.3, is more central to the focus of this chapter. The data showed that 64% of American Indian/Alaska Native only, 50% of Black only, and 55% of Hispanic science and engineering bachelor's and master's degree recipients in 2004 and 2005 attended community college (Tsapogas 2007).

To date, one of the most comprehensive treatments of the community college pathway to engineering careers is the 2005 study by Mattis and Sislin (see also chap. 4 in this volume). The National Science Foundation (NSF) funded the study, which was overseen by the Committee on Engineering Education and the Committee on Diversity in the Engineering Workforce, both of the National Academy of Engineering (NAE), along with the Board on Higher Education and the Workforce of the National Research Council (NRC). Mattis and Sislin's (2005) study described the changing role of community colleges in engineering education, identified exemplary practices and partnerships between community colleges and four-year educational institutions, and recommended critical areas for further study.

This chapter builds on the 2005 NAE/NRC report (Mattis and Sislin 2005) and recent research and policy studies to frame a discussion around promising and best practices in facilitating the movement of African American students from the community college to successful completion of the bachelor's degree in engineering. Specifically, this discussion will (1) define the "new" American dilemma as conceptualized by the National Action Council for Minorities in Engineering (NACME 2008a); (2) review key challenges and exemplary approaches in pre-engineering and recruitment, student support services and retention, innovations in teaching and learning, the special case of developmental mathematics, and transfer and articulation; and (3) suggest further issues for research, practice, and policy.

## The "New" American Dilemma

In 2008, NACME released a report titled *Confronting the "New" American Dilemma: Underrepresented Minorities in Engineering: A Data-Based Look at Diversity* (NACME 2008a). Gunnar Myrdal, a Swedish social scientist whose two-volume work was commissioned by the Carnegie Corporation during the Second World War, originally coined the term "American dilemma," with results published near the close of the war in 1944. While the United States had no shortage of social scientists with the requisite skills to complete this study, racism was embedded so deeply into US society that a foreign analyst, as an outsider, was commissioned to study the status of race relations in the United States.

The dilemma was this: despite a strong ethos of equality, African Americans were subjected to searing inequality. On any measure of life chances, including health care, schools, jobs, housing, and social facilities, the 10% of Americans of African descent were far worse off than their non-Latino White compatriots were. Myrdal's (1944) work highlighted the gap between the rhetoric of equality on the one hand and the reality of inequality on the other.

Since the 1940s, there have been sweeping changes in the legal structures associated with the segregated society Myrdal (1944) studied. Despite the civil and women's rights movements' success in dismantling the legal basis for discrimination, the dilemma persists, but it has taken on new dimensions. The Latino population is the fastest-growing segment of the total US population, yet wide educational gaps between Latinos and other race/ethnic categories have been persistent. At the same time, the race to innovate has become a global engineering imperative, suggesting an urgent need to develop new ways of preparing students for engineering study.

Just as the dilemma described by Myrdal in the 1940s was a pressing social problem, the dilemma of increasing underrepresented minority (URM) participation in engineering has moved to the fore. Solving this dilemma is a matter of increasing national significance as we have become more diverse in a flatter world. At stake are the following questions: What kind of country do we wish to live in? Do we wish to live in one in which all have access to good schools, jobs, health care, and a high quality of life? Or will some people continue to be left behind? Engineering provides its practitioners with a relatively prosperous level of resources for their families and avenues of upward mobility within work settings. More CEOs have engineering degrees than any other major does (Stuart 2005). The research literature makes it clear that diversity can be a key strength of the United States, but the discipline of engineering has yet to utilize our rich cultural tapestry fully (Alfred P. Sloan Foundation 1974).

## Underrepresented Minorities in Engineering: A Data-Based Look at Diversity

NACME's vision is an engineering workforce that looks like America. In total, members of the three URM categories in engineering (African Americans, Latinos, and American Indians) constitute 30% of the US population currently and will account for 40% by 2050.

We are far from having an engineering workforce that looks like America. While women represent 47% of the US labor force and an even greater share of persons in professional and related jobs—the larger category into which engineering fits—they account for 11% of those in four types of engineering occupations: engineering technicians, sales engineers, engineers, and engineering managers. While African Americans represent 10.7% of the

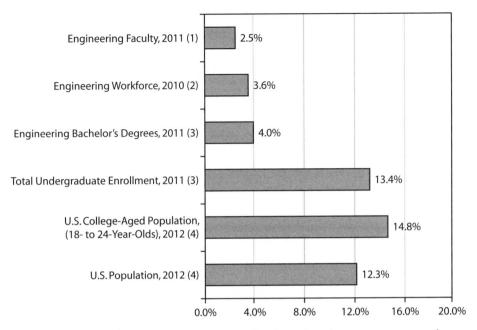

FIGURE 12.1 African American representation in engineering as a percentage of each group
Source: Yoder (2012); Finamore et al. (2013); NACME (2013).

total US labor force, they only constitute 2.5% of engineering managers, 4.5% of engineers, and 7.0% of engineering technicians (NACME 2011b).

According to many analysts, our nation's engineering schools are not producing enough engineers, and even if there were sufficient numbers of new engineers, the lack of diversity among the US engineering workforce poses a significant threat to our nation's ability to maintain an innovative edge in an increasingly competitive world. The representation of African Americans among engineering bachelor's degree recipients has not improved since 2000 (see chaps. 4, 5, and 6 in this volume). The evidence on the positive impacts of diversity on the corporate bottom line is well documented (Herring 2009; Deloitte Touche Tohmatsu 2011; see also chap. 8 in this volume).

The underrepresentation of African Americans in engineering begins at the educational level, as successful graduates of engineering are the necessary building blocks for successful careers. According to the National Center for Education Statistics, 286,000 African American high school completers enrolled in two- or four-year colleges or universities in the fall of 2010, representing just 62% of recent African American completers (US Department of Education 2014). The science, technology, engineering, and mathematics (STEM) fields continue to attract low percentages of African Americans. In 2011, the levels of African American students enrolled in engineering programs remained at 5%, even with new freshmen applications (American Association of Engineering Societies 2011). Underrepresentation at the postsecondary level extends beyond the students. African American faculty in engineering represent only 2.5% of all engineering faculty in the United States.

## The Knowledge Base

As shown in the chapters in this volume, there is evidence that African American students who develop mathematics and science confidence in high school are more likely to consider STEM studies and careers. There is also evidence that exposure to STEM careers at an early age, interventions at the community college level to address gaps in mathematics and science education, and programs developed specifically to attract African American high school students into STEM studies may have potential. More research is necessary in this area to gain an accurate understanding of what the issues are and how to address them to best prepare African American students for engineering study and careers.

Once students have begun their STEM-related studies at the community college level, their lack of educational preparedness becomes apparent through assessment and performance in basic skills courses. This issue is often addressed through student support services and retention efforts implemented at the community college level. Peer-reviewed studies focused specifically on African American students and the effectiveness of student support services and retention efforts as related to community college engineering success are limited. The articles reviewed in this chapter explored the issue from a variety of perspectives, including recruiting and retaining URM students in STEM at community colleges, recruiting and retaining Hispanic students in community college STEM studies, recruiting and retaining all STEM students in two-year and four-year programs, and recruiting and retaining community college students in general. The articles identified some common themes for successful recruitment and retention of all students that may be applicable to African American and other URM students.

As African American students continue on their path to an engineering career, classroom instruction must be transformed to support the acquisition of the requisite knowledge for success in engineering studies. The

"Innovations in Teaching and Learning" section identifies the current state of instructional design and implementation, discusses the characteristics of effective instructional reform strategies, introduces culturally relevant mathematics pedagogy, and discusses a holistic-based practice approach to teaching. NACME's seminal innovation, the Beyond the Dream initiative, is discussed here. Although the studies identified for this section do not focus on the special case of African American students in engineering, they describe approaches intended to enable faculty to teach all students successfully.

This leads to a discussion of instructional practices in developmental mathematics. Mathematics is an effective barrier to African American students' ability to move forward in community college engineering studies. The "Special Case of Developmental Mathematics" section begins with a statistical description of African American and URM student performance in mathematics. Various efforts to address the issue of providing effective developmental mathematics courses are introduced, although, again, studies focusing specifically on successful approaches to teaching African American community college engineering students are limited. The studies identified as useful for this review introduce innovations in developmental mathematics instruction, including contextualized mathematics instruction, modularization, acceleration and compression, alternative instructional delivery methods, and several other initiatives in mathematics education.

As students continue to progress in their community college engineering program, they are headed toward either earning an associate's degree or transferring to a four-year university. The "Transfer and Articulation" section discusses the need for collaboration between faculty at community colleges and faculty at four-year universities to establish effective articulation agreements and the need for formal support systems for transfer students at both the community college level and the four-year university level. Again, the studies focusing specifically on African American students matriculating from community college to engineering programs at four-year universities are limited, but most focused on URM students, and African American students were part of the population studied.

## Pre-engineering and Recruitment

For African Americans, community colleges are a significant gateway into engineering, with 44% of African American students who go on to complete degrees in science and engineering beginning their studies at community colleges. Austin's (2010) study of African American high school students in a pre-engineering program sheds some light on the factors that influence students' decisions to enter into engineering programs, but more studies are needed to gain a solid understanding of what attracts African American high school students into engineering. Austin surveyed 396 African American high school students in two schools with pre-engineering programs in the southeastern United States and found that students' confidence in their ability to do mathematics and science and their interest in mathematics and science were the most significant variables influencing their intention to enter engineering. A moderate correlation was found between family involvement (family's interaction with the student about education and career decisions) and students' intentions to enter engineering (Austin 2010). Austin's study suggested that if students are able to develop confidence in their mathematics ability, they are more likely to consider engineering careers. The issue of student exposure to mathematics and efforts to overcome lack of mastery in basic mathematics are discussed more fully in the "Special Case of Developmental Mathematics" section.

Although their study is not focused on African American or URM students, Hagedorn and Purnamasari (2012) provided information about the state of community college STEM program pre-engineering and recruitment in general, which may be useful when considering the special issue of how to increase African American and URM student enrollment in and completion of STEM programs. Their review of the literature examined ways community colleges can participate in solving the problem America faces regarding workforce shortages in STEM careers. They concluded that community colleges could improve remedial instruction to help students overcome educational deficits, expose students to STEM-related careers, offer extended orientation programs to prepare students for the rigors of college, provide them with information about occupational pathways and career planning, and offer targeted early-college programs designed specifically to lead students to consider STEM majors or careers (Hagedorn and Purnamasari 2012). These recommendations should be considered when designing, redesigning, or refocusing pre-engineering and recruitment programs targeted at African American students.

Mosley et al. (2010) described a collaborative effort between a community college and two four-year universities that used a pre-engineering robotics program to attract 11th and 12th grade African American students into STEM studies and careers and to improve the STEM educational process. The program incorporated a "2 + 2 + 3" model with two years of robotics-based instruction at the high school level, two years at the community college level, and three years at the four-year university level. The curriculum provided students an opportunity to test the results of abstract concepts through hands-on robotics experiences. Students could either pursue an associate's degree in applied science or transfer to the local university in pursuit of a bachelor's degree (Mosley et al. 2010). At the time the article was published, the program had not been evaluated.

## Student Support Services and Retention

Student support services include intervention, reform, and retention at the community college level and transfer support from community college to four-year engineering program studies. There are four significant points in students' educational engineering pathway when interventions are typically implemented: prior to college entry, during community college studies, prior to beginning their studies at four-year institutions, and after transferring to four-year institutions. The following studies focus on interventions implemented at the community college level and beyond.

Craft and Mack (2001) described the impact of reforms implemented at a two-year technical college in South Carolina intended to increase enrollment and retention rates of STEM students. The reforms were intended to increase the quantity, quality, and diversity of the college's graduates and to improve student retention. The reforms incorporated an integrated problem-based curriculum, collaborative teaching strategies, extensive active learning techniques, pre-engineering bridge courses in developmental mathematics and communications, projects based on core courses (mathematics, physics, communications, and technology), and faculty team teaching. The results were impressive. Craft and Mack (2001) reported that retention rates improved to 94% in the fall of 1998 and 100% in the spring of 1998 compared with a nationwide retention rate for associate's-degree-granting colleges of 50%. Enrollment doubled overall, with enrollment of minorities and women tripling between 1998–1999 and 1999–2000.

Kane et al. (2004) reported on the impact of reforms intended to increase enrollment, persistence, and success in computer science, engineering, mathematics, and related technological programs at a community college in California whose student population was 67% minority, 52% Latino, and 41% nonnative English speaking. Using a variety of interventions, including outreach and recruitment; orientation activities; a Mathematics, Engineering, and Science Achievement (MESA) student center; student cohort clustering; academic excellence workshops; academic planning; counseling support; professional development opportunities; university campus tours;

and hands-on experience, the reforms had a dramatic impact on student success. A total of 42% of Hartnell College's students were retained through completion of the targeted degree, and 21% continued their education at four-year institutions (Kane et al. 2004).

Tsui (2007) conducted a review of the literature on the effectiveness of intervention strategies in higher education to increase minority participation in STEM fields. The study included 10 intervention strategies implemented at both two- and four-year institutions, and the critical findings were as follows:

- Participants in summer bridge programs were more likely than nonparticipants to persist into their second year of college.
- Students who are mentored experience better college adjustment, higher GPAs, lower attrition rates, increased self-efficacy, and better-defined academic goals.
- Experience with hands-on research impacts positively the number of minority students who pursue degrees in STEM fields.
- Studies on the impact of tutoring on student performance report inconsistent results; some studies found no significant impact on student achievement outcomes, while others reported positive influences on persistence, attitudes, and grades.
- Exemplary interventions use an integrated approach that employs institutional leadership, targeted recruitment, faculty engagement, personal attention, peer support, student research experiences, bridging, and continuous evaluation (see also Landis 1985; Fleming 2012).

Brock (2010) studied intervention programs at community colleges, although the focus of this study was not on STEM programs specifically. Brock concluded that the effectiveness of various interventions and programs is largely unknown. Some of Brock's findings were that

- contextualized instruction, compared with regular remedial education, results in higher persistence, increased earning of credits and certificates, and greater increases in remedial education scores;
- acceleration and immersion strategies, supplemental instruction, and success centers should be researched;
- performance-based scholarship programs improve student retention rates; and
- a combination of scholarships and counseling has a significant impact on grades and persistence, especially among women.

To overcome the lack of quality education many African American students receive at the PK–12 level, community college STEM education must be taught by highly qualified teachers who understand how to design and implement quality curricula. These teachers must consider the community of learners they are charged with teaching so that African American and other URM students are given a meaningful opportunity to gain the foundational knowledge and understanding necessary to complete STEM studies successfully.

## Innovations in Teaching and Learning

Community college engineering instructional practices must be transformed to meet the needs of African American students. Because of the important role community colleges play in educating African American students and preparing them to transfer to four-year institutions and complete baccalaureate degrees, it is imperative that quality teaching occur at the community college level. Moore (2005) suggested using a culturally relevant K–16 pedagogical approach to teaching mathematics. Moore investigated undergraduate mathematics education for a group of primarily African American engineering students in a four-year institution. Although Moore's study is not community college specific, it informs instructional practices designed to improve the performance of African American students in mathematics and may inform instructional practices in mathematics at the community college level. Moore described culturally relevant mathematics pedagogy as "a student-centered social constructivist pedagogy that recognizes, appreciates, incorporates, and builds upon the knowledge that students bring to the classroom to facilitate the school mathematics achievement of all students. It acknowledges and accounts for the psychological, sociological, historical, political, and cultural factors that influence educational experiences in America. The theory purports that learning is enhanced when it occurs in contexts that possess sociocultural and cognitive meaning for the learner. . . . It calls for an emphasis on student discourse, real-life problem solving curricula, and learning environments which include collaborative group work and student-teacher interaction" (2005, 531).

Moore (2005) found that students in a supplemental collaborative learning calculus program that incorporated a student-centered constructivist/culturally relevant mathematics pedagogy had significantly higher rates of success in the primary calculus courses than other students did. The retention rate for students in the program was 100%, while retention rates for minority students outside of the program were below 20%. The average calculus GPA for the program participants was 3.07, and 100% of the participants studied were still pursuing their engineering degree and on track for graduation at the time of publication of the article.

Hollins also believed that teachers must have a deep understanding of their students to construct effective instructional practices:

> Teachers need to know learners as individuals; as members of social and cultural groups; as learners with particular characteristics; and as learners at a particular point in their academic, emotional, psychological, and social development. The specific ways of facilitating and scaffolding *learning* depend on our understanding of the learning process and knowledge of the background experiences, perceptions, and values of the particular learners. Findings from the new learning sciences emphasize the idea that existing understandings form the basis for new learning (Donovan and Bransford 2005). This finding illuminates the importance of the teacher's knowledge of the learner's background experiences and prior knowledge (2011, 397).

According to Hollins (2011), to teach students from cultures different from one's own effectively, teachers must have a well-integrated theoretical perspective and a clearly articulated philosophical stance to develop powerful pedagogy that supports certain academic and social outcomes. Teachers must also design and use meaningful assessments that ensure that students develop a deep understanding of the content, are able to apply what they have learned, and can identify when misunderstanding occurs and design effective interventions to address the misunderstood issues.

Hollins (2011) charged teachers with taking full responsibility for their ability to teach URM students, advocating a holistic-based practice approach that integrates academic knowledge of theory, pedagogy, and curriculum in the design of authentic contextual experiences and the application of focused inquiry, directed observation, and guided practice to inform and improve instruction. Teachers must understand how students learn, the purpose for their learning, how students are expected to apply the knowledge once it is learned, how to use what students already know in order to facilitate learning, how to assess the effectiveness of the teaching, and how to adjust the teaching if students have not achieved the desired outcome.

Finally, McPhail (2002) argued that current learning theory has not improved the academic performance of underserved learners, especially African Americans, and is inappropriate because it does not consider the influence of culture on learning. All learners deserve high-quality instruction linked appropriately to their experiences and backgrounds. The theory of culturally mediated instruction extends the learning paradigm by demonstrating how the relationship among culturally mediated cognition, culturally valued curriculum content, and culturally appropriate social discourse can result in increased opportunities for learning. Moore (2005) and Moses and Cobb (2001), among other educators, have validated the relevance of the theory of culturally mediated instruction in teaching mathematics to African American students.

The transformation of instructional practices at the community college level will result in an increase in the numbers of African American and URM students entering engineering studies and careers only if it facilitates a change in the teaching of developmental mathematics. This, as discussed earlier, is the biggest obstacle to African American and other URM student successes in community college pre-engineering studies.

## The Special Case of Developmental Mathematics

Teaching developmental mathematics (lower-level mathematics courses) effectively to African American and other URM students in the community college remains a conundrum (Bahr 2010; Hagedorn and DuBray 2010; Bragg 2012; Pearson and Miller 2012). Resolving this conundrum has been a major focus of the national Achieving the Dream: Community Colleges Count initiative (Achieving the Dream 2014). The goal of Achieving the Dream is success for more community college students, especially students of color and students from low-income families. Success is defined by the rates at which students

- complete remedial or developmental instruction and advance successfully to credit-bearing courses;
- enroll in and complete successfully the initial college-level or gateway courses in subjects such as mathematics and English;
- complete the courses they take with a grade of C or better;
- persist from one term to the next; and
- attain a certificate or degree.

MDRC—a nonprofit, nonpartisan education and social policy research organization—and the Community College Research Center (CCRC) are evaluating the work of the first 26 colleges to join the initiative from Florida, New Mexico, North Carolina, Texas, and Virginia. Rutschow and Schneider concluded,

> While research on best practices in developmental education abounds, little rigorous research exists to demonstrate the effects of these reforms on

students' achievement. Programs that show the greatest benefits with relatively rigorous documentation either mainstream developmental students into college-level courses with additional supports, provide modularized or compressed courses to allow remedial students to more quickly complete their developmental work, or offer contextualized remedial education within occupational and vocational programs. These strategies show the most promise for educators and policymakers who must act now, but they should also continue to receive attention from researchers. Many of the strategies have not yet been evaluated using more rigorous and reliable methods, and/ or early promising results have not been replicated in other settings (2011, iii).

While the evidence is still limited, a body of rigorous research exists on the success of *contextualized learning programs* (Bragg and Barnett 2009; Jenkins, Zeidenberg, and Kienzl 2009). Of particular interest to this discussion are *contextualized instructional models* that may be used in specific academic disciplines to promote students' integration of course concepts with mathematical skills by infusing the mathematics curriculum with real-world problems and scenarios (also see Rattan and Klingbeil 2011; Bragg 2012).

NACME's collaboration with Springfield Technical Community College (MA) in the Beyond the Dream: From Developmental Mathematics to Engineering Careers initiative is an example of such a *contextualized instructional model* in engineering education (NACME 2011b). With support from Lumina Foundation for Education, NACME identified 19 Achieving the Dream community colleges that have formal or informal transfer and articulation agreements with one or more of the 51 partner universities in the NACME national network. NACME convened a National Roundtable of 37 distinguished mathematics and engineering science faculty from Achieving the Dream community colleges, engineering faculty and deans from NACME partner universities, and workshop facilitators from ConnectEd: The California Center for College and Career, to explore the possibilities of problem-based learning (PBL) in intermediate algebra and precalculus courses that integrate engineering awareness, concepts, and skills. The National Roundtable is intended as the first step in a multiyear effort to develop curriculum resources for all levels of developmental mathematics courses that engage students in learning and raise academic achievement through PBL approaches in mathematics and engineering; increase minority students' awareness of engineering careers and opportunities for transfer to four-year engineering colleges, beginning in developmental mathematics courses; and pilot test these new PBL curriculum materials and approaches to introduce minority students and others to engineering careers at a group of leading two-year institutions.

Springfield Technical Community College participated in the National Roundtable and began implementation of PBL approaches in developmental mathematics and engineering in fall 2009/spring 2010 in pre-algebra, algebra 1, and elementary algebra 2. The PBL classes met for five full periods (Monday to Friday) each week for six contact hours per week. Faculty incorporated PBL projects emphasizing the application of mathematical principles and concepts in engineering design projects. In contrast, the traditional developmental mathematics classes met three days a week.

Figure 12.2 compares students' end-of-semester grades in the two PBL courses with the average grades of students in all traditional sections offered in the same semester. The key metric of student performance was achieving a grade of C or better. Figure 12.2 shows that in both pre-algebra and algebra 1, a larger percentage of students were successful in the PBL classes than in the sections with traditional pedagogy and scheduling.

Table 12.4 presents another way to look at these classes by using the instructors as their own controls. The table compares the PBL classes with traditional three-day sections taught by these same instructors in previous

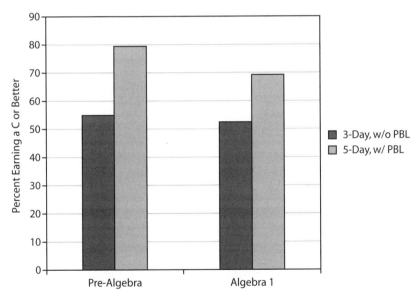

FIGURE 12.2 PBL vs. traditional sections/end-of-semester grades

| Success Rate | Instructor A<br>Pre-algebra (%) | Instructor B<br>Algebra 1 (%) |
|---|---|---|
| With three-day format without using PBL | 46 | 74 |
| With five-day format using PBL | 80 | 68 |

years. Students appeared to do better with the five-day PBL model in Instructor A's pre-algebra classes, but the outcomes were not as clear for Instructor B's algebra 1 class. The disparity in these results may result from variations in how instructors implemented the PBL approach, differences in instructional materials, or differences in the preparation and other characteristics of the individual classes.

Although it is impossible to disentangle whether the additional time, the PBL approach, or both factors together were associated with these findings, mathematics and engineering science faculty are convinced that the approach can be important for student success. Not only were students in the PBL sections more actively engaged in learning, but also many of the PBL / extended hours / algebra 1 students continued on to the algebra 2 course in spring 2010. To unravel these questions, NACME and Springfield Technical Community College are pursuing joint grant funding to deepen the engagement of mathematics and engineering faculty in introducing engineering concepts into developmental mathematics courses through PBL and to expand the model to other community colleges.

The innovations under way at both the instructional design level and the instructional delivery level in developmental mathematics are promising. It remains clear, however, that we need more rigorous research on learning outcomes. We also need more radical changes in developmental education programs to maximize the number of students prepared to be successful in college-level courses. NACME's Beyond the Dream initiative is a prototype for this new level of academic strategy.

## Transfer and Articulation

Given the significant role that community colleges play in providing URM matriculants to four-year institution engineering programs, there is a clear need for policies that eliminate barriers to African American and other URM students transferring successfully from two-year colleges to four-year universities. Zinser and Hanssen note, "Ideally, there would be universal acceptance of the associate's degree as the first two years of university study" (2006, 29).

At the undergraduate level, 47% of African American undergraduates enrolled in community college, and only 15% of those students transferred to four-year institutions; 56% of Hispanic undergraduates enrolled in community colleges, and 15% transferred; 40% of Asian American undergraduates enrolled in community college, and 40% transferred; 65% of American Indian undergraduates had a 10% transfer rate; while 37% of White undergraduate students enrolled in community colleges, with a 25% transfer rate to four-year institutions (Wilson 2000). According to Rendon and Nora (1994), to improve retention and increase the number of students who transfer successfully to STEM education at four-year institutions, community colleges must improve faculty development to strengthen curricula and teaching, develop articulation committees to improve the transition process from two-year to four-year institutions, improve student assessment to ensure that students learn what they are intended to learn, and engage in program evaluation to assess the effectiveness of student support programs and transfer articulation committees. Four-year universities must also do their part by providing effective bridge programs and mentors for incoming students, publishing transfer catalogs that explain the transfer process, accepting community college credits as agreed on in their articulation agreements, offering campus visits, providing financial aid information, diversifying their faculty, and creating successful minority engineering programs (Rendon and Nora 1994).

Packard et al. (2011) analyzed data from a sample of 30 female community college students, 23% of whom were ethnic minorities, to determine what supported and what hindered their intentions to transfer to four-year institutions in the pursuit of a bachelor's degree in a STEM major. At the community college level, students reported that faculty inspiration to pursue STEM careers, peer academic support, helpful transfer advising, family support, and work schedule flexibility supported their intentions, while ineffective initial advising and limited finances hindered their progress. At the four-year university level, students reported that helpful professors and advisors and support from peers supported their ability to persist, while negative course experiences, poor experiences with advisors, unwelcoming campus environments, financial pressures, and work challenges hindered their ability to persist (Packard et al. 2011).

Jain et al. (2011) focused on a four-year institution's perspective of community college transfer students in an effort to understand how critical race theory (CRT) informs a transfer receptive culture (TRC) that increases successful transfer for community college students who are first-generation college attendees, underrepresented, from low-income families, or students of color. They concluded that when viewing partnerships between community colleges and four-year institutions through the CRT lens, five elements are necessary to establish a TRC: (1) establish transfer as an institutional priority, (2) provide outreach and resources specifically for transfer students, (3) offer financial and academic support geared toward transfer students, (4) create an environment that welcomes the family and community of transfer students, and (5) develop a reflective and analytic process for measuring success, rather than measuring success according to standards used for traditional students (Jain et al. 2011).

Zinser and Hanssen (2006) analyzed data collected from 700 institutions participating in a program sponsored by NSF and implemented under the Advanced Technological Education (ATE) program. The purpose of ATE was to foster collaboration between community colleges and four-year universities in the development of formal articulation agreements for associate's degree students in various technological areas. Zinser and Hanssen reported that the successful creation and implementation of articulation agreements between community colleges and four-year universities benefit students, the educational institutions, and the broader community. Students benefit from reduced educational costs, flexibility in scheduling their courses, broader access, and additional student services, such as career counseling. Institutions benefit from marketing a pipeline to new students and collaboration between faculties at different institutions. The community benefits by the increase of trained available workers for local businesses.

## The NACME Community College Strategy

These aforementioned studies indicate that effective articulation agreements, together with quality institutional support, are imperative if we are going to increase the number of African American and other URM students pursuing engineering degrees to completion.

NACME has long recognized the significance of the community college pathway to engineering careers for African American and other students of color and has developed a number of innovative intervention strategies in this area. For example, in 2006 NACME collaborated with Qualcomm and the University of California, San Diego (UCSD), to create the NACME/Qualcomm Community College Pre-Engineering Studies Transfer Scholarship Program. The program was intended as a pilot project to gather information on effective practices for moving more pre-engineering students at the community college into the Jacobs College of Engineering at UCSD. NACME commissioned a program evaluation study of the program in 2008, conducted by Dr. Bevlee A. Watford (NACME 2008b).

Watford framed her program evaluation around the four exemplary practices that contribute to the successful transfer of students from community colleges to four-year institutions identified in the 2005 NAE/NRC report (Mattis and Sislin 2005). The four exemplary practices are (1) accessible, clearly defined articulation agreements; (2) frequent communication between institutional partners; (3) equivalency and uniformity of curricula; and (4) financial support for transfer students. Eight URM community college students were selected annually for a period of two years (2006–2007 and 2007–2008). Of the 16 URM students who received the NACME/ Qualcomm Scholarship in the 2006–2008 time frame, 11 earned their bachelor's degrees in engineering. This performance level exceeded both the national and UCSD retention rates for all students in engineering. Consistent with the research and policy recommendations on effective transfer and articulation in engineering education, Mattis and Sislin's (2005) study pointed to the critical importance of frequent communication between all institutional partners.

With generous funding from the Motorola Foundation, NACME compared data for 1,688 NACME Scholars—including 355 community college transfers—at 29 NACME Partner Universities and 17 NACME Affiliate Universities (NACME 2010). These data were supplemented with interviews with key informants at three NACME Partner Universities about the reasons for transfer and the ways that institutions could increase the success of transfer students. Figures 12.3 and 12.4 document the following findings: (1) transfer scholars' overall GPAs were higher than those of traditional NACME Scholars who began their engineering education as freshman at the university, (2) transfer scholars were more likely to be retained as of the study date, and (3) slight variations across race/ethnic groups require additional analysis. Among the reasons given for why NACME Scholars chose to begin their engineering education at the community college were (1) lack of financial aid for a four-year college; (2) family circumstances—desire to be close to family; (3) affordability—can live with family and save on room and board; and (4) smaller first- and second-year classes than at major research universities. Table 12.5 reveals that 21% of NACME Scholars are community college transfers, a metric consistent with national data.

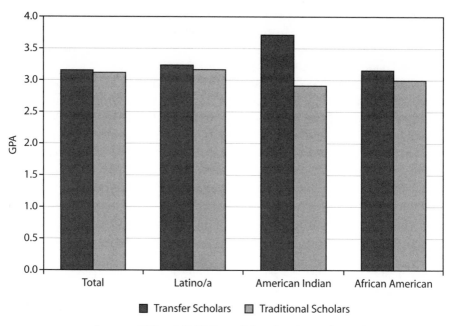

FIGURE 12.3 Average GPAs—NACME traditional and transfer students

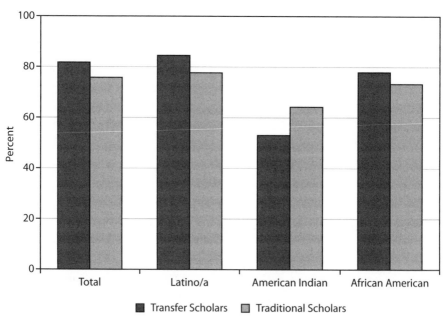

FIGURE 12.4 Retention of NACME traditional and transfer scholars still enrolled or graduated by 2009

TABLE 12.5 NACME Scholar demographics and institutional characteristics

| Parameter | Total | | Traditional | | Transfer | |
|---|---|---|---|---|---|---|
| | N | % | N | % | N | % |
| Sex | | | | | | |
| Female | 524 | 31.0 | 422 | 31.7 | 102 | 28.7 |
| Male | 1,164 | 69.0 | 911 | 68.3 | 253 | 71.3 |
| Race/ethnicity | | | | | | |
| Latino/a | 791 | 46.9 | 586 | 44.0 | 205 | 57.7 |
| African American | 776 | 46.0 | 652 | 48.9 | 124 | 34.9 |
| American Indian | 83 | 4.9 | 64 | 4.8 | 19 | 5.4 |
| Other | 38 | 2.3 | 31 | 2.3 | 7 | 2.0 |
| Institutional control | | | | | | |
| Public | 1,434 | 85.0 | 1,106 | 83.0 | 329 | 92.7 |
| Private | 253 | 15.0 | 227 | 17.0 | 26 | 7.3 |
| Type of institution | | | | | | |
| Research university | 1,157 | 68.5 | 928 | 69.6 | 229 | 64.5 |
| Minority-serving institution | 525 | 31.1 | 391 | 29.3 | 134 | 37.7 |
| Grand total | 1,688 | 100 | 1,333 | 79.0 | 355 | 21.0 |

SOURCE: (1) US Census Bureau, 2011. Statistical Abstract of the United States, 2010. (2) NACME Research, Evaluation and Policy analysis of 2009 American Community Survey Public Use Microdata Sample (PUMS) data.

NOTES: "Female" includes females of all race/ethnic categories. Likewise, the race/ethnic category data include females and males within those categories.

## Issues for Further Research, Practice, and Policy

This discussion makes clear the need for more focused research on African American community college students in engineering. The peer-reviewed literature in this area is scant. Rigorous research in the key areas and exemplary approaches identified in this chapter would permit community college leaders in engineering education to design more effective intervention strategies, redesign courses, develop more effective transfer agreements and articulation policies, and implement student support programs that result in increased movement of African American learners from engineering science transfer programs at the community college to successful completion of bachelor's degrees and beyond in engineering and increased entry to engineering careers.

The following are some of the critical issues for further research, practice, and policy which emerge from this discussion:

*Pre-engineering and Recruitment*

- Rigorous program evaluation of reforms that seek to establish a correlation between reforms and educational attainment.
- Impact of engineering awareness programs in middle schools on academic motivation and progression in engineering education.
- Impact of high school pre-engineering programs—such as the Academy of Engineering, developed as a partnership between the National Academy Foundation (NAF), Project Lead the Way (PLTW), and NACME—on academic motivation and retention in engineering education.
- Effectiveness of bridge programs in engineering education.
- Strategies to increase public awareness about the role of the community college as a gateway to bachelor's degree completion in engineering.
- Impact of targeted financial aid at the community college on recruitment and retention of high school students in engineering science transfer programs.

*Student Support Services and Retention*

- Disaggregation of student data to understand better and close gaps in student success at the community college.
- Use of noncognitive factors as predictive indices for African American student performance in engineering study (see the Vanguard Engineering Scholarship Program; NACME 2006).
- Identification of explicit factors in the culture, student services, and learning environments of community colleges which facilitate the recruitment, enrollment, education, associate's degree attainment, and successful transfer of African American women and men in engineering.

*Innovations in Teaching and Learning*

- Impact of culturally congruent pedagogy in mathematics and science education.
- Effectiveness of online courses with African American learners in the community college engineering science program.
- Impact of faculty diversity in two- and four-year engineering programs.
- Reform in teacher education programs at community colleges to address the shortage in the quality of STEM teachers.

*The Special Case of Developmental Mathematics*

- Effectiveness of the Beyond the Dream approach to contextualized learning in developmental mathematics and engineering applications.

*Transfer and Articulation*

- Effectiveness of documented exemplary practices: accessible, clearly defined articulation agreements; frequent communication between institutional partners; equivalency and uniformity of curricula; and financial support for transfer students.
- Role of institutional leadership at community colleges in negotiating the creation of partnerships between STEM faculty at community colleges and four-year universities and in transforming institutional and/or faculty perceptions about community college transfer students in engineering.
- Analysis and broad dissemination of student success data on community college transfer students in engineering at four-year universities.
- Effectiveness of state-wide policies that foster the implementation of transfer and articulation agreements in engineering which are both vertically aligned and coherent.

## Summary and Conclusion

This chapter began with a discussion of why the United States must turn to African Americans and other URMs in order to maintain American competitiveness in STEM subjects. Activating the hidden workforce of young African American men and women is at the core of the competitiveness solution. Resolving the "new" American dilemma requires the active engagement of K–12 education, higher education, government, and business. The community college is a key stakeholder in this process.

Community college students constitute a significant percentage of all US undergraduates. African Americans choose the community college to begin their postsecondary education in record numbers, and this number is expected to grow as US demographics continue to change at a rapid rate. African American community college students already constitute 43% of all US undergraduates (American Association of Community Colleges 2012).

This is a propitious moment to connect four strands that relate directly to the concerns about US competitiveness in the flat world: (1) the fact that diversity drives innovation and that its absence imperils our designs, our products, and, most of all, our creativity—all components of competitive-ness; (2) African Americans remain one of the most underrepresented minority groups in engineering-related fields; (3) African American students are well represented in the community college sector, although not in the STEM disciplines; and (4) community colleges are already essential to the education of engineers in the United States.

Positioning the community college as an even more prominent and effective gateway to engineering careers for African American students will require bold action and academic strategy in a number of key areas:

- Introduce STEM education and career options at an early age to African American students by providing access to academic support programs, after-school tutoring for ACT/PSAT preparation, and STEM-integrated curricula to increase the ability for high school graduates to enter the college arena prepared for the academic rigor of engineering study.
- Provide research-based academic support services, financial support, and trained and competent faculty.
- Transform instructional practices at the community college to incorporate student-centered, social constructivist pedagogy in mathematics and science.
- Introduce contextualized instructional models that infuse the developmental mathematics curriculum with real-world engineering problems and scenarios.
- Implement exemplary practices in transfer and articulation in engineering education.

How, then, might community colleges meet this seemingly Herculean challenge? This chapter provides actionable recommendations and examples of promising and best practices in the critical areas of pre-engineering and recruitment, student support services and retention, innovations in teaching and learning, the special case of developmental mathematics, and transfer and articulation. NACME's seminal work in contextualized instructional models has also been discussed. A concrete set of recommendations for further research, practice, and policy has been offered. Given that the number of college-age African American and other minority students will grow dramatically over the next decade, and that many of these students will elect to begin their postsecondary education at a community college, now is the time to strengthen the pathway from community college to engineering careers for these students. In our quest to confront the "new" American dilemma, the time for *doing*—for taking action—is *now*.

The dream is real, my friends.
The failure to make it work is the unreality.

TONI CADE BAMBARA (1939–1995)

## Acknowledgments

I wish to acknowledge Kimberly Hollins, Esq., Hollins Law Firm, for her assistance in the review of related literature, and Christopher Smith, PhD, Director of Research and Program Evaluation, NACME Inc., for his technical assistance in editing the final draft of the manuscript.

## References

Achieving the Dream. 2014. "Overview." www.achievingthedream.org.

Adelman, Clifford. 1998. *Women and Men of the Engineering Path: A Model for Analysis of Undergraduate Careers*. Washington, DC: US Government Printing Office.

Alfred P. Sloan Foundation, The Planning Commission for Expanding Minority Opportunities in Engineering. 1974. *Minorities in Engineering: A Blueprint for Action*. New York: Alfred P. Sloan Foundation.

American Association of Community Colleges. 2012. *Community College Facts at a Glance*. Washington, DC: American Association of Community Colleges.

American Association of Engineering Societies. 2011. *Engineering and Technology Enrollments*. Reston, VA: American Association of Engineering Societies.

Austin, Chandra Y. 2010. "Perceived Factors That Influence Career Decision Self-Efficacy and Engineering Related Goal Intentions of African American High School Students." *Career and Technical Education Research* 35 (3): 119–135.

Bahr, Peter R. 2010. "Preparing the Unprepared: An Analysis of Racial Disparities in Postsecondary Mathematics Remediation." *Journal of Higher Education* 81 (2): 209–237.

Bragg, Debra D. 2012. "Two-Year College Mathematics and Student Progression in STEM Programs of Study." In *Community Colleges in the Evolving STEM Education Landscape: Summary of a Summit*, edited by Steve Olson and Jay B. Labov, 81–106. Washington, DC: National Academies Press.

Bragg, Debra D., and Elisabeth Barnett. 2009. "Lessons Learned from Breaking Through." http://occrl.illinois.edu/files/InBrief/Breaking-Thru4-09.pdf.

Brock, Thomas. 2010. "Young Adults and Higher Education: Barriers and Breakthroughs to Success." *Future of Children* 20 (1): 109–132.

Carnegie Commission on Higher Education. 1970. *The Open-Door Colleges: Policies for Community Colleges*. New York: McGraw-Hill.

Cohen, Arthur M., and Florence B. Brawer. 2003. *The American Community College*. San Francisco: Jossey-Bass.

Craft, Elaine L., and Lynn G. Mack. 2001. "Developing and Implementing an Integrated Problem-Based Engineering Technology Curriculum in an American Technical College System." *Community College Journal of Research and Practice* 25 (5–6): 425–439.

Deloitte Touche Tohmatsu. 2011. *Only Skin Deep? Re-examining the Business Case for Diversity*. Sydney: Deloitte Touche Tohmatsu.

Donovan, M. Suzanne, and John D. Bransford. 2005. *How Students Learn: History, Mathematics, and Science in the Classroom*. Washington, DC: National Academies Press.

Finamore, John, Daniel J. Foley, Flora Lan, Lynn M. Milan, Steven L. Proudfoot, Emilda B. Rivers, and Lance Selfa. 2013. "Employment and Educational Characteristics of Scientists and Engineers." http://www.nsf.gov/statistics/inbrief/nsf13311/nsf13311.pdf.

Fleming, Jacqueline. 2012. *Enhancing Minority Student Retention & Academic Performance: What We Can Learn from Program Evaluations*. San Francisco: Jossey-Bass.

Hagedorn, Linda S., and Daniel DuBray. 2010. "Math and Science Success and Nonsuccess: Journeys within the Community College." *Journal of Women and Minorities in Science and Engineering* 16 (1): 31–50.

Hagedorn, Linda S., and Agustina V. Purnamasari. 2012. "A Realistic Look at STEM and the Role of Community Colleges." *Community College Review* 40 (2): 145–164.

Herring, Cedric. 2009. "Does Diversity Pay? Race, Gender, and the Business Case for Diversity." *American Sociological Review* 14:208–224.

Hollins, Etta R. 2011. "Teacher Preparation for Quality Teaching." *Journal of Teacher Education* 62 (4): 397.

Jain, Dimpal, Alfred Herrera, Bernal Santiago, and Daniel Solorzano. 2011. "Critical Race Theory and the Transfer Function: Introducing a Transfer Receptive Culture." *Community College Journal of Research and Practice* 35 (3): 252–266.

Jenkins, Paul D., Matthew Zeidenberg, and Gregory S. Kienzl. 2009. *Building Bridges to Postsecondary Training for Low-Skill Adults: Outcomes of Washington State's I-BEST Program*. New York: Columbia University, Teachers College, Community College Research Center.

Kane, Michael A., Chuck Beals, Edward J. Valeau, and M. J. Johnson. 2004. "Fostering Success among Traditionally Underrepresented Student Groups: Hartnell College's Approach to Implementation of the Math, Engineering, and Science Achievement (MESA) Program." *Community College Journal of Research and Practice* 28 (1): 17–26.

Landis, Raymond B. 1985. *Improving the Retention and Graduation of Minorities in Engineering*. New York: NACME.

Mattis, Mary C., and John Sislin, eds. 2005. *Enhancing the Community College Pathway to Engineering Careers*. Washington, DC: National Academies Press.

McPhail, Irving P. 2002. "Culture, Style, and Cognition: Expanding the Boundaries of the Learning Paradigm for African American Learners in the Community College." In *Making It on Broken Promises: African American Male Scholars Confront the Culture of Higher Education*, edited by Lee Jones, 107–131. Sterling, VA: Stylus.

Moore, J. 2005. "Undergraduate Mathematics Achievement in the Emerging Ethnic Engineers Programme." *International Journal of Mathematical Education in Science and Technology* 36 (5): 531.

Moses, Robert P., and Charles E. Cobb. 2001. *Radical Equations: Civil Rights from Mississippi to the Algebra Project*. Boston: Beacon Press.

Mosley, Pauline H., Yun Liu, S. Keith Hargrove, and Jayfus T. Doswell. 2010. "A Pre-engineering Program Using Robots to Attract Underrepresented High School and Community College Students." *Journal of STEM Education* 11 (5–6): 44–54.

Myrdal, Gunnar. 1944. *An American Dilemma: The Negro Problem and Modern Democracy*. 2 vols. New York: Harper & Brothers.

NACME (National Action Council for Minorities in Engineering). 2006. *Vanguard Engineering Scholars Program: A Report on Students and Institutional Participation and Progress*. White Plains, NY: NACME.

NACME (National Action Council for Minorities in Engineering). 2008a. *Confronting the "New" American Dilemma: Underrepresented Minorities in Engineering: A Data-Based Look at Diversity*. White Plains, NY: NACME.

NACME (National Action Council for Minorities in Engineering). 2008b. *The NACME/Qualcomm Community College Pre-Engineering Studies Transfer Scholarship Program: An Analysis of the Current Efforts*. White Plains, NY: NACME.

NACME (National Action Council for Minorities in Engineering). 2010. "Community College Transfers and Engineering Bachelor's Degree Programs." *NACME Research and Policy Brief* 1 (1): 1–2.

NACME (National Action Council for Minorities in Engineering). 2011a. "Beyond the Dream: From Developmental Mathematics to Engineering Careers." *NACME Research and Policy Brief* 1 (5): 1–2.

NACME (National Action Council for Minorities in Engineering). 2011b. *2011 NACME Data Book*. White Plains, NY: NACME.

NACME (National Action Council for Minorities in Engineering). 2013. *2013 NACME Data Book*. White Plains, NY: NACME.

National Science Foundation, National Center for Science and Engineering Statistics. 2012. "Characteristics of Recent Science and Engineering Graduates: 2008." www.nsf.gov/statistics/nsf12328/.

National Student Clearinghouse Research Center. 2012. "The Role of Two-Year Colleges in Four-Year Success." http://research.studentclearinghouse.org.

Packard, Becky W., Janelle L. Gagnon, Onawa LaBelle, Kimberly Jeffers, and Erica Lynn. 2011. "Women's Experiences in the STEM Community College Transfer

Pathway." *Journal of Women and Minorities in Science and Engineering* 17 (2): 129–147.

Pearson, Willie, Jr., and Jon D. Miller. 2012. "Pathways to an Engineering Career." *Peabody Journal of Education* 87:46–61.

Rattan, Kuldip S., and Nathan W. Klingbeil. 2011. *Introductory Mathematics for Engineering Applications: Preliminary Edition*. Hoboken, NJ: John Wiley & Sons.

Rendon, Laura I., and Amaury Nora. 1994. "Clearing the Pathway: Improving Opportunities for Minority Students to Transfer." In *Minorities in Higher Education*, edited by Manuel J. Justiz, Reginald Wilson, and Lars G. Bjork, 120–138. Washington, DC: American Council on Education/Oryx Press.

Rutschow, Elizabeth Z., and Emily Schneider. 2011. *Unlocking the Gate: What We Know about Improving Developmental Education*. New York: MDRC.

Stuart, Spencer. 2005. "2004 CEO Study: A Statistical Snapshot of Leading CEOs." https://bus.wisc.edu/~/media/bus/mba/why%20wisconsin/statistical_snapshot _of_leading_ceos_relb3.ashx.

Townsend, Barbara K. 2001. "Redefining the Community College Transfer Mission." *Community College Review* 29 (2): 29–42.

Tsapogas, John. 2007. "The Role of Community Colleges in the Education of Recent Science and Engineering Graduates." Handout presented at STEM Conference, Montgomery College, MD, October.

Tsui, Lisa. 2007. "Effective Strategies to Increase Diversity in STEM Fields: A Review of the Research Literature." *Journal of Negro Education* 32 (3): 2–13.

US Department of Education, National Center for Education Statistics. 2014. "The Condition of Education 2014" (NCES 2014-083). Immediate Transition to College.

Wilson, Reginald. 2000. "Barriers to Minority Success in College Science, Mathematics, and Engineering Programs." In *Access Denied: Race, Ethnicity, and the Scientific Enterprise*, edited by George Campbell Jr., Ronni Denes, and Catherine Morrison, 193–206. New York: Oxford University Press.

Yoder, Brian L. 2012. "Engineering by the Numbers." http://www.asee.org/papers -and-publications/publications/college-profiles/2011-profile-engineering-sta tistics.pdf.

Zinser, Richard W., and Carl E. Hanssen. 2006. "Improving Access to the Baccalaureate: Articulation Agreements and the National Science Foundation's Advanced Technology Education Program." *Community College Review* 34 (1): 27–43.

# CHAPTER THIRTEEN

# Spelman's Dual-Degree Engineering Program

## A Path for Engineering Diversification

CARMEN K. SIDBURY, JENNIFER S. JOHNSON,
*and* RETINA Q. BURTON

O VER THE PAST 20-PLUS YEARS, the nation has maintained a central focus on increasing and diversifying the field of engineering. Spelman College maintains an important role in this effort. Spelman has a strong record of preparing African American female students for degrees in engineering. Since its founding in 1881, Spelman has held the distinction of being America's oldest historically Black college for women. From fall 2008 to fall 2012, Spelman's enrollment averaged approximately 2,200 students per year. Of the students declaring a major, 31% selected a major in biology, chemistry, mathematics, computer science, physics, or environmental science (Spelman College 2013). The college has a solid track record with regard to awarding African American women bachelor of science degrees in science, technology, engineering, and mathematics (STEM)—setting them on the track for success in engineering.

Infrastructural developments in the past decade reflect Spelman's strong commitment to building a research-active environment necessary to sustain innovative science- and engineering-oriented curricular and training resources on par with other Carnegie baccalaureate institutions. For instance, in 2001, Spelman College completed construction of the $34 million Albro-Falconer-Manley Science Center, a 154,000-square-foot training facility equipped with state-of-the-art science equipment to support comprehensive STEM research and training.

In addition to providing students with state-of-the-art equipment and graduating a large group of students with STEM degrees, the college currently employs 54 full-time faculty in STEM disciplines, of which 83% are racial and ethnic minorities and 52% are women. Of the 28 women faculty members, 64% are African American. This represents a diverse group of research-active STEM faculty who provide a unique support system of role models and undergraduate research training for African American women in STEM.

## History of the Dual-Degree Engineering Program

Engineering is a field plagued with barriers for African American women and other women of color. Yet, it is a disciplinary field that provides the opportunity for these same women to make profound contributions once given the opportunity to do so. In 1969, as a strategy to reduce barriers that denied access to engineering, Spelman College, as a member of the Atlanta University Center (AUC), became a partner in the formation of the Dual-Degree Engineering Program (DDEP). The purpose of the DDEP was to expose African American students to engineering curricula and place them in engineering environments. The idea of the DDEP originated with Monte Jacoby, a college relations officer at the Olin Corporation, a leading North American producer of chlorine, caustic soda, and ammunition. Jacoby recognized the benefits of exposing African American students attending colleges in the AUC to engineering education (Olin Corporation 1973). Conversations began between Monte Jacoby and Louis Padulo, a Georgia Tech alumnus and chair of the Morehouse College Mathematics Department. Padulo wanted to implement an engineering program in collaboration with the Georgia Institute of Technology (Georgia Tech) (see also chap. 1 in this volume).

The conversation expanded to include Dr. Charles Meredith (a chemistry professor at Morehouse College), the late Dr. Prince Wilson (executive director of the AUC Corporation), and Dr. William Schutz (assistant dean of engineering at Georgia Tech), resulting in a plan to expose AUC schools to engineering education and employment opportunities. With the support of Dr. Albert E. Manley, then president of Spelman College, and the AUC schools, the DDEP was established with an agreement between the AUC schools and Georgia Tech. By successfully completing the DDEP, students would be able to receive two degrees, a bachelor of science or bachelor of arts degree awarded by one of the AUC-affiliated colleges and a bachelor of science in engineering degree awarded by Georgia Tech. Although Georgia Tech had similar arrangements with other liberal arts institutions, the AUC DDEP was the first one with a focus on increasing the number of African American students in engineering (Georgia Tech 1969).

By 1972, the total DDEP enrollment was 108 students, of which 10% were Spelman College students (Olin Corporation 1972). The AUC DDEP increased its staff to include a full-time director, Charles Meredith, and coordinators at AUC and Georgia Tech. Under Meredith's direction, the enrollment numbers increased to 164 students, with approximately 70 first-year students entering each year. He recruited 70% of the program participants, and the remaining 30% learned of the program once on the AUC's campuses. Female student enrollment in the DDEP increased from 10% to 15.2%, compared with the national average of 3% female enrollment in engineering schools (Olin Corporation 1973).

## Structure of the DDEP Program

The DDEP was implemented through a 3/2 format in which students matriculate at their home AUC institution for three years, to complete required courses in English, economics, history, social science or a modern language, chemistry, physics, and mathematics, and then attend Georgia Tech for two years, to complete the required courses for the engineering major. During the beginning phases of the program, the mathematics, chemistry, and physics courses were taught at Morehouse College, with the vision to extend these courses to the remaining AUC schools (Georgia Tech 1969).

Over the past 45 years, the structure of the DDEP has remained intact, with the overarching goal to continue to increase the number of women of color in the engineering workforce. As initially planned, the AUC schools have increased their curriculum offerings to include mathematics, chemistry, physics, and computer science on their respective campuses. The pre-engineering courses (engineering graphics and engineering mechanics) are also offered by the AUC schools.

Given the success of the partnership between the AUC schools and Georgia Tech, the Spelman College DDEP has expanded to include 14 additional participating engineering schools: Auburn University, California Institute of Technology, Clarkson University, Columbia University, Dartmouth College, Indiana University–Purdue University Indianapolis, Missouri University of Science and Technology–Rolla, North Carolina A&T State University, Rensselaer Polytechnic Institute of Technology, Rochester Institute of Technology, University of Alabama–Huntsville, University of Michigan–Ann Arbor, University of Notre Dame, and University of Southern California.

## Atlanta University Center Consortium

The AUC—now AUCC (Atlanta University Center Consortium)—DDEP still plays a critical role in the development and financial support of the students in the program. A significant number of Spelman College students are first-generation STEM majors and are unfamiliar with the resources available to support their career development. Through the AUC DDEP, students gain access to information regarding scholarships, resume writing, summer internships, permanent hire positions, and many other academic support services. Despite the fact that each institution now maintains its own DDEP, the AUC DDEP serves as a common rallying point for all DDEP students. The annual AUC DDEP awards banquet and the tri-campus National Society of Black Engineers (NSBE) chapter are two examples of formal AUC DDEP support services in which the consortium unites to celebrate and acknowledge the academic achievements of its participants. These and other cooperative arrangements were commonplace prior to the establishment of individual programs. In past years, AUC DDEP students were enrolled in the same statics, dynamics, and pre-engineering courses, depending on which campus offered the class during that term. Currently, those courses are offered at each of the campuses, though students can elect to take them at any of the consortium institutions. The structure of the AUC DDEP serves to promote a culture of communal, knowledge-based sharing among its students.

## Historical Pathway Programs to Engineering

To better support students' development as scientists and engineers, Spelman College faculty conceptualized the Pre-freshman Summer Science Program and established it as a bridge to help improve students' chances for access to science and engineering (S&E). Spelman College president Albert Manley agreed that an increase in numbers also requires cocurricular programs that help students to "overcome the psychological fears of science instilled in Black women through unequal educational opportunity and sexual discrimination" (Scriven 2006). During its first year of operation in 1972, the Pre-freshman Summer Science Program accepted 16 students. Eventually, it supported between 40 and 50 students annually. The program was funded with institutional and external grants (including support from the Jessie Smith Noyes and Rockefeller Foundations and the federal government). Courses were taught by faculty from Spelman College and other AUC institutions. The six-week program offered both basic and accelerated instruction in biology, precalculus, reading, chemistry, and computer science, with students' schedules dependent on their declared majors. Freshman students earned up to three college credits in mathematics and four in biology, and they could also satisfy the college's reading requirement through their participation in the program (Scriven 2006).

In addition to granting students college course credits, the Pre-freshman Summer Science Program included counseling and enrichment components. As chronicled in a 1989 *SAGE* article, Falconer (1989) noted that the counseling and enrichment components were particularly important as they helped students develop a greater awareness of what it means to be a scientist. Professionals in the field were invited to campus to give lectures and career talks. These practicing scientists, many of whom were female and African American, did more than simply talk about their jobs. They served as critical role models to students who had never encountered an African American woman scientist (Scriven 2006). The Pre-freshman Summer Science Program provided a model of student development which would be adapted for other science and health programs, including Spelman College's DDEP.

Fast-forward 18 years. In 1987, the National Aeronautics and Space Administration (NASA) and Spelman College partnered to increase the number of African American female students receiving advanced degrees in engineering and in science, technology, and mathematics. The partnership resulted in the formation of Women in Science and Engineering (WISE), operating as a complement to the DDEP. The WISE program supported NASA's goal of strengthening the agency's and the nation's future workforce. From its inception, the WISE program sought to recruit strong, science-focused students. The program provided students with scholarships and unique opportunities such as NASA internships. The late Dr. Etta Falconer, a Spelman College mathematics professor and one of the first African American women in the country to receive a doctorate in mathematics, initiated the WISE program (NASA 2009).

In addition to meeting its primary goal, the WISE program led to the expansion of Spelman College's science offerings and created a critical mass of STEM scholars who would become outstanding graduates. During peak enrollment periods (2000–2004), there were approximately 100 WISE scholars annually. The program provided

- prefreshman summer curricular enhancement;
- research experiences during the academic year;
- summer internships at NASA or other sites;
- research training with the professional staff at NASA centers during the summer;
- scholarships to cover the cost of tuition, fees, books, supplies, and room and board; and
- a stipend during the summer research experience at a designated NASA center.

In 1994, the National Science Foundation (NSF) and NASA selected Spelman College as one of only six institutions in the country to be designated as Model Institutions for Excellence (MIE) in undergraduate science and mathematics education. The MIE program was a joint venture program designed to increase the number of African Americans and other underrepresented minorities in STEM through funding to a select group of minority-serving institutions (MSIs; Rodriguez, Kirshstein, and Hale 2005).

The MIE program represented an unprecedented strategic investment by NSF and NASA. The program was unprecedented because it (1) specifically targeted a small cadre of MSIs and (2) recognized that success in attracting and retaining underrepresented students to STEM fields required the development of a STEM infrastructure. Funding for the planning and implementation of the MIE program from NSF alone exceeded $88 million over a 10-year period. Combined with the NASA funding, the entire project represented an investment of nearly $120 million. The success of the MIE and WISE programs is evident in the quality of Spelman graduates and provides definitive evidence on the impact of strategic partnerships on diversifying the STEM workforce. Of the more than 320 women who participated in the WISE program, more than half received graduate degrees and at least 40 earned doctorate degrees in a STEM discipline (NASA 2009).

Through the DDEP, a modest 4% of women graduates (from 2008 to 2012) receiving the bachelor of science degree in any of the aforementioned fields (except biology) pursue dual majors. Once they complete the course of study at Spelman College, they go on to receive an engineering degree from one of Spelman's partnering institutions. Spelman College's history presents a valuable opportunity to examine the development of the DDEP, which currently plays a unique role in bolstering students' participation for future engineering education and training opportunities.

## Data and Methods

### RESEARCH QUESTIONS

Through programs like MIE, NASA WISE, and various federal and private grants, Spelman College has received a significant amount of funding to enhance its STEM education programs. The increase in funding for STEM education from multiple sources allowed Spelman College to gradually achieve its original goal of increasing the number of African American women receiving degrees in engineering. The impact of programs like the DDEP, MIE, and NASA WISE is reflected in the high percentage of survey respondents who participated in these programs while at Spelman College and who subsequently pursued postbaccalaureate degrees after graduating from Spelman. Strong indicators of success in postgraduate education—leadership, scholarly development, research, and mentorship—characterize the collective voice of Spelman STEM alumnae and provide the rationale for the phenomenon of a high percentage of graduate school enrollment among Spelman's STEM students. Additionally, the increase in funding levels and subsequent development of programs provided Spelman administrators with an important opportunity—to ask and answer two overarching research questions using case study methodology:

1. What has been the historical impact of the DDEP and other science- and engineering-focused programs on Spelman alumnae's preparation and entrée into the workforce?
2. What are the perspectives of Spelman alumnae regarding their experiences in the DDEP and other STEM-focused programs?

COLLECTION OF EMPIRICAL DATA

With support from the Spelman DDEP office, which provided the names and e-mail addresses of DDEP alumnae, online survey invitations were e-mailed to over 1,300 DDEP alumnae from the classes of 1985–2012. Results from survey participants, self-identified as DDEP alumnae, were sorted for further analysis. A total of 215 DDEP alumnae responded to the survey, and their responses are included in the case study.

Qualitative and quantitative data were collected to respond to the research questions. The data are presented in a case study format. Numerous documents were analyzed to understand the historical development of programs like the DDEP, MIE, and WISE. A historical narrative is presented to provide context for alumnae who responded to the DDEP alumnae survey. An analysis of survey responses is included in this case study. Survey questions were modeled after the National Survey of College Graduates and the Assessing Women & Men in Engineering (AWE) self-efficacy instruments to examine the success of Spelman graduates and their experience at Spelman and partnering engineering institutions (AWE 2013). Graduates of the DDEP were invited to respond to a brief survey so that researchers could learn more about their individual experiences as STEM majors at Spelman and at the partnering engineering schools, their graduate school pursuits and motivations for attending, and their participation in the engineering workforce. The DDEP alumnae e-survey was divided into four sections delineating the transition from first-year undergraduates to industry professionals. Each section contained questions to ascertain those influences leading to major decisions about attending Spelman, choosing a major, selecting the partner engineering school, pursuing a graduate degree, and ultimately choosing their employment. Survey participants were also asked multiple-choice questions about their campus involvements (professional societies, enrichment programs, scholarships awarded, research opportunities, etc.) at Spelman College and at the partnering engineering school.

## DATA ANALYSIS

Surveys were analyzed for descriptive statistics—graduation information, employment in S&E, and campus involvement. Spelman alumnae were compared with a nationally representative sample of STEM degree recipients. The results provide significant feedback about aspects of the DDEP and other science- and engineering-based programs, along with student involvement while at Spelman College, at the partnering engineering institutions, and ultimately in the workforce. Analysis of data reveals the context and underpinnings of Spelman alumnae success beginning with student enrollment at Spelman College. Almost immediately emerging from the data are three key factors that account for the high percentage of graduate school attendance, as reported by DDEP alumnae survey respondents: desire for career advancement, family encouragement, and availability of fellowships and financial aid. Each of these factors provides insight into the influences not only for opting to attend graduate school but also for the larger aspects of decision making which characterize this unique population of African American women in engineering.

## STEM AT SPELMAN COLLEGE

A closer look at bachelor of science degrees earned by African American women at Spelman offers insight that can shift the balance in the distribution of STEM degrees. From 2001 to 2010, African American women earned, on average, 84,520 bachelor of science degrees per year across all fields. Approximately 24,643 were in S&E, representing 29% of degrees granted.

Drilling down to specific S&E disciplines, 0.5% and 1.2% of the degrees earned by African American women were in mathematics and engineering, respectively. At Spelman approximately 3.7% and 2.5% of degrees awarded are in mathematics and engineering through the DDEP, respectively. The fact that a small liberal arts women's college has found a way to encourage student persistence in S&E, where African American women are sig-

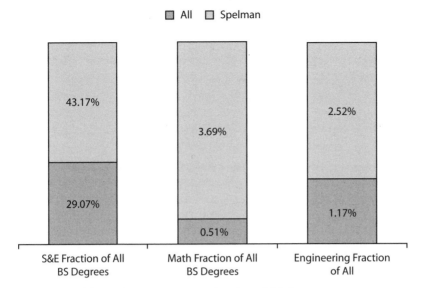

All ☐ Spelman

43.17%

3.69%

2.52%

29.07%

0.51%

1.17%

S&E Fraction of All
BS Degrees

Math Fraction of All
BS Degrees

Engineering Fraction
of All

FIGURE 13.1 African American women fraction of BS degrees, average
(2001–2010)

nificantly underrepresented, is encouraging with respect to addressing the
nation's desire to achieve racial, ethnic, and gender parity in S&E (see
fig. 13.1).

## Contemporary Voice of Spelman STEM Alumnae

Historical perspectives were presented at the beginning of this case study.
Responses from STEM alumnae provide a contemporary voice for an often-
neglected group—Black women mathematicians and engineers. Of the
more than 200 survey respondents, over 40% were mathematics majors (see
fig. 13.2). The findings presented build on the work of Perna et al. (2009)
by following the degree attainment of Spelman STEM alumnae and exploring the factors supporting their success. These identified factors align
squarely with recommendations cited in a report from the National Academies (2011) on increasing minority participation in STEM. The results
from the survey show that the success of Spelman STEM majors can be attributed to several key factors:

1. extended family support structures and learning communities;
2. financial assistance; and
3. desire for career advancement.

Based on the results of the survey, the majority (41%) of dual-degree engineering participants major in mathematics. Chemistry majors (23%) and computer science majors (15%) combine with math majors to represent approximately 80%. The other 20% of DDEP participants were physics majors (8.7%), general science majors (7.7%), or other (4.1%).

## EXTENDED FAMILY SUPPORT STRUCTURES AND LEARNING COMMUNITIES

On the topic of factors in choosing their STEM major, surveyed Spelman DDEP alumnae are very diverse in their responses, but an overarching theme is the impact of family in guiding their choices. Respondents indicate that family (parents and other family members) and high school teachers were most influential in their decision to choose STEM as a major at Spelman (36% and 31%, respectively). This is an interesting finding and contradicts the common perception that parents of color do not encourage their children to pursue STEM degrees. Most of the literature on students of color in STEM talks about family members and teachers discouraging their participation in STEM. Hence, these findings are encouraging and warrant further study.

Several DDEP survey respondents provided specific insight about why they chose their majors, while at the same time revealing their beliefs about themselves and likelihood of success in engineering:

- "Math seemed to be the backbone of engineering, so I went with that."
- "I was good in math and liked it."
- "I excelled in mathematics in high school."
- "My own giftings in math."
- "My love for the subject."
- "Always had a love for math and science."
- "Enjoyment and its relationship to engineering."
- "Prior knowledge and success in the subject area."
- "Personal assessment of my strongest areas."
- "Physics was the first science I enjoyed learning and helping others to learn."
- "I actually hated physics in high school, but chose it as a college major b/c there were so few women who majored in it (5 during my freshman year)."
- "I just had a love for mathematics."
- "I always like math."

- "Love for Math."
- "Interest in Math."
- "I liked math from an early age."

These alumnae responses reveal the extent to which self-efficacy, as described by Bandura (1977, 1989), plays a role in STEM attainment. Bandura (1977) described self-efficacy as "the beliefs in one's capacity to organize and execute the courses of action required to produce given attainments." In this context, self-efficacy specifically addresses students' beliefs in the likelihood that they can complete an undergraduate or graduate degree in a STEM discipline. Pajares (1996) states that self-efficacy affects the choices that one makes, how hard one works at a task, and how long one perseveres when the task is problematic and difficult. One's self-efficacy regarding a task or situation also has implications for the affective domain. Bandura (1977) indicated four sources from which we assess the likelihood that we can successfully reach a given attainment: (1) mastery experiences, (2) social persuasion, (3) vicarious experiences, and (4) emotional and physiological states.

Mastery experiences refer to the opportunities that one has had in accomplishing similar tasks or goals. It also encompasses the discrete knowledge that one possesses which is deemed necessary for completion of the attainment. This includes not only having an appropriate background in key academic areas but also having experienced a measure of academic success. The findings of Besterfield-Sacre et al. (2001) and Arnette (2004) suggest that simple preparation in math and science fields is insufficient for the positive mastery experience that is important for high self-efficacy in the completion of degrees in STEM fields. Instead, these authors indicate that another of Bandura's sources, social persuasion, may more fully explain the challenges of STEM disciplines to attract and retain women and other underrepresented students. Social persuasion refers to the influence that others have in a particular context. In the academic context it describes, for example, faculty encouraging and offering students support. Parents, friends, and significant others can also serve as effective social persuaders in that they may provide the social support necessary for students to thrive in an academic environment where they are underrepresented and sometimes overlooked.

The literature on the importance of social support, particularly for the academic success of ethnic minority and female students in college and STEM disciplines, is extensive. Much of the literature focuses on the lack of support that ethnic and gender minority students perceive in their relationships with faculty (Solorzano, Ceja, and Yosso 2000; Zeldin and Pajares 2000). This is not the case at Spelman, as Spelman's WISE program and designation as an MIE were developed to increase self-efficacy and to promote social persuasion of participants. Of those surveyed, 82% were affiliated with an engineering society, 79% participated in activities sponsored by their department or major, 23% attended an early college program, and 52% had a summer bridge experience. The prefreshman summer curricular enhancements, research experiences during the academic year, and summer internships at NASA or other sites provided valuable student development opportunities for each participant. Additionally, the faculty at Spelman served as influential role models who provided students with vicarious experiences that motivated them to pursue and complete a course of study which led to a STEM-related career. As evident in the survey results, the environment at Spelman had a positive impact on the emotional and physiological state of students and empowered women to persist and fulfill their academic endeavors.

FINANCIAL ASSISTANCE

Another commonly mentioned factor in undergraduate and graduate school attendance is financial aid. Financial support from government agencies and the private sector is directly correlated with the number of scientists and engineers who complete their undergraduate studies and subsequently enroll in and complete graduate school. Historically, scholarships and financial aid packages have been DDEP recruitment tools. Of those surveyed, approximately 44% report that their scholarship offer influenced their decision to attend Spelman College. During their matriculation, 81% received institutional, national, or corporate scholarships and 61% received similar financial support at the partnering engineering institutions. The value of these educational scholarships extends beyond the payment of tuition, fees, room and board, and books. These scholarships provide opportunities for immersion in the field of study through advanced academic courses, scientific research, and corporate internships. Through scholarship programs such as NASA WISE, ExxonMobil WISE, Xerox Corporation, and Georgia Power Women in Engineering (WIE), engineering students have maximized their academic and professional options. It is not a surprise, therefore, that one of the most influential factors in deciding to attend graduate school is the availability of fellowships and financial aid. Funding of and participation in the various aspects of these scholarship and fellowship programs enable students to persist in engineering through graduate-level education into the engineering workforce. To continue with the diversification of the engineering workforce, therefore, it is important to amplify support of proven scholarship models and foster new education-industry partnerships.

When asked which three factors were most influential in their decision to attend graduate school, the following common themes emerged: career advancement (73%), family encouragement (47%), and fellowship/financial aid (38%).

Approximately 78% of the survey respondents completed graduate degrees, which is well above the 2006 national average of 26.3% for female graduates, as reflected in figure 13.2. These trends hold for completed graduate STEM degrees as well. The rate at which DDEP alumnae are earning STEM master's degrees and PhDs relative to the national average is encouraging as it reflects student preparedness and a potential pipeline of future African American women STEM faculty. Further, encouraging data reveal that of those DDEP alumnae who are employed at an educational institution, nearly 47% work at a four-year college or university (as opposed to medical schools). Future work must be done to continue to strengthen pathways to graduate schools and access to financial aid, which could yield

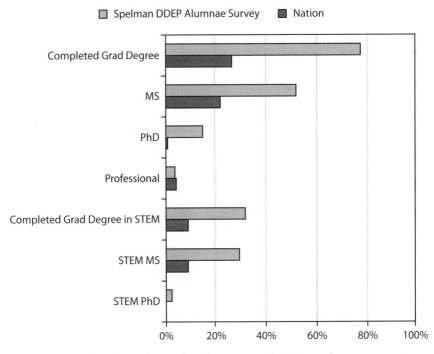

FIGURE 13.2 Female graduate education: surveyed Spelman alumnae vs. nation

increasing numbers of African American women STEM professionals with Spelman College baccalaureate origins.

Responses from Spelman alumnae reveal that there is a 91% employment rate among Spelman DDEP alumnae, and nearly 64% of Spelman DDEP alumnae survey respondents are currently working in an S&E occupation.

Results based on S&E categories defined as computer and mathematical sciences; biological, agricultural, and other life sciences; physical and related sciences; social and related sciences; and engineering sciences are summarized in figures 13.3 and 13.4. Figure 13.3 illustrates the comparison of national data, as reported in the 2011 Education and Statistics Administration (ESA) Issue Brief, for college-educated, female workers with a STEM degree and the DDEP surveyed group (US Department of Commerce 2011). In this figure, there is a striking difference between employment trends of Spelman survey respondents and those of women in general with STEM degrees. Spelman women are 2.5 times more likely to work in a STEM occupation than women in general as reported in the 2011 ESA Issue Brief. This divergence can likely be attributed to factors such as female role models and the encouragement students receive while at Spelman to pursue engineering careers (Tate and Linn 2005).

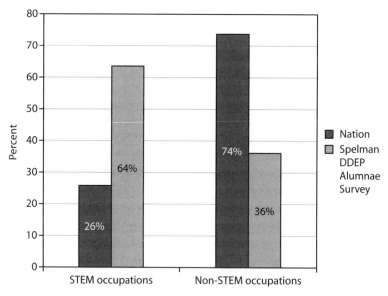

FIGURE 13.3 College-educated female workers with a STEM degree by STEM occupation (nation vs. Spelman DDEP)

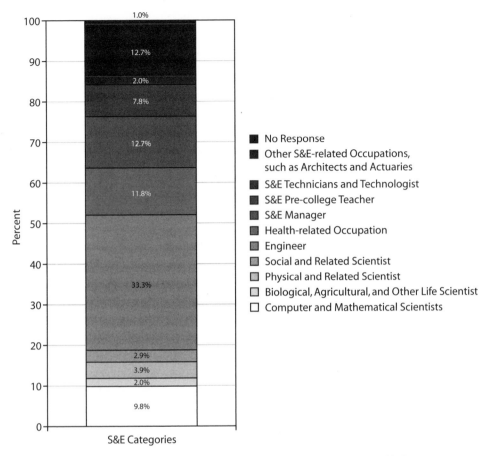

FIGURE 13.4 Spelman College DDEP alumnae survey participants: S&E occupations

As recently reported in an NSF InfoBrief (2013), the engineering occupation continues to be dominated by white males, with women engineers representing only 13% of S&E occupations. The vast majority of women are employed in the physical and life sciences sectors (NSF 2013). However, the majority (33%) of Spelman dual-degree engineering alumnae are currently employed as engineers, while only 11.8% work in health-related occupations. The diversity in occupations, as shown in figure 13.4, reflects the versatility of an engineering background and the wide range of alumnae interests.

Table 13.1 includes a representative sample of the S&E employment of Spelman DDEP alumnae survey participants (as of October 1, 2012).

TABLE 13.1  Spelman College DDEP alumnae S&E employment

| Employer | Department/Division | Job Title |
|---|---|---|
| Dept. of Navy—Naval Facilities Engineering Command | Environmental | Chemical Engineer— Environmental Program Manager |
| Baxter Healthcare | Packaging Engineering | Engineer II |
| ExxonMobil | Refining & Supply | Process Contact Engineer |
| TEKSystems | Federal Consulting | Senior Architect |
| Kraft Foods | Process Engineering/ Product Development for Mainstream/Premium Coffee | Associate Engineer II |
| US Department of Transportation | Federal Highway Administration | Major Projects Highway Engineer |
| Deloitte Consulting | Technology | Consultant |
| State of Georgia Dept. of Natural Resources | Environmental Protection/ Air | Permit Engineer |
| Rockwell Collins | Commercial Systems Design Support | Mechanical Engineer |
| JC Penney | Engineering | Senior Industrial Engineer |
| Norfolk Southern | Information Technology | Senior Technology Engineer |
| Department of Defense | Office of the Secretary of Defense/Acquisition, Technology and Logistics | Senior Program Management Analyst |
| General Electric | Energy | Lead System Engineer |
| Federal Aviation Administration | Atlanta Aircraft Certification Office | Aerospace Engineer |
| ExxonMobil Chemical Company | Global Polyolefin Products Technology | Research Engineer for Product Development |
| Accenture | Systems Integration and Technology | Senior Manager |
| Air Force Office of Scientific Research | Physics and Electronics | Program Manager |
| Raytheon Co. | Mixed Signals and Power | Senior Electrical Engineer I |
| Medtronic Cardiovascular | Coronary Product Development | Associate R&D Engineer |
| Southern Company | Engineering Design | Senior Engineer |
| Kaiser Permanente | Supply Chain | Product Manager |
| General Dynamics IT | National Homeland Security Division | Deputy Program Manager |
| NASA—Marshall Space Flight Center | Materials and Processes Laboratory/Environmental Effects Branch | Assistant Branch Chief |
| Iatric Systems Inc. | Application Software | Developer |
| Clorox | Cleaning | Senior Scientist |

TABLE 13.1 *continued*

| Employer | Department/Division | Job Title |
|----------|---------------------|-----------|
| Raytheon Company | Performance Excellence | Senior Quality Assurance Engineer II |
| Marathon Petroleum Company | Pipeline Integrity | Project Engineer |
| Jackson and Tull | Aerospace Engineering Division | Chief Systems Engineer |
| National Institute of Standards & Technology | Office of Information Systems Management | Supervisor |
| The Mitre Corporation | CAASD | Associate Department Head |
| Sandia National Laboratories | Program & Test Integration | Senior Member of Technical Staff |
| Caterpillar Inc. | IMOD | Manufacturing Engineer |
| USPTO | Commerce | Chemistry Patent Examiner |

## Conclusion and Recommendation

The US STEM workforce is crucial to America's innovative capacity and global competitiveness. Yet women are vastly underrepresented in STEM jobs and among STEM degree holders despite making up nearly half of the US workforce and half of the college-educated workforce. This has been the case throughout the past decade, even as college-educated women have increased their share of the overall workforce.

An analysis of the historical record shows that DDEP graduates are more likely to work in STEM jobs than their female counterparts (13% vs. 33%). Data show three key factors that account for the high percentage of graduate school attendance for Spelman alumnae: desire for career advancement, family encouragement, and availability of fellowships and other financial aid. This historical narrative serves as the context for understanding the contemporary voice and experiences of Spelman STEM alumnae and represents a case analysis of the formation of DDEP at Spelman College in an effort to diversify engineering.

A close examination of the success with African American females at a liberal arts college offers promise that the model can be extended to other institutions and result in a significant increase in the diversification of the engineering workforce. The close-knit Spelman DDEP community and their desire to inspire future generations afford the opportunity to expand programs to develop a strategic communication and education campaign (set of synchronized activities) to diversify the field of engineering. The results of this work will allow us to articulate the experiences of women of color for the educational advancement of younger women of color interested in becoming engineers. The findings illustrate the success of African American women and provide a platform through changing behavior to disrupt existing and often outdated paradigms and to create a twenty-first-century paradigm shift that acknowledges that young girls of color can and will successfully do engineering because those that came before them did.

## References

Arnette, Robin. 2004. "Perspectives: The Gender Gap in Math and Science." http:// sciencecareers.sciencemag.org/career_magazine/previous_issues/articles/2004 _01_30/nodoi.11840235497843220998.

AWE (Assessing Women & Men in Engineering). 2013. www.engr.psu.edu/awe /misc/about.aspx.

Bandura, Albert. 1977. "Self-Efficacy: Toward a Unifying Theory of Behavior Change." *Psychological Review* 84:191–215.

Bandura, Albert. 1989. "Social Cognitive Theory." In *Annals of Child Development*, edited by R. Vasta, 6:1–60. Greenwich, CT: JAI Press.

Besterfield-Sacre, Mary, Magaly Moreno, Larry Shuman, and Cynthia Atman. 2001. "Gender and Ethnicity Differences in Freshman Engineering Student Attitudes: A Cross-Institutional Study." *Journal of Engineering Education* 90:477–490.

Falconer, Etta Z. 1989. "A Story of Success: The Sciences at Spelman College." *SAGE* 6 (2): 36–39.

Georgia Tech. 1969. *Large Grant Gets New Program with Negro Colleges Under Way*. Atlanta: Georgia Tech Alumni Association.

NASA (National Aeronautics and Space Administration). 2009. "A Wise Choice." www.nasa.gov/audience/foreducators/postsecondary/features/a-wise-choice .html.

National Academies. 2011. *Expanding Underrepresented Minority Participation: America's Science and Technology Talent at the Crossroads*. Washington, DC: National Academies Press. www.nap.edu/catalog/12984/expanding -underrepresented-minority-participation-americas-science-and-technology -talent-at.

NSF (National Science Foundation). 2013. *Employment and Educational Characteristics of Scientists and Engineers*. NSF 13-311. Washington, DC: NSF National Center for Science and Engineering Statistics.

Olin Corporation. 1972. *Atlanta U's 3-2 Program Feature of D.C. Conference*. Stamford, CT: Olin Corporation.

Olin Corporation. 1973. *Black Students Ride Dual Track to Engineering*. New York: Olin Corporation.

Pajares, Frank. 1996. "Self-Efficacy Beliefs and Mathematical Problem-Solving of Gifted Students." *Contemporary Educational Psychology* 21:325–344.

Perna, Laura, Valerie Lundy-Wagner, Noah Drezner, Marybeth Gasman, Susan Yoon, Enakshi Bose, and Shannon Gary. 2009. "The Contribution of HBCUs to the Preparation of African-American Women for STEM Careers: A Case Study." *Research in Higher Education* 50 (1): 1–23.

Rodriguez, Carlos, Rita Kirshstein, and Margaret Hale. 2005. *Creating and Maintaining Excellence: The Model Institutions for Excellence Program*. Washington, DC: American Institutes for Research.

Scriven, Olivia. 2006. "The Politics of Particularism: HBCUs, Spelman College, and the Struggle to Educate Black Women in Science, 1950–1997." PhD diss., Georgia Institute of Technology.

Solorzano, Daniel, Miguel Ceja, and Tara Yosso. 2000. "Critical Race Theory, Racial Microaggressions and Campus Racial Climate: The Experiences of African American College Students." *Journal of Negro Education* 69:60–73.

Spelman College. 2013. *Spelman College Fact Book 2012–13: A Profile in Facts & Figures*. Atlanta: Division of Academic Affairs, Office of Institutional Research, Assessment & Planning. www.spelman.edu/docs/oirap/fact-book-final-10-29-13rCACBAB95F21D.pdf?sfvrsn=2.

Tate, Erika D., and Marcia C. Linn. 2005. "How Does Identity Shape the Experiences of Women of Color Engineering Students?" *Journal of Science Education and Technology* 14 (5/6).

US Department of Commerce. 2011. ESA Issue Brief #04-11. Washington, DC: Economics and Statistics Administration.

Zeldin, Amy, and Frank Pajares. 2000. "Against the Odds: Self-Efficacy Beliefs of Women in Mathematics, Scientific, and Technical Careers." *American Educational Research Journal* 37:215–246.

# CHAPTER FOURTEEN

# Enhancing the Number of African Americans Pursuing the PhD in Engineering

## Outcomes and Processes in the Meyerhoff Scholarship Program

KENNETH I. MATON, KAREN M. WATKINS-LEWIS,
TIFFANY BEASON, *and* FREEMAN A. HRABOWSKI III

## The Challenge

Professional engineers merge knowledge from a complex array of scientific fields to develop technology that sustains, protects, and improves the quality of human life (Capobianco et al. 2011). At the doctorate level, engineers develop new technology by advancing research ideas and product design. Countries with research talent in engineering not only are well positioned to develop innovative, life-enhancing technology for its citizens but also reap tremendous social and economic benefits worldwide. Historically, the United States has produced talented engineering PhDs whose research has placed America at the global helm of applied science. However, in recent decades there have been growing fears of a shortfall of PhDs in the United States, which in large part appears to be resulting from a smaller pipeline of undergraduate students (Congressional Commission 2000; National Science Board 2006). Of major concern is the impact that this shrinking pool of engineering resources will have on America's standard of living and its leadership position in the technological race toward innovation (Carter, Mandell, and Maton 2009).

Why is there a threat of a shortfall among American PhD engineers? Mounting evidence cites two demographic factors that give rise to the concern. First, White males, who have traditionally dominated the engineering labor force, are retiring without similar numbers of the same demographic to replace them. The situation is further exacerbated by a long-standing history of limited diversity among college students who persist in engineering programs (Sowell 2008). Similar to the "leaky pipeline," a concept used to describe female attrition in science and technology (Huyer 2002), a systematic leak has occurred among African American and other underrepresented minority (URM) students in engineering. Research on the pipeline phenomenon has identified a number of barriers to increasing diversity in engineering, including (1) adverse sociocultural attitudes experienced by URM students at the K–16 level which undermine their academic success, (2) a lack of educational resources at the K–12 level which foster learning and study skills development, (3) little or no diversity among teaching and research faculty, and (4) departmental requirements and expectations of faculty (e.g., research, large course load) which compete with their interests in monitoring individual student performance (Roach 2006; Borman, Tyson, and Halperin 2010; Nelson and Brammer 2010; Pittman 2010).

To remain competitive in a global economy, the United States must develop its technological talent and expertise across all sectors of society. Despite over 50 years of attempts by political and research organizations to raise awareness about the critical imperative to increase diversity in engineering, barriers persist for African American students, the focus of the current chapter (Maton et al. 2009; National Action Council for Minorities in Engineering 2011). Such factors have countered efforts to expand the engineering pathway for African American students (Walter and Austin 2012). In response, numerous organizations have launched initiatives to reverse this trend by applying education and support strategies that cultivate and preserve engineering talent among African American students and other URMs (Congressional Commission 2000; Committee on Underrepresented Groups 2010).

## Intervention Programs

Prior to enrolling in a doctoral program, students must complete engineering programs at the undergraduate level with a solid academic record. African American engineering students are more likely to advance in this way with the assistance of intervention programs offered by their respective colleges and universities, especially at traditionally White institutions (American Society for Engineering Education 2007). By recognizing the strength of a holistic strategy, minority engineering and science programs strive to provide support in diverse ways with some combination of the following: (1) enrolling freshman students in summer bridge orientation programs, (2) awarding or identifying student tuition assistance, (3) assigning student mentors, (4) monitoring student academic progress, (5) promoting multicultural faculty hiring practices across engineering disciplines, (6) encouraging supportive faculty–minority student relations, and (7) establishing peer academic and social support groups (Chubin, May, and Babco 2005; Committee on Underrepresented Groups 2010; MacLachlan 2012). In so doing, more African American students are provided opportunities to develop strong study skills, achieve academic efficacy, and experience feelings of social inclusion, intellectual respect, and high expectations within their collegiate community.

Model programs using these modes of intervention retain high percentages or graduate large numbers of African American engineering students (Reichert and Absher 1997; American Society for Engineering Education 2007; Diverse Issues in Higher Education 2012). For instance, the Minority Engineering Program at North Carolina State University currently ranks third in graduating African American engineers (Diverse Issues in Higher Education 2012). This program combines a variety of strategies that promote high achievement among African American and other URM students. Starting in year 1, incoming minority freshmen are invited to attend a summer bridge program and are partnered with upperclassmen for peer mentoring. The program also values faculty diversity, encourages faculty–minority student outreach, and assists students with finding tuition assistance. Similar programs exist at several other non–historically Black colleges and universities, including Georgia Tech; University of Maryland, College Park; the City College of New York (CCNY); Michigan State University; and the University of Houston (Cooper 2012; Journal of Blacks in Higher Education 2012). While these programs have experienced success at the undergraduate level, few have been successful in helping these students go on to pursue and complete engineering doctoral degrees.

## The Meyerhoff Scholars Program

One example of success in sending students on to pursue and complete doctorates is the Meyerhoff Scholars Program (MSP). Generally, the rigor of engineering courses, along with the attractive lure of other majors, necessitates a collage of social and academic support for engineering undergraduates. Adding to the academic challenge for African American students is the scarcity or absence of peer and sociocultural support. In addition to providing financial, academic, and peer mentoring support for African American students, some programs place special emphasis on creating a culture of student peer collaborations, close student-faculty relations, and student involvement in real-world applications of complex theoretical principles. The MSP at the University of Maryland, Baltimore County (UMBC), has successfully done so since its inception in 1988 (Maton et al. 2012). The MSP initially only accepted African American students, but beginning in 1996 the program was opened to students of all backgrounds committed to increasing diversity in the science, technology, engineering, and math (STEM) fields. In terms of engineering outcomes for its African American students, UMBC ranked third (tied with University of Michigan–Ann Arbor, and behind the Massachusetts Institute of Technology and Georgia Tech) in the nation among primarily White baccalaureate-origin institutions for producing African American engineering graduates who proceeded to receive PhDs in engineering during 2007–2011 (National Science Foundation 2013).

MSP is a comprehensive intervention program that is designed to address four critical sets of factors that affect minority success in science: (1) knowledge and skills, (2) motivation and support, (3) monitoring and advising, and (4) academic and social integration (Maton and Hrabowski 2004). MSP provides ongoing monitoring, persistent advising, and early feedback to its minority STEM students. Specifically, MSP immerses African American students in social and academic support across the following domains:

- *Financial aid*: The MSP provides students with a financial package ranging from $5,000 to $22,000. This support is contingent upon maintaining a B average in a STEM major.
- *Recruitment weekend*: The top applicants and their families attend one of the two recruitment weekends on the campus.
- *Summer bridge*: Meyerhoff students attend a mandatory prefreshman summer bridge program and take courses in math, science, and "Race, Science and Society." They also participate in STEM-related cocurricular activities and attend social and cultural events.
- *Study groups*: Group study is strongly and consistently encouraged by the program staff, as study groups are viewed as an important aspect of success in STEM majors.
- *Program values*: Program values include support for academic achievement, seeking help from a variety of sources, peer supportiveness, high academic goals (with emphasis on PhD attainment and research careers), and giving back to the community.
- *Program community*: The Meyerhoff program provides a family-like social and academic support system for students. Students live together in the same residence hall during their first year and are required to live on campus during subsequent years.
- *Staff academic advising, staff personal counseling*: The program employs full-time advisors who monitor and support students on a regular basis. The staff focus not only on academic planning and performance but also on any personal problems students may have.

- *Tutoring*: The program staff strongly encourage students to either tutor others or be tutored to maximize academic achievement (i.e., to get As in difficult courses).
- *Summer research internships and academic year research*: Each student participates in multiple summer research internships, often at leading sites around the country, as well as some international locations. Many students also participate in academic-year research, including a subset of students who participate in UMBC's Minority Access to Research Careers (MARC) program.
- *Faculty involvement*: Key STEM department chairs and faculty are involved in the recruitment and selection phases of the program. Many faculty further provide opportunities for student laboratory experience during the academic year to complement summer research internships.
- *Peer mentoring*: Freshmen and sophomores receive mentoring from selected upperclassmen.
- *Community service*: Meyerhoff students are encouraged to volunteer to help at-risk youth.
- *Administrative involvement*: The MSP is supported at all levels of the university, including ardent support from the university's president (the program cofounder).

## The Current Study

The current study extends previous research on the MSP, with a focus on the subset of African American Meyerhoff students who graduated with engineering majors. Specifically, entry into engineering graduate programs, program experience, individual predictors of outcome, and experiences in graduate school for this subpopulation are examined. Four research questions are addressed: (1) Are African American engineering MSP students (first 16 cohorts) more likely to enter engineering PhD programs than a comparison sample, with covariates controlled? (2) Which MSP components do African American engineering students rate as especially helpful? (3) Do various precollege and college variables predict which of these students are more likely to pursue an engineering PhD, with covariates controlled? (4) What are the supports and barriers that African American MSP engineering alumni experience in graduate engineering programs, and how might they relate to ethnic minority status? Taken together, these four research questions provide insight into the effectiveness of the program in enhancing PhD pursuit for African American engineering students, enhance understanding of both program-level and individual-level factors that appear to contribute to program impact, and contribute to our understanding of the graduate school experience for MSP engineering alumni.

## Method

### RESEARCH PARTICIPANTS

Table 14.1 presents characteristics of the samples for each of the four aspects of the study.

### POSTCOLLEGE OUTCOME

The 67 African American Meyerhoff students who graduated with engineering degrees from the first 16 entering classes (fall 1989–fall 2004) with full baseline and outcome data composed the Meyerhoff sample for the outcome analyses (an additional 13 African American Meyerhoff students were excluded who lacked full baseline or outcome data). The "Declined" comparison sample consisted of 49 African American students who graduated with engineering majors with full baseline and outcome study data and were offered admission to the MSP between 1989 and 2004 but declined the offer (an additional 17 African American Declined comparison sample students were excluded who lacked full baseline or outcome data). In the vast majority of cases the Declined comparison sample students attended other institutions, mostly highly selective or selective universities.

Preliminary analyses revealed several statistically significant differences between samples. The Declined students in the study sample, compared with the MSP students, had a later year of college entry, $t(114) = 2.0$, $p < .05$, had

TABLE 14.1   Sample characteristics for each aspect of study

| Characteristic | Outcomes | | Program Components | Individual Predictors | Grad School Experience |
| | Meyerhoff | Declined | | | |
|---|---|---|---|---|---|
| Sample size (N) | 67 | 49 | 28–62[a] | 51–67[b] | 20 |
| Male (%) | 68.7 | 49.0 | 69.4 | 68.7 | 75.0 |
| Math SAT | 658.1 | 679.8 | 654.5 | 648.6 | 662.1 |
| Verbal SAT | 619.4 | 644.1 | 623.2 | 620.7 | 622.6 |
| High school GPA | 3.70 | 3.80 | 3.70 | 3.89 | 3.67 |
| Year of entry[c] | 1997.4 | 1998.9 | 1997.4 | 1997.4 | 1995.5 |

[a]Sample size varies depending on missing data on the program component item.
[b]Sample size varies depending on missing data on the predictor variable.
[c]Year of entry is considered a continuous covariate for current study purposes.

higher SAT math scores, $t(224) = 2.3$, $p < .05$, had higher SAT verbal scores, $t(114) = 2.1$, $p < .05$, and were more likely to be female, $X^2(1) = 4.6$, $p < .05$ (means are reported in the first two columns of table 14.1). High school GPAs did not differ significantly. All of these variables (including high school GPA) were included as covariates in the outcome analysis.

## PROGRAM COMPONENTS

The 62 African American Meyerhoff engineering students from the first 16 entering classes (fall 1989–fall 2004) who completed most or all of the survey items assessing program components composed the Meyerhoff sample for the descriptive analysis of the program component ratings. The third column of table 14.1 presents the sample characteristics.

## PREDICTORS OF POSTCOLLEGE OUTCOME

The 67 African American Meyerhoff engineering students from the third through the 16th entering classes (fall 1991–fall 2004) who provided information on most or all of the predictor variables composed the sample for the analyses linking predictor variables to outcomes. Students from the first two entering classes were not included since we did not begin collection of research excitement data until summer 1991.

## GRADUATE SCHOOL EXPERIENCE INTERVIEWS

Twenty African American graduates of the program who (a) entered the MSP between 1989 and 1998 and (b) were pursuing or had already obtained a doctorate degree ($n = 12$) or master's degree ($n = 8$) in engineering at the time of the interviews were included. At the time of interview, eight of the participants were still enrolled in a PhD program. Twelve individuals, including all of the participants with an MS in engineering, had already received their degree. Participants ranged in age from 24 to 31, with an average age of 27.

## PROCEDURE

### Informed Consent

The research study received expedited review by the UMBC institutional review board. Students completed an informed consent form at the time they applied to the program.

## Postcollege Outcomes

When applying to the MSP, students completed a form providing permission to obtain college and future graduate school transcripts from registrar offices at other institutions they might choose to attend. Information on postcollege destination was obtained from multiple sources, including program records (in the case of Meyerhoff students), the students, family members, or through Internet or paid searches. Information was confirmed (or clarified) in the vast majority of cases through review of transcripts or by phone calls to graduate school registrar offices.

## Program Components

Program component items were included as part of process evaluation surveys administered periodically to students. Identifying information was obtained on the cover page. The survey was administered during program-wide meetings in the early years of the program, in later years it was directly sent to students, and most recently it was included as part of an exit survey e-mailed to graduating students (see Maton et al. 2008).

## Predictors of Postcollege Outcome

Father and mother education levels, research excitement, and math/science intrinsic motivation at college entry were collected from the Meyerhoff students in a group meeting during the six-week summer bridge orientation program. The measures were included in a larger set of survey items and administered by graduate student research assistants. Identifying information was obtained on the survey cover page. The number of summer research internships was obtained from MSP staff. Undergraduate transcripts were obtained from the university registrar's office.

## Graduate School Experience

As part of a larger study on the graduate student experience of Meyerhoff students conducted in 2006, recruitment of potential participants included multiple e-mails and telephone calls from research and Meyerhoff staff. There were various reasons potential participants were not interviewed, including outdated contact information or students who did not have the time to participate. Individuals that responded to recruitment efforts were interviewed by phone. As this study is an extension of existing research on the MSP, the consent forms participants previously signed extended to this interview. Students were informed that their participation was voluntary and that the information they provided would be fully confidential. Interviews, averaging two hours in length, were audiotaped. Participants were paid $20 at the completion of the interview.

Demographic and Academic Background Variables
Ethnicity, gender, university entrance date, SAT scores (both math and verbal), and high school GPA were obtained from university application records.

Graduate Education
The engineering graduate education outcome variable contained four post-college categories: (1) entered an engineering PhD program, (2) entered an engineering master's program, (3) entering a nonengineering STEM graduate program (e.g., including PhD, master's, MD/PhD, and MD), or (4) no postcollege education in STEM (includes students who did not complete college, those who graduated in a non-STEM major, STEM graduates who did not pursue graduate or professional education, and those who attended non-STEM graduate or professional school programs). If a student entered an engineering master's program upon graduation but later entered an engineering PhD program, the latter (i.e., the higher degree program) was coded. However, if a student entered an engineering PhD program but left the program with a terminal engineering master's degree, then the degree actually received was counted. For nonengineering graduate programs, STEM was defined to include programs in the physical and life sciences, computer science, and mathematics; the behavioral and social sciences were excluded.

Program Components
Survey items were developed by the research team to assess student perceptions of the majority of the program components. Items included (1) the summer bridge program, (2) financial support, (3) being part of the Meyerhoff community, (4) study groups, (5) staff academic advising, (6) staff personal counseling, (7) faculty involvement in the program, (8) academic tutoring services, (9) UMBC administrators' involvement and support, (10) Baltimore/DC-assigned mentor, and (11) summer research experience (this item was added later). The wording of items and the anchors on the five-point Likert scale were somewhat modified over the years (see Maton et al. 2008). In the most recent version, the question stem reads, "Please indicate the degree to which the following aspects of the Meyerhoff Pro-

gram have been helpful in your experience as a student." The most recent rating scale includes five anchors: $1 = $ not at all helpful; $2 = $ a little helpful; $3 = $ somewhat helpful; $4 = $ helpful; $5 = $ very helpful. For the subset of students who completed surveys on more than one occasion, responses on the most recent survey were used. Although the items have face validity, their psychometric properties (reliability, validity) are not known.

### Father and Mother Education

The highest education level completed by each parent was assessed using a nine-point scale ($1 = $ grade school; $2 = $ some high school; $3 = $ high school diploma or equivalent; $4 = $ business or trade school; $5 = $ some college; $6 = $ associate or two-year degree; $7 = $ bachelor's or four-year degree; $8 = $ some graduate or professional school; $9 = $ graduate or professional degree).

### Precollege Research Excitement

Precollege research excitement was assessed with a single item developed by the research team. The item read, "I am excited about the idea of doing scientific research." Students respond on a five-point Likert-type scale, indicating how accurate a statement this is for them ($1 = $ not at all accurate; $3 = $ somewhat accurate; $5 = $ completely accurate). Although the item has face validity, its psychometric properties (reliability and validity) are not known.

### Math/Science Intrinsic ("Effectance") Motivation

Intrinsic ("effectance") motivation assesses the extent to which students feel positive about their math/science problem-solving abilities, actively seek out challenges in math/science, and are not easily discouraged by difficult problems in math/science. A slightly modified version of the 12-item Fennema-Sherman Effectance Motivation in Mathematics Scale (Fennema and Sherman 1976) was used in the current study. For each item in the original version of the scale, the term "mathematics" was changed to "mathematics and science" to broaden the STEM-related relevance of the scale. Six of the items are positively worded, and six are negatively worded. A sample item is "When a math or science problem arises that I can't immediately solve, I stick with it until I have the solution." The original scale has demonstrated strong reliability and validity. The alpha reliability of the modified scale used in the current study is 86.

### Summer Research Internships

Summer research internships were assessed by the number of summers that students took part in a research internship.

Undergraduate GPA

Undergraduate GPA was obtained from student transcripts. For students who graduated college, this was their final GPA; for the small number of students who dropped out of college, this was their cumulative GPA before leaving.

Graduate School Experience Interviews

The major interview questions were aimed at identifying supports and barriers students encountered during their graduate education. Participants were asked whether they received any encouragement or special support that impacted their graduate education. Participants were then asked to indicate whether and how a list of factors, including faculty mentors and/ or advisors, course instructors, departmental climate, peers in program, family, and the MSP, supported their success during graduate school. After discussing supports for success, barriers to success in graduate school were investigated. Participants were asked whether they faced any challenges, barriers, and/or unpleasant experiences that impacted their graduate education. Then, participants were asked to indicate whether and how the above-mentioned factors acted as barriers to success during graduate school. Participants were given an opportunity to list additional supports and barriers to success which may not have been addressed during the interview. Both supports and barriers were investigated to determine how they may relate to ethnic minority status.

DATA ANALYSIS PROCEDURES

The four research questions were examined using a variety of analytical procedures. Differences in postcollege outcomes between Meyerhoff and Declined students were examined using multinomial logit regression. The relative rankings of student perceptions of the helpfulness of program components, an exploratory descriptive analysis, were based on calculation of mean scores. The independent contributions of predictors of program outcome were examined using analyses of variance and multinomial logit regression analysis.

All interviews were transcribed and imported into NVivo, a qualitative software package, to be coded. Codes are short rubrics that indicate specific qualitative content—themes, patterns, or processes. Codes used for this data set were identified through past literature, theory, the researchers (their insights on the topic and the research question being asked of the data), and the data themselves. Iterative, consensus coding of interviews was conducted using a team approach. Multiple coders, led by their own experiences (including conducting interviews and knowledge of the literature), read through interview transcripts and discussed how they viewed the data. These deliberations led to the addition and modification of the initial coding format to incorporate not only the spontaneous conversations of participants but also the input of all members of the team. Consequently, the resulting final coding format included codes from existing literature, the interview questions, and emergent content from the data. Eventually, consensus on a coding format, which included a set conceptualization and consistent assignment of the codes, was attained.

## Results

### POSTCOLLEGE OUTCOMES

The first research question addressed program outcomes. To examine trends in outcomes over time, the sample was divided into 1989–1995 and 1996–2004 subgroups (table 14.2; as noted earlier, 1996 was the year that the program was opened to students who were not URMs). The 1989–1995 African American Meyerhoff engineering students entered engineering PhD programs at a rate comparable to their African American Declined sample counterparts (15.7% vs. 16.7%); in contrast, the 1996–2004 African American Meyerhoff students were 2.5 times more likely to enter engineering PhD programs than their African American Declined sample counterparts (46.3% vs. 18.9%). Of note, the percentage of Meyerhoff students entering engineering PhD programs tripled from 1989–1995 to 1996–2004. Equally striking is that the percentage of Meyerhoff engineering students not entering any graduate STEM program declined greatly from the earlier to the later time period (34.6% to 4.9%), while the percentage of Declined comparison students increased (33.3% to 45.9%). Although both Meyerhoff and Declined comparison samples show a notable decrease in the percentage of students entering master's engineering programs across the two time periods, only the Meyerhoff students appear to be attending engineering PhD programs instead. Across the entire study period, African American Meyerhoff students were almost two times more likely to attend engineering PhD programs than the African American Declined comparison students (34.3% vs. 18.4%).

Multinomial logit regression analyses were conducted for each time period comparing engineering PhD degree pursuit, master's engineering degree pursuit, and nonengineering STEM degree pursuit, respectively, with no graduate STEM degree pursuit, to determine whether the observed

TABLE 14.2  Postcollege STEM outcomes for African American Meyerhoff and Declined comparison sample students: 1989–1995, 1996–2004, and 1989–2004

| Postcollege Outcome | 1989–1995 Entering Cohorts | | 1996–2004 Entering Cohorts | | 1989–2004 Entering Cohorts | |
|---|---|---|---|---|---|---|
| | Meyerhoff | Declined | Meyerhoff | Declined | Meyerhoff | Declined |
| Engineering PhD | 15.4% (4) | 16.7% (2) | 46.3% (19) | 18.9% (7) | 34.3% (23) | 18.4% (9) |
| Engineering MS | 46.2% (12) | 41.7% (5) | 24.4 % (10) | 29.7% (11) | 32.8% (11) | 32.7% (16) |
| Nonengineering STEM PhD, MS, MD/PhD, or MD | 3.8% (1) | 8.3% (1) | 24.4 % (10) | 5.4% (2) | 16.4% (11) | 6.1% (3) |
| No graduate STEM | 34.6% (9) | 33.2% (4) | 4.9% (2) | 45.9% (17) | 16.4% (11) | 42.9% (21) |
| Total | 100.0% (N=26) | 100.0% (N=12) | 100.0% (N=41) | 100.0% (N=37) | 100.0% (N=67) | 100.0% (N=49) |

NOTES: Findings from multinomial regression analyses with six covariates indicated that Meyerhoff students were more likely than Declined comparison sample students to enter engineering PhD programs (1996–2004; 1989–2004), engineering MS programs (1996–2004; approached significance for 1989–2004), and nonengineering STEM graduate programs (1996–2004; 1989–2004) than not to attend graduate school (No graduate STEM). See text for details.

differences achieved statistical significance. High school GPA, SAT math and verbal scores, gender, and year of entry were included as covariates. For 1989–1995, African American Meyerhoff engineering students did not differ significantly from the Declined students in any of the three comparisons. However, with covariates controlled, for 1996–2004 the MSP students were significantly more likely than Declined students to enter (1) engineering PhD programs (odds ratio = 47.4, wald (1) = 13.7, $B$ = 3.8, $p$ = .000), (2) master's engineering programs (odds ratio = 12.2, wald (1) = 6.6, $B$ = 2.5, $p$ = .01), and (3) nonengineering STEM graduate programs (odds ratio = 169.9, wald (1) = 13.9, $B$ = 5.3, $p$ = .000) than not to enter a STEM graduate program (reference group). Finally, with covariates controlled, for the entire 1989–2004 study period, Meyerhoff students were significantly more likely than Declined students to enter (1) engineering PhD programs (odds ratio = 8.9, wald (1) = 11.1, $B$ = 2.2, $p$ = .001) and (2) nonengineering STEM graduate programs (odds ratio = 16.3, wald (1) = 10.1, $B$ = 2.8, $p$ = .001) and (3) marginally significantly more likely than Declined students to enter master's engineering programs (odds ratio = 2.8, wald (1) = 3.5, $B$ = 1.0, $p$ = .063) than not to enter a STEM graduate program (reference group).

In terms of engineering degree receipt, of the 23 Meyerhoff students who entered doctoral engineering programs, 15 (65.2%) have completed their PhDs to date (most of the remainder are still enrolled). Of the nine Declined sample students who entered engineering PhD programs, five (55.6%) have completed their PhDs to date (most of the remainder are still enrolled). Of the 22 Meyerhoff students who entered master's engineering programs, 21 (95.5%) completed their master's degree. Of the 16 Declined sample students who entered master's engineering programs, 14 (87.5%) completed their master's degree. There are no statistical differences in terms of degree receipt to date between Meyerhoff and Declined students at either the doctoral or master's degree level.

## PROGRAM COMPONENTS

The second research question addressed student perceptions of the relative helpfulness of the different MSP components. The mean score of African American MSP engineering students from 1989–1995 and 1996–2004 for each program component item was calculated (table 14.3). Across both time periods, the five most highly rated components (receiving ratings of 4.0 or higher on a five-point scale) were: striving for outstanding academic achievement (program value), financial scholarship, being part of the MSP community, summer bridge, and study groups. Of these five, only striving

TABLE 14.3 Perceived benefit of Meyerhoff program components for African American Meyerhoff engineering students, 1989–1995 and 1996–2004

| Program Component | 1989–1995 | | | 1996–2004 | | |
|---|---|---|---|---|---|---|
| | Mean | Standard Deviation | N | Mean | Standard Deviation | N |
| Strive for outstanding academic achievement | 4.2 | 1.0 | 6 | 4.9 | 0.3 | 22 |
| Financial scholarship | 4.6 | 0.9 | 23 | 4.7 | 1.0 | 35 |
| Being part of the MSP community | 4.0 | 0.9 | 23 | 4.7 | 0.5 | 35 |
| Summer Bridge | 4.2 | 1.1 | 23 | 4.6 | 0.6 | 35 |
| Study groups | 4.0 | 1.2 | 23 | 4.3 | 1.0 | 39 |
| Staff academic advising | 3.1 | 1.2 | 21 | 4.3 | 1.0 | 39 |
| Summer research | 3.2 | 1.2 | 13 | 4.2 | 0.9 | 39 |
| Faculty involvement | 3.4 | 1.1 | 19 | 4.1 | 1.1 | 35 |
| Staff personal counseling | 2.9 | 1.3 | 21 | 3.9 | 1.4 | 34 |
| Academic tutoring services | 3.6 | 1.6 | 10 | 3.6 | 1.0 | 39 |
| UMBC administrators' involvement and support | 2.9 | 1.0 | 21 | 3.4 | 1.1 | 39 |

for outstanding academic achievement and being part of the MSP community showed an increase of more than half a point from the earlier to the later time period (4.2 to 4.9 and 4.0 to 4.7, respectively). Four additional components showed increases of more than half a point from the earlier to the later time period (in each case receiving a rating of 3.9 or higher in the more recent time period): staff academic advising (3.1 to 4.3), summer research, (3.2 to 4.2), faculty involvement (3.4 to 4.1), and staff personal counseling (2.9 to 3.9). The two components with the lowest ratings in each time period were academic tutoring services and UMBC administrators' involvement and support. These findings focused on change over time, however, should be interpreted with caution and were not subjected to statistical testing, given changes in item wording and scaling and means and timing of administration of the process evaluation surveys from earlier to later years, as described earlier.

PREDICTORS OF POSTCOLLEGE OUTCOME

The third research question examined individual precollege and college predictors of postcollege outcome for the Meyerhoff African American engineering students. The top portion of table 14.4 presents the mean values (or percentages) for the precollege predictor variables: gender, father education level, mother education level, math and verbal SAT scores, high school GPA, research excitement, math/science intrinsic motivation, and year of entry. The bottom portion of the table presents the corresponding information for the two college predictors: number of summer research internships and college GPA. Univariate analyses of variance with post hoc analyses were conducted for each precollege and college variable, with graduate outcome serving as the independent variable. In the table, differing superscripts indicate statistically significantly differences across the specified graduate outcome categories (post hoc analysis results).

A number of significant findings emerged. Students who entered engineering PhD programs reported higher levels of father education (mean = 7.6) and mother education (mean = 7.5) than those who entered master's engineering programs (5.8 and 5.9, respectively). Students who entered engineering PhD programs (3.8) and nonengineering STEM graduate programs (3.8) achieved higher high school GPAs than those who entered a master's engineering program (3.6) or did not pursue graduate STEM education (3.5). Meyerhoff engineering students who entered PhD programs were from later entering cohorts (mean = 1999.6) than those who entered non-PhD programs (1996.2) or did not pursue a graduate degree (1994.0). Students

TABLE 14.4  Precollege and college predictors of postcollege outcome: African American engineering Meyerhoff students, 1989–2004

| | Postcollege Outcome | | | |
| Predictor | Engineering PhD Program | Engineering MS Program | Other Graduate STEM Program | No Graduate STEM Program |
|---|---|---|---|---|
| Precollege | | | | |
| Male (%) | 70 [23] | 77 [22] | 55 [11] | 64 [11] |
| Father's education | 7.6$^a$ (1.4) [21] | 5.8$^b$ (2.4) [19] | 6.1 (2.5) [10] | 6.9 (2.3) [11] |
| Mother's education | 7.5$^a$ (1.7) [22] | 5.9$^b$ (2.1) [22] | 6.4 (2.5) [10] | 6.0 (1.6) [11] |
| Math SAT | 657.4 (54.5) [23] | 649.6 (41.3) [22] | 682.7 (33.8) [11] | 651.8 (49.4) [11] |
| Verbal SAT | 607.0 (48.8) [23] | 623.2 (52.0) [22] | 634.6 (55.4) [11] | 622.7 (52.6) [11] |
| High school GPA | 3.8$^a$ (0.3) [23] | 3.6$^b$ (0.3) [22] | 3.8$^a$ (0.4) [11] | 3.5$^b$ (0.4) [11] |
| Research excitement | 4.3 (1.3) [15] | 4.0 (1.2) [19] | 3.8 (1.1) [11] | 3.6 (1.2) [8] |
| Math/science intrinsic motivation | 4.5$^a$ (0.4) [18] | 3.9$^b$ (0.6) [17] | 4.0$^b$ (0.7) [8] | 4.4 (0.3) [8] |
| Year of entry | 2000.6$^a$ (4.2) [23] | 1995.5$^b$ (4.2) [22] | 1998.9$^a$ (3.1) [11] | 1994.0$^b$ (2.6) [11] |
| College | | | | |
| No. summer research internships | 2.3 (0.8) [23] | 1.9 (1.1) [22] | 2.5 (1.2) [11] | 1.6 (1.1) [11] |
| College GPA | 3.51$^a$ (0.2) [23] | 3.28$^b$ (0.3) [22] | 3.47$^a$ (0.3) [11] | 2.93$^c$ (0.4) [11] |

NOTES: Number in parentheses to the right of means is standard deviation. Number in square brackets under the means (or percentages) is sample size. Differing superscripts indicate statistically significant differences for one-way analyses of variance, Duncan post hoc test ($p < .05$).

who attended engineering PhD programs reported higher levels of math/science intrinsic motivation (4.5) than those who entered a master's engineering program (3.9) or a nonengineering graduate program (4.0). Students who entered engineering PhD programs (2000.6) and nonengineering STEM graduate programs (1998.9) were from later entering MSP cohorts than those who entered a master's engineering program (1995.5) or did not pursue graduate STEM education (1994.0). Finally, students who pursued a PhD program in engineering (mean = 3.51) and who pursued a nonengineering STEM graduate program (3.47) achieved higher college GPAs than those who pursued a master's in engineering (3.28), and all three of these groups achieved higher college GPAs than those who did not pursue graduate STEM (2.93).

A multinomial logit regression analysis was conducted to determine the unique contributions of each of the six variables that achieved significance in the univariate analyses, with graduate outcome the criterion variable. In five of the six possible contrasts across categories of outcome, one variable emerged as a significant predictor, independent of the other five variables. In the sixth contrast, none of the six variables were significant.

Specifically, with all six variables entered, students entering engineering PhD programs reported (1) significantly higher math/science intrinsic motivation than students entering master's engineering programs (odds ratio = 14.6, wald (1) = 4.3, $B$ = 2.7, $p$ = .038), (2) significantly higher math/science intrinsic motivation than students entering nonengineering STEM graduate programs (odds ratio = 12.5, wald (1) = 4.2, $B$ = 2.5, $p$ = .041), and (3) a significantly later year of college entry than students who did not pursue graduate STEM education (odds ratio = 2.6, wald (1) = 4.2, $B$ = 0.9, $p$ = .040). Students entering nonengineering STEM graduate programs reported significantly higher college GPAs (4) than students entering master's engineering programs (odds ratio = 28.3, wald (1) = 5.2, $B$ = 5.6, $p$ = .023) and (5) than students who did not pursue graduate STEM education (odds ratio = 141.8, wald (1) = 5.6, $B$ = 7.3, $p$ = .018). Finally, with all six variables entered, none emerged as significant in comparisons between students who entered master's engineering programs and those who did not pursue graduate STEM education.

ENGINEERING GRADUATE SCHOOL EXPERIENCE

Participants were questioned about the impact of six major factors on their success in graduate school. These factors include faculty mentors and advisors, course instructors, departmental climate, peers in program, family, and the Meyerhoff Scholars Program. Overall, most factors were described as both specific supports for success and specific barriers to success. The MSP is the single factor that was mentioned solely as a support for success in graduate school. Peers in the program and family were most frequently described as resources that encouraged academic success in graduate school. Faculty mentors and advisors served as supports for academic success in most cases. However, in some cases, interactions or relationships with mentors and advisors (or lack thereof) inhibited academic success for participants. Course instructors were equally regarded as both supports and barriers to success in graduate school. Finally, departmental climate was most often described as a barrier to academic success for the engineering graduate students and alumni sampled in the study.

Below, key themes related to supports from each of these sources, along with representative quotes, are provided first, followed by key themes and illustrative quotes related to barriers.

## SUPPORTS

### Faculty Mentors and Advisors

Participants identified their relationships with both official and unofficial mentors and advisors as important to their success in graduate school. The guidance students received from their mentors helped prepare them for careers in science. Participants reported that the most helpful mentors were open-minded about the research questions students pursued and affirmed students' abilities. A master's student expressed appreciation for their mentor: "He wanted you to think independently, he's not a micromanager . . . if I came to him and asked for help, he'd help, but for the most part he'd leave us alone to do our own work." A doctoral student echoed this sentiment, saying, "I was encouraged by the fact that [my advisor] trusted my intuition and my judgment, and he was very willing to let me sort of find my own path." Participants also expressed appreciation for mentors who were accessible and who nurtured personal relationships with students. For example, one advisor had "an open door policy where you can go in there at any time and talk to him about whatever. . . . It was essentially camaraderie with his grad students . . . we'd pick a Friday and go out with everybody and hang out after the evening hours."

Most participants did not associate the support they received from their mentors with being students of color, but a couple of participants did make this association. One participant recalled how an unofficial mentor, a Black faculty member, advocated for him and ultimately secured him research funding that his official mentor was unable to offer to him. The participant found the response from Black faculty to be the result of personal, not just business, relationships. He stated, "[I] came to the realization that one of the greatest challenges to making it through this process is having someone on the other side who can act as an advocate for you but also takes on some level personally . . . the reality for me was that the only people in the department . . . who were willing to act as my advocate at that level were people of color."

### Course Instructors

Course instructors were named as sources of support for graduate students as well. Participants expressed that instructor enthusiasm, empathy, and patience were characteristics that encouraged them to succeed. Participants also said they benefited from having professors who demonstrated genuine concern for them as students. A participant described how a course instructor made extra efforts to help him when he was having difficulty in a course. The student described his professor as "very patient when I struggled in his class. . . . He gave me extra work to help make up a bad exam score."

Participants reported that course instructors were helpful in pointing students to labs that may be a good fit, funding opportunities, and other professors. Instructors also provided social support by being available to students to provide feedback and advice on coursework, papers, presentations, and students' plans for the future.

### Departmental Climate

Departmental climates described as collegiate, friendly, and/or collaborative were often listed by participants as influential to their success in graduate school. Participants felt it encouraging when their department placed emphasis on producing good scientists. Participants felt supported when their departments provided adequate resources for them to get the most out of their experience, as opposed to weeding out people with tactics like graduating only the top 30%–50% of students. Participants reported that environments that were receptive to student input and fostered communication and collaborative efforts between labs yielded a sense of support and belonging. One doctoral student noted, "The professors in my department always had an open-door policy. [Students] could always go and talk to them and ask them questions and they were always willing to give [students] support. . . . Just always being able to have an environment where you could always talk to the professors and get some feedback and get information helps [students] improve."

### Peers in the Program

One of the encouraging aspects of participants' graduate experiences is the support they received from peers. Participants reported that their peers shared similar experiences, studied together, mentored each other, and offered their expertise on projects when needed. One participant recalled that "classmates were there to study with me. . . . We had our own departmental student board . . . to become organized and things of that nature." For many participants, graduate school is/was a new and stressful experience, so knowing that others shared the same experience and went through the same process was reassuring. A participant exclaimed, "We were all just going through the same hell together and we helped each other with coursework. . . . On weekends we would go out . . . we just kind of stuck together. And [it] helped a lot to know there were other people going through the same."

Some participants associated the support they received from peers with being a student of color. One doctoral student noted, "When you . . . represent a minority group, there's a tendency to support other members of that group." A group for minority engineers helped to encourage students through discussions of issues related to being a student of color. A participant reported that "[the minority student organization] had different sessions in which they discussed issues that come up among [minority] graduate students . . . and they try to work out those issues that we as newer students could expect along the way." Involvement in minority-focused groups on campus helped students overcome challenges they associated with being minority graduate students, helping them to achieve academic success.

### Family

Family was often mentioned as a source of support, providing both emotional and financial support to students. Families served as outlets for students to discuss the graduate school experience, and they helped students retreat from the stress of graduate life. Participants felt encouraged when family members engaged in their success by attending important academic events and praying for students' academic success. Family members who had already received an advanced degree, or were in the process of obtaining one, encouraged participants to pursue their academic goals by serving as role models to participants. A doctoral student exclaimed, "My wife, she's kinda trying to push me. . . . And the fact that she's trying to finish school herself, kinda motivated me."

Some participants associated the encouragement they received from their family with being a student of color. A student talked about how his family members recalled when it was impossible for African Americans to get doctoral degrees, and they encouraged the student to strive for success in achieving this formerly unattainable goal. Another family was helpful by regularly checking up on a student to ensure the student's well-being, because the student moved to a city that they believed was not supportive of African Americans.

Family helped participants put their academic struggles in perspective by offering students unconditional support regardless of their academic letdowns. One participant stated, "I just remember . . . feeling really bad, you know, I'd gone from being a 4.0 student in undergraduate school to, you know, doing horribly. . . . I just realized why this isn't what defines me as a human being. . . . Even if I failed . . . they were always there for me." Many participants explained that family members encouraged their continuation in graduate school by pushing them to success and reminding them that they were not quitters. One participant exclaimed, "Without my family, I probably would have quit a long time ago. I mean, my parents really wanted me to just keep pushing. . . . In the end, that's what I did."

### MSP

The MSP provided encouragement to participants through its staff, mentors, and scholars. Participants reported that Meyerhoff staff and mentors provided support by sending students words of encouragement, notifying students of relevant talks and conferences, and directing students to various resources. For some students, the MSP helped them make the decision to pursue graduate degrees and persist in that goal despite challenges. One student said, "[Meyerhoff] first exposed me to graduate school and encouraged us to pursue Ph.D.'s. . . . I have had conversations with [Meyerhoff staff] about graduate school and [they] have always encouraged me to stick to it." Participation in the MSP helped participants build networks of peers in undergraduate school that they maintained in graduate school. One participant said, "Going through the Meyerhoff Program really helped get me ready and running when I hit the ground in graduate school, but then, even afterwards, some of my peers in my particular program were also in graduate school so we kept up our network."

Several participants associated the support they received from the MSP with being a student of color. Participants noted that during their time in the program it was limited to students of color. Participants were encouraged by former Meyerhoff scholars who completed their doctoral degrees, specifically because those scholars were African American students who accomplished the challenge of obtaining a PhD. A former doctoral student said, "I was looking at them, very proud of them. These were young African Americans, people of color who are very smart and doing wonderful things and ambitious and that in itself made me want—I wanted to be a part of that group."

### Faculty Mentors and Advisors

Although in many cases participants' relationships with their mentors helped them during the graduate experience, in some cases mentor relationships were barriers to success. Students reported poor communication skills and overinvolvement in student research projects as mentor-related barriers to success. One participant referred to her advisor as "very hard [and] so we had to work very hard so that's great on paper but I think that, you know, I think the general . . . morale of my research group was rather low." Being assigned to mentors who were not knowledgeable about students' preferred area of research presented as a challenge for students as well. A participant reported working with an advisor whose expertise did not match the student's research area. The participant explained that the advisor "wasn't always the most knowledgeable person. Because it was a multi-disciplinary program, he had to know about all the different engineering disciplines . . . and he might say one thing and that might not be exactly true so you just had to make sure and double-check." Working with advisors who lacked knowledge of their students' areas of research placed students at a disadvantage in which they were unable to fully capitalize on the advisor's expertise.

### Course Instructors

Course instructors who lacked enthusiasm for teaching and did not appreciate student questions during class periods were mentioned as barriers to academic success by participants. Participants felt that unenthusiastic and/ or disgruntled course instructors dampened the learning experience and made it difficult for them to seek out answers to questions in their courses. One doctoral student's report of an experience with an unpleasant professor included, "He really absolutely hated teaching the class and he was just very, very unpleasant. Many times during the class, he mentioned his dislike for other faculty, his dislike for the department, his dislike for the university. . . . Most of the class was okay until I actually had some questions on some things. He [got] annoyed that I was asking so many questions and yeah, so that just sort of made a strain on everything."

## Departmental Climate

Many participants discussed the challenges they encountered related to the climate of their department or program. Some participants remarked that programs that lacked structure and organization served as a barrier to success for students. For example, the interdisciplinary nature of one program translated into a student having difficulty communicating messages between faculty members from different departments. Additionally, programs that emphasized advisor autonomy and lack of administrative oversight led to students lingering in a program and not graduating.

Minority issues that served as challenges involved departmental climate and minority recruitment. Participants stated that administrators appreciated minority representation, particularly because high minority student numbers were required for National Institutes of Health (NIH) grants. However, when it came time to recruit students of color, administrators needed to be convinced of the extra efforts to secure highly qualified minority students. One participant said, "There was a general frustration in my school which I think that all Black grad-students go through . . . in that of minority recruitment. We make a lot of effort to recruit minorities and we were told that yeah, sure, we'll support you and then when the time came the department turned a blind eye." Occurrences such as this sent the message to students that the department ultimately did not truly value diversity.

## Peers in the Program

Peers were mentioned as a challenge by several participants for a variety of reasons, including cultural and language barriers, feelings of isolation, and differences in personality. Several participants associated being a person of color with the challenges they encountered with their peers. One participant believed that White peers were less culturally sensitive than African American peers, and conflict was created because of a lack of exposure to students of color. A female participant reported that peers in her lab "weren't very receptive or welcoming to having a female come in and then an African American female to be sitting there in the office with them, and so I think that the environment was rather hostile in the beginning simply because they just weren't used to it." Foreign peers were mentioned by a few participants as being especially challenging, because they felt that foreign peers held negative stereotypes about African Americans. Another former doctoral student said, "In grad school . . . there are a lot of students . . . straight from India, straight from China and so their perception of African Americans was really just what they saw in the news . . . that was sort of how they had categorized me."

Peers also posed a challenge because of differences in personality or background which resulted in some participants being excluded from social groups in their program. Speaking of his former peers, one participant explained, "They buddied together, they were over at each other's house socializing, and they had a social structure that I was completely left out of." Social isolation from peers resulted in some students having difficulty finding study groups for courses. A former master's student remarked, "They basically worked together and I was just left hanging to do work by myself . . . for some of my classes I just didn't have a study group."

Some participants associated this type of isolation with being students of color. One participant said, "There's definitely a sense among some students that people of color were there because [the university] made special exceptions for them." They felt that their non–African American peers naturally bonded over their shared backgrounds and did not feel particularly connected to African American students. As a participant explained it, "I think it's just being cliquish. You know, the Chinese students hung together. The Greek students hung together. The Indian students hung together. . . . I just think that's the way it was set up. That's just the way it is."

### Family

The primary concern of participants in this study involving family and their impediment in the pursuit of graduate education was that family members did not understand the process and the obstacles faced by graduate students. As put by one PhD student, "I think my family sort of generically is supportive of my efforts but on the flipside, there is a lack of . . . the ability to imagine what it is I'm doing in the sense that I'm engaged in something that no one in my family has done . . . the Ph.D. process is a black box and [they don't] know what my struggle is inside of it." Participants felt that family members may not comprehend the desire or the benefits of attaining a doctoral degree, as family members would argue that obtaining an undergraduate degree is adequate to begin a career. One doctoral student said that family members did not easily understand the need for "such a long education, you know spending nine years in school before actually going out and getting a real job." The student also felt that this struggle of dealing with the lack of understanding of the graduate experience is specific to African American students. He explained, "The family problems . . . not understanding the Ph.D. process. That could be from lack of exposure because of [African American] history of not having these opportunities before."

## Discussion

The current findings suggest that the MSP is an effective intervention, enhancing the number of African American students who pursue and complete graduate education in engineering PhD programs. The number of program graduates pursuing their engineering PhD has tripled in recent years, and UMBC is now one of the top baccalaureate-origin institutions for African American PhDs in engineering. Based on the broad array of core program components rated highly by students, it appears that it is the combination of components rather than any one or two individual components which is responsible for program effectiveness. Furthermore, multinomial regression findings reveal that intrinsic motivation in math/science at college entry is the only predictor variable that uniquely distinguishes Meyerhoff students who pursue an engineering PhD program from those who pursue a master's in engineering or other graduate STEM programs. Interviews with African American Meyerhoff alumni who completed or were pursuing graduate degrees in engineering revealed that faculty mentors, course instructors, fellow graduate students, family, and the Meyerhoff program are key sources of support and, in some cases, sources of barriers. Each of the study findings is discussed below.

PROGRAM OUTCOMES

The current findings support earlier published findings of positive MSP impact on African American students in the entire range of STEM majors (e.g., Maton et al. 2012). Meyerhoff students who entered the program in 1989–2004 were more likely to attend engineering PhD programs than comparison sample students who were accepted into the program but declined the offer and initiated engineering coursework elsewhere. The Meyerhoff students did not have higher SAT scores or high school GPAs than comparison students, suggesting that their higher levels of engineering PhD pursuit were not due to greater precollege preparation or capability. Nonetheless, in the absence of random assignment, it cannot be ruled out that the groups differed, at least to some extent, in other characteristics (e.g., motivation) that may have contributed in part to outcomes.

Of special note, the percentage of Meyerhoff students in engineering PhD programs tripled between 1989–1995 and 1996–2004. This dramatic increase in Meyerhoff students attending engineering PhD programs may be due to a number of factors, including the program attracting more PhD-focused, better-prepared students in more recent years; improvements over time in faculty involvement and program quality; the program's enhanced national reputation; and the program's increased emphasis on the importance of completing PhDs. Surprisingly, the percentage of comparison sample students not attending any graduate program increased from 34.6% to 45.9%. It is not clear what explains this negative trend; to some extent this may reflect lower success rates in engineering majors, a retreat from affirmative action, enhanced competition (e.g., international students), and increased opportunities to pursue other career paths.

PROGRAM COMPONENTS

Striving for outstanding academic achievement, financial support, being part of the Meyerhoff community, summer bridge, study groups, staff academic advising, summer research opportunities, and faculty involvement were aspects of the program rated by students in the 1996–2004 time period as providing the greatest benefit. Each addresses an important challenge facing minority students pursuing undergraduate STEM degrees (Chubin, May, and Babco 2005; Committee on Underrepresented Groups 2010). The program value of striving for outstanding academic achievement helps motivate students to work extremely hard to achieve at the highest level. Financial support allows students to afford college and to focus exclusively on their studies rather than working during college. Being part of the Meyerhoff community, with its critical mass of high-achieving African American peers, contributes to both academic and social integration, which are important for student retention and enhancing student academic success. Similarly, the summer bridge program helps students become adjusted to college life prior to their first full semester, while developing peer support networks and friendships that can help them handle the demands of freshman year (see also Stolle-McAllister 2011). Study groups give students the opportunity to provide and receive academic help in difficult engineering courses. Staff academic advising is important to help students make informed, strategic decisions about the number and type of courses to take (and retake) and which possible research opportunities to pursue. Finally, participation in research provides students with critical experience and knowledge, as well as opportunities to develop personally and professionally rewarding relationships with leading researchers (see also Pender et al. 2010).

In our view, the multifaceted, comprehensive nature of the program, addressing multiple areas of African American student need and challenge, explains its high levels of success over time.

Meyerhoff engineering students with higher levels of intrinsic motivation in math/science at college entry were more likely to pursue engineering PhDs than engineering master's degrees or nonengineering STEM degrees, independent of other predictor variables. Higher levels of intrinsic motivation likely help students develop and maintain an interest in a research career, overcome hurdles, and achieve academically in difficult math/science coursework. It also may link them to a like-minded community of similarly motivated scholars— who further encourage, support, and prepare them for successful entry into a PhD program. Students who entered engineering PhD programs also were more likely to be in later Meyerhoff cohorts than those who did not enter any graduate STEM program, independent of other predictor variables. This finding likely reflects, as noted above, improvements over time in program quality and in its ability to attract high-quality students.

Furthermore, undergraduate GPAs were higher for students who entered nonengineering STEM graduate programs than those who entered engineering master's programs or who did not pursue graduate work in STEM. This finding is not surprising, as a strong qualification for PhD programs is a strong GPA (Millett 2000), and the majority of the Meyerhoff students entering nonengineering graduate programs entered either PhD or MD/PhD programs.


GRADUATE SCHOOL EXPERIENCE

The analysis of interviews focused on supports and barriers experienced by African American Meyerhoff students in master's and doctoral engineering programs revealed both substantial supports and substantial barriers to success. Key sources of support included faculty mentors and advisors, course instructors, departmental climate, peers in the program, family, and the MSP. Conversely, faculty mentors and advisors, course instructors, department climate, and peers in the program for some students were also barriers, and in some cases students linked these to their ethnicity. The sources of support and barriers appear similar to those in the existing literature on the STEM graduate school experience of URM students (MacLachlan 2006; Mwenda 2010). It is noteworthy, though, that the undergraduate MSP experience, as well as ongoing relationships in some cases with MSP staff and peers, continued to be identified as a source of support for a number of those interviewed.

LIMITATIONS

This study has a number of limitations. Possible self-selection differences between Meyerhoff and comparison students limit the strength of the conclusions that can be drawn about the outcome findings. That is, African American engineering students who opted to attend the MSP may be more committed initially to obtaining a PhD than those who declined the admissions offer. It should be noted, though, that all students accepted into the program had expressed a strong interest in pursuing the PhD—both before and during the on-campus interview weekend. Only a random assignment design would provide a definitive means to overcome this design weakness, a design that is difficult to implement in this research area. A second limitation relates to the unknown reliability and validity of the program component measures. Such ratings are generally of concern methodologically given their subjectivity. Nonetheless, the items do possess strong face validity.

A limitation of the predictors of outcome analysis was the lack of analysis of multiple pathways of influence—that is, both direct and indirect relationships among covariates, precollege predictors, and college predictors. Yet another limitation was the lack of inclusion of measures specific to engineering, such as college GPA in engineering and other STEM courses.

One limitation of the graduate experience interviews is the relatively small number of students who took part. A second limitation is the relatively brief and one-time nature of the interviews—more in-depth interviews, over time, would provide more in-depth information about the nature of the graduate school experience in engineering and evolving experiences in terms of mentoring, research experience, and career focus—especially after students have completed doctorates and have had more time to reflect on their experiences.

Finally, the generalizability of findings to programs in different universities and with differing arrays of program components is likely limited. The MSP is relatively unique in its focus, its comprehensiveness, its high level of resources, and the high levels of commitment of the university administration to its success. As other universities (e.g., University of North Carolina at Chapel Hill and Pennsylvania State University) are replicating the MSP—with the assistance of UMBC faculty and program staff—we should be able to compare findings on different campuses over time.

The limitations notwithstanding, the current study represents one of the few systematic examinations of a college-based intervention program designed to increase PhDs among URM students, including African American students, in engineering. Future research should include systematic comparisons of different intervention approaches, use of established measures of known reliability and validity, in-depth examination of the student experience (in the classroom, lab, and social settings), the role of funding for program staff and students, the role and perceptions of faculty (engineering and nonengineering) regarding student performance, and longitudinal tracking of outcomes through receipt of the PhD and beyond (i.e., postdoctoral experiences and engineering career options including academic research, teaching, corporate opportunities, and policy).

In conclusion, enhancing the academic success of African American (and other URM) students in engineering is a pressing national priority. It represents both an economic necessity, so that our nation can stay competitive in the global economy, and a critical part of our nation's larger social justice agenda. Increased understanding of the effectiveness of intervention programs, including the program components and individual student predictors that contribute to positive outcomes, represents a critical priority for future work. The current study represents one contribution to this important research agenda.

## Acknowledgments

This project is supported by grant no. 5R01GM075278-3 from the National Institute of General Medical Sciences (NIGMS). The content is solely the responsibility of the authors and does not necessarily reflect the official views of NIGMS or NIH. The authors acknowledge and are extremely appreciative of the cooperation and support of the MSP staff and students and the UMBC faculty and staff who have been involved in the program over the years.

## References

American Society for Engineering Education. 2007. "Start: A Formal Mentoring Program for Minority Engineering Freshmen." http://search.asee.org/search /fetch;jsessionid=33x5gnlwisn4n?url=file%3A%2F%2Flocalhost%2FE%3A %2Fsearch%2Fconference%2F14%2FAC%25202007Full2785.pdf&index

=conference_papers&space=1297467972036057917166676178&type
=application%2Fpdf&charset=.

Borman, Kathryn M., Will Tyson, and Rhoda H. Halperin, eds. 2010. *Becoming an Engineer in Public Universities: Pathways for Women and Minorities.* New York: Palgrave Macmillan.

Capobianco, Brenda M., Heidi A. Diefes-Dux, Irene Mena, and Jessica Weller. 2011. "What Is an Engineer? Implications of Elementary School Student Conceptions for Engineering Education." *Journal of Engineering Education* 100: 304–328.

Carter, Frances D., Marvin Mandell, and Kenneth I. Maton. 2009. "The Influence of On-Campus, Academic Year Undergraduate Research on STEM Ph.D. Outcomes: Evidence from the Meyerhoff Scholarship Program." *Educational Evaluation and Policy Analysis* 31:441–462.

Chubin, Daryl E., Gary S. May, and Eleanor L. Babco. 2005. "Diversifying the Engineering Workforce." *Journal of Engineering Education* 94:73–86.

Committee on Underrepresented Groups and the Expansion of the Science and Engineering Workforce Pipeline. 2010. Expanding Underrepresented Minority Participation: America's Science and Technology Talent at the Crossroads. Washington, DC: National Academies Press. https://grants.nih.gov/training /minority_participation.pdf.

Congressional Commission. 2000. "Land of Plenty: Diversity as America's Competitive Edge in Science, Engineering and Technology." Report of the Commission on the Advancement of Women and Minorities in Science, Engineering and Technology Development. www.nsf.gov/pubs/2000/cawmset0409/cawmset _0409.pdf.

Cooper, K. J. 2012. "Turning the Corner." Diverse: Issues in Higher Education 29:12–13.

Diverse Issues in Higher Education. 2012. "Top 100 Bachelor Degree Producers to Students of Color." http://diverseeducation.com/top100/BachelorsDegree Producers2012.php.

Fennema, Elizabeth, and Julia A. Sherman. 1976. "Fennema-Sherman Mathematics Attitudes Scales: Instruments Designed to Measure Attitudes toward the Learning of Mathematics by Females and Males." *Journal for Research in Mathematics Education* 7:324–326.

Huyer, Sophia. 2002. "The Leaky Pipeline: Gender Barriers in Science, Engineering and Technology." The World Bank: Information and Communication Technologies and Gender Seminar, February 5. http://go.worldbank.org/H88 FNC5MN0.

Journal of Blacks in Higher Education. 2012. "Universities Honored for Their Efforts to Increase Retention of Minority Engineering Students." www.jbhe .com/2012/04/universities-honored-for-their-efforts-to-increase-retention-of -minority-engineering-students/.

MacLachlan, A. 2006. "Developing Graduate Students of Color for the Professoriate in Science, Technology, Engineering and Mathematics (STEM)." *University of California: The Center for Studies in Higher Education*. www.cshe.berkeley.edu/sites/default/files/shared/publications/docs/ROP.MacLachlan.6.06.pdf.

MacLachlan, A. 2012. "Women and Students of Color as Non-traditional Students: The Difficulties of Inclusion in the United States." In *Issues in Higher Education*, edited by Tamsin Hinton-Smith, 263–279. Basingstoke, UK: Palgrave Macmillan.

Maton, Kenneth I., and Freeman A. Hrabowski III. 2004. "Increasing the Number of African American Ph.D.s in the Sciences and Engineering: A Strengths-Based Approach." *American Psychologist* 59:547–556.

Maton, Kenneth I., Freeman A. Hrabowski, Metin Özdemir, and Harriette Wimms. 2008. "Enhancing Representation, Retention, and Achievement of Minority Students in Higher Education: A Social Transformation Theory of Change." In *Toward Positive Youth Development: Transforming Schools and Community Programs*, edited by Marybeth Shinn and Hirokazu Yoshikawa, 115–132. New York: Oxford University Press.

Maton, Kenneth I., Shuana Pollard, Tatiana McDougall Weise, and Freeman A. Hrabowski III. 2012. "Meyerhoff Scholars Program: A Strengths-Based, Institution-Wide Approach to Increasing Diversity in Science, Technology, Engineering, and Mathematics." *Mount Sinai Journal of Medicine* 79:610–623.

Maton, Kenneth I., Mariano Sto Domingo, Kathleen Stolle-McAllister, J. Lynn Zimmerman, and Freeman A. Hrabowski III. 2009. "Enhancing the Number of African Americans Who Pursue STEM PhDs: Meyerhoff Scholarship Program Outcomes, Processes, and Individual Predictors." *Journal of Women and Minorities in Science and Engineering* 15:15–37.

Millett, C.M. 2000. "Race Matters in Access to Graduate School." Paper presented at the Annual Meeting of the American Educational Research Association, April, New Orleans, LA.

Mwenda, Margaret Nkirote. 2010. "Underrepresented Minority Students in STEM Doctoral Programs: The Role of Financial Support and Relationships with Faculty and Peers." PhD diss., University of Iowa.

National Action Council for Minorities in Engineering. 2011. 2011 NACME Data Book: A Comprehensive Analysis of the "New" American Dilemma. www.nacme.org/publications/data_book/NACMEDatabook.pdf.

National Science Board. 2006. Science and *Engineering* Indicators 2006. Vol. 1. www.nsf.gov/statistics/seind06/pdf/volume1.pdf.

National Science Foundation. 2013. "Survey of Earned Doctorates." www.nsf.gov/statistics/srvydoctorates/.

Nelson, Donna J., and Christopher N. Brammer. 2010. *A National Analysis of Minorities in Science and Engineering Faculties at Research Universities.*

http://faculty-staff.ou.edu/N/Donna.J.Nelson-1/diversity/Faculty_Tables _FY07/07Report.pdf.

Pender, Matea, Dave E. Marcotte, Mariano Sto. Domingo, and Kenneth I. Maton. 2010. "The STEM Pipeline: The Role of Summer Research Experience in Minority Students' Ph.D. Aspirations." *Education Policy Analysis Archives* 18:1–34.

Pittman, David. 2010. "Understanding Diversity." *Chemical and Engineering News* 88:26. http://cen.acs.org/articles/88/i47/Understanding-Diversity.html.

Reichert, Monty, and Martha Absher. 1997. "Taking Another Look at Education: The Importance of Undergraduate Retention." *Journal of Engineering Education* 86:241–253.

Roach, Ronald. 2006. "Under Construction: Building the Engineering Pipeline." *Diverse: Issues in Higher Education* 23:24–27.

Sowell, Robert. 2008. "Ph.D. Completion and Attrition: Analysis of Baseline Data." www.phdcompletion.org/resources/cgsnsf2008_sowell.pdf.

Stolle-McAllister, Kathleen. 2011. "The Case for Summer Bridge: Talented Minority STEM Students and the Meyerhoff Scholarship Program." *Science Educator* 20:12–22.

Walter, A. M., and S. J. Austin. 2012. "Expanding the Engineering Pathway for Underrepresented Minorities." Paper presented at American Society for Engineering Education Annual Conference and Exposition, July, San Antonio, TX.

# Future Directions

Similar to part I, part V consists of a single chapter (chap. 15), by Daryl E. Chubin. Chapter 15 summarizes the substantive findings of the previous parts and discusses their implications for future research, program development, and policies. Chubin concludes with proposed policies and practices, linked to different actors and stakeholders that are derived from the analyses featured in the collection. Moreover, he appropriately acknowledges that many of the authors in this volume have experienced the glacial pace of cultural change which frames the history of minority participation in engineering education and the engineering workforce.

# CHAPTER FIFTEEN

# Challenges and Opportunities
# for the Next Generation

## DARYL E. CHUBIN

S OME SCHOLARLY BOOKS are a "who's who" of a particular field. This is
a "what's what"—a compilation of the history and impact of minority
participation in engineering. As a compendium of experiences and perspectives, it encompasses analysis both retrospective and contemporary to reveal
a legacy of accomplishment of which all of science and technology, not just
engineering, should be justly proud.

Moreover, this collection is a testament to the human spirit—the dignity
and dedication of many educators and practitioners who contributed through
their careers to constructing the fabric of this nation and of a specific discipline. The result is engineering and technology as we know it, though largely
taken for granted today.

We begin with the realities that the preceding chapters elucidate. From
first-person testimony to the mining of institutional and national data sets
(with chapter numbers noted below), we know with confidence that

- the US engineering workforce does not yet look like America
  (chap. 12);
- there is a thoroughgoing lack of awareness about the benefits of a
  career in engineering (to oneself and society) (chap. 2);
- mentors are necessary but not sufficient. Young engineers need
  sponsors and champions who not only know the rules of the game
  but also have clout and the contacts that can be used to advance the
  career of the protégé (chap. 9).

How did we acquire such wisdom? It was not easy. Surely the history told
in the preceding chapters features organizational collaborations that were
heroic in increasing minority, and especially African American, participation

in engineering, notably the Alfred P. Sloan Foundation, the American Society for Engineering Education, General Electric, historically Black colleges and universities (HBCUs), the National Academy of Engineering, the National Action Council for Minorities in Engineering (NACME), the National Scholarship Fund for Minorities in Engineering, and the National Society of Black Engineers (chap. 1).

## A Tainted History

The journey recounted here seems remote from engineering and is not always uplifting. It reflects the ugly racial legacy of the United States both long before and after the civil rights legislation of the mid-1960s (Coates 2014).

Indeed, the Supreme Court's 1896 *Plessy v. Ferguson* decision established a "separate but equal doctrine in public education." A half century later, the racial and social climate of the United States ensured that even with the passage of the GI Bill, Black and White veterans would benefit inequitably. For example, only 12% of Black veterans compared with 28% of White veterans used the GI Bill for higher education (Ciment 2007).

By 1950, public HBCUs applied three other Supreme Court decisions—each reflecting the separate but equal doctrine—to graduate and professional education: *Sinuel v. Board of Regents of University of Oklahoma* in 1948 stipulated that a state must offer schooling for Blacks as soon as it provided it for Whites; *MacLaurin v. Oklahoma State Regents* stated in 1950 that Black students must receive the same treatment as White students; and with *Sweatt v. Painter* in 1950, a state had to provide facilities of comparable quality for Black and White students. Black students were increasingly admitted to traditionally White graduate schools if their program of study was unavailable at HBCUs (Office for Civil Rights 1991). Arguably, desegregation of higher education began at the postbaccalaureate level.

In 1954, the Supreme Court in *Brown v. Board of Education* overturned *Plessy*, ruling that racially segregated public schools deprive Black children of equal protection under the Fourteenth Amendment of the US Constitution. But after *Brown*, most HBCUs remained segregated, with poor facilities (libraries, research equipment) and minuscule budgets relative to traditionally White institutions.

The economic clout of the civil rights movement was also demonstrated in 1958, when the Reverend Leon Sullivan organized 400 other ministers and launched "a 'selective patronage' program whose main purpose was to boycott the Philadelphia-based companies that did not practice equal oppor-

tunity in employment" (www.casey.senate.gov/imo/media/doc/Leon%20H
.%20Sullivan%20Biography.pdf).

A decade later, Arthur Fletcher was appointed by President Nixon as assistant labor secretary for employment standards, which included supervision of the Office of Federal Contract Compliance Programs (OFCCP). Golland (2011) notes, "Fletcher now had the power to revoke federal contracts and debar contractors from bidding on future work. On June 27, 1969, Fletcher implemented the Revised Philadelphia Plan, the nation's first federal affirmative action program, which required federal contractors to meet specified goals in minority hiring for skilled jobs in the notoriously segregated construction industry."

Thus, the passage of Title VI of the Civil Rights Act of 1964, spurred by slow progress in desegregating educational institutions despite *Brown*, sanctioned penalties on organizations receiving federal funding which discriminated on race, color, or national origin. Furthermore, the Office for Civil Rights, established in what is now the US Department of Education, would monitor such illegal discrimination and enforce necessary penalties. Equal-opportunity lawsuits followed: *Adams v. Richardson* in 1977 challenged disparate state support for historically Black institutions in formerly segregated higher education systems, and in the landmark 1992 *United States v. Fordice* case the Supreme Court ordered 19 states to take immediate action to desegregate their public higher education systems.

Today, there is arguably a more level playing field for African Americans and other disadvantaged minorities competing for the higher-paying jobs in American corporations and indeed in all sectors of the economy. This is one prominent legacy of Sullivan, Fletcher, and countless civil rights leaders who created opportunities for many who had been systematically disadvantaged and undereducated, particularly African Americans who succeeded in becoming engineers.

While we remain ever mindful that law and policy, in principle, eliminate the discretion to discriminate, human behavior often evades the strictures of enforcement. Scholars have long recognized that antidiscrimination laws governing the selection process by which a pool of applicants was constituted "did not address issues of isolation and injustice embedded within the institutional climate which impeded the successful matriculation of many minority students. It soon became clear that access-focused laws were necessary but not sufficient, because long-term policies, programs, and permanent infrastructure were also needed" (chap. 6).

Remnants of historical wrongs thus remain in more subtle ways, unconscious biases and cultural habits deeply ingrained even in the most educated among us. Today underrepresentation in science, technology, engineering, and mathematics (STEM) fields is a global issue of economic and national development rather than merely a reflection of social injustice and inequality. Recall, too, that while "minority" was virtually synonymous in the twentieth century with "African American," today a diversity of groups is underrepresented in engineering. While focusing in this volume on one group, the authors have not ignored others. Some trends generalize, but disaggregation is the key to understanding and supporting disparate populations (chap. 6).

## Interpreting This Collection's Empirical Record

The contributions presented above are like the windows of a house. Beyond first-person narratives, secondary analysis of national data sets, new survey and interview data, and institutional case studies, we can see the "rooms" of engineering demarcated by specialty and the "corners" where they articulate. Who populates that house, how they arrived there, and what they have wrought constitute a story of many lifetimes.

The "story" has been told here in 14 chapters that can be bifurcated into (1) educational pathways to leadership forged through programs and other interventions and (2) institutional contexts, models, and evidence of collective success. Highlights of each generate an agenda of future possibilities (presented at the end) nestled between prescriptions and injunctions for succeeding generations of scholars and other practitioners to ponder.

### PATHWAYS TO LEADERSHIP

A simple truth motivates any chronicling of the growth of the national minority engineering effort: the small pool of minority technology professionals makes it less likely that many minority youth will have meaningful, life-changing interactions or encounters with people who look like them in STEM fields (chap. 10). This makes dissecting the pathways to engineering an even more noble undertaking. Researchers have been relentless in charting those paths, demarcating the alternatives, and filling gaps in our knowledge of career choices and destinations that cross institutional and disciplinary lines.

The diversity of ways of becoming an engineer is underappreciated. The lacunae in our empirical knowledge of the intersection of race, ethnicity, and gender hold the key to increasing minority participation in engineering. We ignore segments of the path at our peril. Concerted action focusing on the personal characteristics of minority students, particularly African Americans, was clearly required. Research has continued to shine a bright light on those who participate and how the workforce lags demographic changes despite the gallant efforts to prepare students desirous of a career in engineering. Consider this: the 2012 US Census shows that African Americans constituted slightly less than 13% of the US population, while data reported in chapter 3 estimate that African Americans are 12% of the total workforce, 5% of the engineering workforce, and 2.5% of the tenured or tenure-track engineering faculty.

For those who excel, the role models have been many, and the recognitions delayed but well deserved. They remind us that honorific awards are not just for the awardees. Rather, they reflect a community of professionals who reinforce their commitment to excellence and exploitation of talent regardless of race, ethnicity, and gender.

Championed through alliances among federal, corporate, philanthropic, and nonprofit organizations, these alliances, fortified by Nixon-era "affirmative action" in contracting, led to the formation of councils, consortia, and other national and regional intervention programs that persist to this day. Prominently recurring in these chapters is NACME, a nonprofit creation of the National Academy of Engineering which, with a Fortune 100 board of directors, has provided scholarships to minority youth—often acting on "non-cognitive factors," e.g., motivation and perseverance (identified and measured by Sedlacek 1989), in addition to test scores and grades. Without such intervention, these students would have remained ignorant of engineering as a career option and deprived the United States of untold creativity and innovation.

NACME has illustrated empirically how creativity transforms challenges into opportunities (chap. 12). Today NACME is part of a cadre of nonprofits targeted to various minority groups and the pursuit of engineering degrees. The most renowned are the National Society of Black Engineers (NSBE), the Society of Hispanic Professional Engineers (SHPE), and the American Indian Science and Engineering Society (AISES). Without these and a slew of other group-centered rather than discipline-centered associations, the minority gains in degrees and workforce entry would simply not have occurred. For the mainstream educational establishment—federal agencies and professional associations alike—failed to meet the financial and cultural needs of these students, and they either did not recognize or were slow to act on the accumulated evidence that "critical mass" is a phenomenon that erodes isolation and doubt, providing peer support and expectations for professional success (Olivas 2013).

Chapter 11 finds that out-of-school and precollege programs "designed, funded, and executed for African American student groups increase their content knowledge and awareness of careers in the STEM disciplines; however, the experiences appear to have a significant negative impact on students' motivation and interest in these areas. It could be concluded that STEM-based programs designed by educators and engineers in the absence of experts in youth development, social cognitive theorists, and/or educational psychologists are not attending to the nuanced experiences necessary to increase motivation and interest."

Fortunately, certain programs, both local and national, have proved to have lasting and sustainable impacts: "The opportunities provided by MESA, SECME, and DAPCEP are invaluable to students who otherwise would not have access to quality resources, role models, and supportive learning environments" (chap. 11). Thus, "precollege preparation can broaden or narrow the postsecondary options for students, affect time to degree, affect cost of degree . . . and thereby affect program persistence and success" (chap. 4).

A related finding by analysts of STEM pathway endpoints is the now 25-year-old principle of "intersectionality" dissected in chapter 3. In brief, "African American men and women do not have the same experiences in STEM. . . . The intersectionality approach argues that race/ethnic and sex effects are not additive. . . . Although they have a double disadvantage involving race/ethnicity and gender in the White male STEM culture, African American women's unique history and gender culture provide them resources to 'swim against the tide.'" It is significantly and plainly apparent that "those with a doctorate experience a more level playing field in the engineering workplace" (chap. 3).

Consider the stark barriers on the one hand and the multiple pathways on the other. If K–12 is a pivotal period of identity formation and attitude toward educational achievement (National Science Board 2014), then two-year colleges are pivotal, yet underutilized, as points of access and skills acquisition for those both underprepared academically and financially constrained. If HBCUs remain disproportionate contributors to baccalaureate engineering production, then the shift in African American enrollment from HBCUs to predominantly White institutions (PWIs)—and in deference to intersectionality—should alarm us. "For African American women in particular, HBCUs play a critical role in broadening participation in engineering subfields that are traditionally dominated by men" (chap. 4). Thus, the female advantage that African American women experience in college enrollment and earned baccalaureate degrees does not translate to engineering. Yet, ironically, the "disappearing Black male" haunts higher education (Esters and Mosby 2007; Frierson, Pearson, and Wyche 2009).

Together these chapters have demonstrated, with data, the synergies among administrators, faculty, and students. As shown in chapter 9, for example, there is a need for "simultaneous or multiple mentors": "Because one person cannot mentor another in terms of every aspect of personal and professional life, it is wise to have multiple mentors at a given point in time. . . . [O]ften, it is necessary to have different mentors at different points/ stages in one's career—'sequential or serial mentors.'"

Mentoring relationships thus lubricate the transition from educational preparation to career entry (http://ehrweb.aaas.org/sciMentoring/mentor awards.php). They engage, inform, promote, and equip students with the tools and career guidance needed to make their way in engineering—and moreover, into leadership positions where they command deference, resources, and clout.

Those who reach the pinnacle of their professions as leaders in engineering—and happen to be minorities—enrich every sector of the US economy. For all sectors benefit from a deep reservoir of talent. While the history of African Americans in the military services is well documented (Moskos and Butler 1996), other organizational settings reflect significant minority impacts—against the odds of success—on federal agencies (with NASA as an exemplar), higher education (especially via academic deans), and business and industry (through executives and practicing engineers).

As former Jet Propulsion Laboratory (JPL) director and Voyager project scientist Edward C. Stone notes, "For a mission to succeed, NASA scientists and engineers must share certain qualities despite their inherent differences, 'qualities like patience, dedication, optimism, faith in colleagues, a willingness to take informed risks, and the capacity to be a team player'" (chap. 7).

From NASA engineers, we have learned here, through eight in-depth interviews, that the guiding principle is to "pay it forward." As one puts it, "I know that my investment of time will pay big dividends down the road when the child that I inspire to become an engineer or scientist actually becomes an engineer or scientist, as well as a productive citizen. . . . If they give back and try to uplift the generation that comes after them, then the cycle of giving will continue. Thus, the impact that I have on my community will last many lifetimes!" (chap. 7).

Such contributions are breathtaking. They put a human face on what is too often systemically described in cold analytical terms as products and practices at the very heart of US innovation, economic competitiveness, and intergenerational responsibility. A subtext is the role of minority-serving institutions and nongovernmental organizations that undergird this "pipeline" from school to work. And the singular role of HBCUs, highlighted below, during the era of legal educational segregation cannot be overestimated.

## INSTITUTIONAL CONTEXTS

Two lessons emerge from legal history. First, even abetted by federal sanctions and in the face of chronic underfunding, programs in expensive majors such as engineering education took root. Today, as this collection amply attests, HBCUs lead in awarding baccalaureate degrees to Black students in STEM disciplines, but especially in engineering. Indeed, HBCUs continue to serve as a resource for PWIs. Second, and perhaps more telling, HBCUs have been unparalleled "as the principal locale for postsecondary degree attainment for Blacks prior to integration in American society, as a mobilizing agent into the middle class, and as the institution primarily responsible for the development of Black leadership" (Minor 2008).

This is a hard-won legacy. Nothing is more revealing than the words of four African American deans serving at majority-serving institutions. These are outliers of supreme accomplishment. Their advice on how to increase minority participation in engineering of students, faculty, and administrators is wise and pragmatic: identify the gatekeepers, use peer-led team learning (a model of teaching undergraduates in peer-led workshops), and "nurture a sense of community and a culture of support among students" (chap. 2). Above all, one interviewee offers this: "To move students forward requires intervention—which starts with undergraduates doing research" (chap. 2). Today, this is a bedrock principle of STEM education.

Making a difference in the life of a single minority student—indicating choices and supporting the movement "from here to there"—is a selfless act that brings credit to all involved. With it, these leaders leave an institutional footprint that is worth tracing and emulating. The footprints of institutional types etched here are "existence proofs" that interventions that work can be documented and refined. The challenge is to adapt and apply them in new settings. By disaggregating the various contexts for engineering education, models are revealed, processes unpacked, degree production demystified, and databases harvested. Scholars renew the knowledge base. They not only extend what is known but also suggest how we might act on those trends and institutional idiosyncrasies that find fidelity with certain students in certain ways (BEST 2004).

Consider the following reality that emerges from the data reported in chapter 12: the ratios of engineering degrees awarded to US minorities relative to foreign students ("nonresident aliens") at the baccalaureate, master's, and doctoral levels are 3:1, 1:2+, and 1:4+, respectively. (Ironically, a fund authorized by Congress in FY 2005 from H-1B petitioner proceeds under-writes National Science Foundation–administered scholarships—$100 mil-lion in FY 2012—for low-income students. This has doubtless boosted com-pletion of BS degrees, with almost 65,000 students—an unknown number of whom are minorities—supported by the fund, now known as Scholar-ships in Science, Technology, Engineering, and Mathematics, or S-STEM; see www.nsf.gov/about/budget/fy2013/pdf/16-EHR_fy2013.pdf.)

Nevertheless, minorities are either not pursuing or not favored in gradu-ate engineering admissions. Lagging college retention explains part of these ratio gaps. What, then, in the undergraduate environment discourages minority engineering students?

A clue lurks in the contributions of community colleges to educating mi-nority engineers. As noted in chapter 12, "community colleges are a signifi-cant gateway into engineering, with 44% of African American students who go on to complete degrees in science and engineering beginning their studies at community colleges." Primary reasons students start at commu-nity colleges are (1) lack of financial aid for a four-year college; (2) family circumstances—desire to be close to family; (3) affordability—can live with family and save on room and board; and (4) smaller first- and second-year classes than at major research universities. In addition, transfer students earn higher GPAs and are more likely to be retained to graduation than stu-dents who matriculate at four-year colleges.

For African Americans, as shown in chapter 5, "five HBCUs have been fairly consistent top producers of African American engineering baccalaure-ates on the whole: North Carolina A&T, Prairie View A&M, Tuskegee, Southern, and Morgan State. But along gender lines . . . Tuskegee ranked considerably higher in producing Black women engineers (#5) compared to Black men engineers (#13) by 2010–2012." Furthermore, "at every degree level HBCUs have closed the engineering education gender gap relative to all institutions. In 2012, 33% of HBCU baccalaureate engineers were women, compared to 19% at all institutions."

But enrollment and degree trends are in flux: "Not only is the share of African American engineering baccalaureates from HBCUs in decline for men and women, but the absolute numbers of African American engineer-ing baccalaureates are down at HBCUs since the 1990s. . . . Undergraduate engineering at HBCUs has grown increasingly racially/ethnically diverse, though at a slower rate among women relative to men. Racial/ethnic diversity has been the norm at the graduate degree level at least since 1990, with about 30%–40% in master's degree cohorts composed of non-Blacks in recent years" (chap. 5).

Fortunately, through case studies and especially interviews with students experiencing engineering education in real time, disparities between minority and nonminority completion (Chubin, May, and Babco 2005) are not just reported and lamented but illuminated. Through comparison group analysis, we begin to understand why some things work and others do not. This is illustrated by evidence of community college influence on eventual BS and MS degree attainment of Hispanic students. Disaggregated by subculture, we find that 69% of Mexican American, 51% of Cuban, and 31% of Puerto Rican engineers began their education in a two-year college. As the Latino population grows more quickly than all others, it is incumbent on analysts to disaggregate as they seek to explain and predict coming trends in engineering participation.

The Spelman College Dual-Degree Engineering Program (DDEP)—initially a collaboration with Georgia Tech and 40-plus years later with 13 other engineering schools as well—underscores the value of disaggregation by gender. Data in chapter 13 "show three key factors that account for the high percentage of graduate school attendance for Spelman alumnae: desire for career advancement, family encouragement, and availability of fellowships and other financial aid. . . . The close-knit Spelman DDEP community and their desire to inspire future generations afford the opportunity to expand programs."

Perhaps the program of distinction in producing minority STEM graduates is the 25-year-old Meyerhoff Scholarship Program (MSP) at the University of Maryland, Baltimore County (UMBC). It is also one of the best documented, fortified by longitudinal and quasi-experimental data (e.g., Maton et al. 2012). In chapter 14, a glimpse of UMBC's record is offered:

> The percentage of Meyerhoff students entering engineering PhD programs tripled from 1989–1995 to 1996–2004. Equally striking is that the percentage of Meyerhoff engineering students not entering any graduate STEM program declined greatly from the earlier to the later time period (34.6% to 4.9%), while the percentage of Declined comparison students increased (33.3% to 45.9%). . . . Across the entire study period, African American Meyerhoff students were almost two times more likely to attend engineering PhD programs than the African American Declined comparison students (34.3% vs. 18.4%). . . .

Participants identified their relationships with both official and unofficial mentors and advisors as important to their success in graduate school. . . . The number of program graduates pursuing their engineering PhD has tripled in recent years, and UMBC is now one of the top baccalaureate-origin institutions for African American PhDs in engineering. Based on the broad array of core program components rated highly by students, it appears that it is the combination of components rather than any one or two individual components which is responsible for program effectiveness.

This is "action" research at its best for it increases our options for implementation and enlarges the policy space within which interventions become "better bets" with greater potential return on investments. As reported in chapter 14, "As other universities (e.g., University of North Carolina at Chapel Hill and Pennsylvania State University) are replicating the MSP—with the assistance of UMBC faculty and program staff—we should be able to compare findings on different campuses over time."

## More Than Looking Forward: Prescriptions for the Next Generation

Collective success is the ultimate aim of all who educate, administer, and oversee workforce development. In this spirit, this chapter concludes with proposed policies and practices implicating different actors and stakeholders derived from the analyses featured in this collection. Building for the future should be the legacy of any profession, and engineering is no exception.

### PRESCRIPTION 1: SELECTIVE PATRONAGE

The federal role in higher education has evolved radically over the past 30 years. Access and affordability have been inextricably linked as the availability of federal aid, especially through Pell Grants, Trio programs, and GEAR UP, has opened opportunities and broadened horizons (Devarics 2014). Alas, this bipartisan support is vanishing.

A conservative Congress bent on debt reduction and "smaller government" is pulling back on research and development, as well as education outlays. Consider the potentially adverse impact on students at HBCUs of US Department of Education eligibility requirements for Parent PLUS loans, which for the moment have been forestalled (Field 2013). Despite budget initiatives of the White House, the advocacy of elected officials such as the

Congressional Black Caucus, and the work of minority-serving professional organizations, the higher education lobbies, and the National Academies, US society has gravitated toward a redefinition of higher education as a private good and of science as an ideological lightning rod. As private sector funding grows, for example, Silicon Valley is displacing—instead of augmenting—federal funding (Broad 2014).

Thus, the funding of institutions skilled at educating minorities but historically underresourced—minority-serving institutions, including community colleges—would be a potent antidote. It would also apply the principle of selective patronage (described above) to higher education, namely, "seller beware" in the competition for students of color.

As observed in chapter 12, even if there were sufficient numbers of new engineers, "the lack of diversity among the US engineering workforce poses a significant threat to our nation's ability to maintain an innovative edge in an increasingly competitive world." This is what corporations call "making the business case for diversity." Their concern is the net profit in the "bottom line." The academic bottom line should demonstrate, both in the composition of its graduates and in the character of the new knowledge it produces, "the educational value of diversity." That is what the University of Michigan argued in the *Grutter* and *Gratz* Supreme Court cases (Malcom, Chubin, and Jesse 2004), but it remains contentious to this day (Kurashige 2014).

While we wait for all institutions to become "minority-serving," we must invest in the 104 historically Black, 250-plus Hispanic-serving, and two dozen Tribal colleges, as well as 1,600-plus community colleges, of the nation to educate legions of minority students (National Research Council 2011). Selective research universities will continue to skim the cream and enroll students of all colors mostly with "means" (if not wealth). Meanwhile, "positioning the community college as an even more prominent and effective gateway to engineering careers for African American students will require bold action and academic strategy" (chap. 12). Self-reliance can only go so far. Incentives—public and private—can mitigate intransigence.

PRESCRIPTION 2: EXPLANATORY, HYPOTHESIS-BASED RESEARCH

National conferences have a vaunted history of convening educators at minority- and majority-serving institutions alike as underrepresented student-centered support agents. Three conferences of vintage, though not exclusively STEM focused, are the QEM (Quality Education for Minorities) Network (http://qemnetwork.qem.org), NAFEO (National Association for Equal Opportunity in Higher Education; www.nafeo.org/community/index.php), and NCORE (the National Conference on Race and Ethnicity in American Higher Education; www.ncore.ou.edu).

A different kind of gathering, organized under the auspices of a National Research Council committee in 2007, was called "Understanding Interventions That Encourage Minorities to Pursue Research Careers" (http:// understanding-interventions.org/2007-report/#sthash.9UumXTdp.dpuf). Today, there is a movement known as "Understanding Interventions That Broaden Participation in Research Careers" which fosters "exchange of hypothesis-based research on interventions and initiatives that broaden participation in science and engineering research careers. Its annual conference is designed to create a dialogue among behavioral/social science and education researchers, evaluators, and faculty in STEM (science, technology, engineering, and mathematics) fields who participate in intervention programs" (http://understanding-interventions.org/about/).

Understanding Interventions (UI) anticipates that the inclusion "of individuals with the sociocultural expertise (e.g., anthropologists, social psychologists) to devise strategies that lead to self-efficacy among precollege students consistently goes untapped" (chap. 11). Indeed, UI research featured at its conferences has explored concepts identified in this collection as central to increasing participation of minorities in STEM careers. Foremost among them are:

- critical mass—the presence of large numbers of a particular minority, as in community colleges and HBCUs, as well as in research-intensive universities with modest minority enrollments "sitting together in the lunchroom" (Tatum 2003);
- self-efficacy—extending Bandura's (1977) findings that students' capacities to organize and execute the courses of action required to achieve relate to students' beliefs in the likelihood that they can complete an undergraduate or graduate degree in a STEM discipline (Chemers et al. 2011); and
- stereotype threat—performance in academic contexts can be harmed by the awareness that one's behavior might be viewed through the lens of racial or gender stereotypes. It can also lead students to choose not to pursue a particular domain of study and, consequently, limit the range of professions they pursue (Steele and Aronson 1995).

These concepts, synthesized in the translation of research findings into practice, have gained explanatory power through replication in different institutional settings and with different populations (Chubin and DePass

2014). Efforts to reduce pernicious effects in the learning environment and grow student confidence to broaden participation in engineering must be redoubled. The UI community—cutting across disciplines and institutional types, sharing problems and formulating solutions—has become a forum for equipping minorities to compete in the future STEM workforce.

PRESCRIPTION 3: DISAGGREGATED ACCOUNTABILITY

At the risk of exacerbating "identity politics," the chronic fringe groups of higher education (and the subset of engineering aspirants) are growing in consciousness, if not in size—the disabled, veterans, Native Americans, and Asian and Pacific Islanders, as well as the fastest-growing categories of first-generation and of Hispanic/Latino students (who, as seen in chap. 12, vary wildly across subculture orientations to pursuing education away from home and the values of the degrees earned). At a time when "race-conscious" college admissions policies are under scrutiny as "undermatching" and "mismatching" (Chubin 2014), the drumbeat for "class-based" affirmative action and "race-neutral alternatives" intensifies (Kahlenberg 2012).

Like race, gender divides more than unites in distributing talent. "Research that does not acknowledge the intersection of race/ethnicity and sex in STEM achievement might also miss important progress that African American and other minority women are making in STEM in catching up with their male counterparts" (chap. 3). Such knowledge must be applied, not just applauded, as insights that move individuals and organizations alike. As African American engineering deans see it, diversifying the faculty and the student body in general, and increasing the participation of African Americans in particular, "becomes one criterion on which the performance of faculty members and administrators is assessed . . . accountability is critical at every level of an institution" (chap. 2). Combine the plight of African American males with the "double bind" suffered by minority women, and no easy solutions appear. To paraphrase, there are several prizes to "keep eyes on."

If accountability starts at the top, there is cause for concern. In the 2014 Survey of College and University Presidents conducted by Inside Higher Ed and Gallup, "90% say that generally speaking the state of race relations on their campus is good or excellent" (Lederman and Jaschik 2014). But the evidence suggests otherwise, whether we consider racial incidents, sexual harassment, or hiring patterns that maintain a faculty that looks unlike its undergraduate student body (Stewart 2014). Who is out of touch here? Or is the composition of faculty and students simply not a high priority? This is a problem not just in engineering or at research universities, but it reflects outmoded perceptions, an antiquated reward system, change-resistant recruitment and hiring procedures, and a campus climate that serves the majority but not the few. Where, then, are the leaders—presidents, provosts, deans—and rank-and-file faculty with a commitment to the next generation of engineers (Chubin 2013)?

PRESCRIPTION 4: INSTITUTIONAL COLLABORATION

It is increasingly apparent that formal, top-down articulation agreements designed to facilitate student transfers, dual-degree programs, and internships (including cooperative education experiences) are of limited effectiveness. They work where third parties like NACME or SECME have cultivated trust for the long term. For most educational institutions, however, partnerships formed by faculty, department to department (and legally sanctioned), hold the key for true research-based collaborations that result in the movement of students and faculty across institutional types (Coleman et al. 2012b). This incubation of relationships holds promise for institutionalizing what universities cannot do alone—connect with the facilities, instruments, and training not readily available at "home," thereby blurring the lines separating education and training, school and work, while expanding students' socialization and professional networks.

In an era of "high debt, no job" for many college graduates (Patel 2014) and uncertainty about the "shelf life" of a STEM degree (Czekalinski 2014), engineering is an antidote. Why should refinancing student loans even be a policy issue? With a robust job market and a receptivity to women and graduates of color (Carnevale, Smith, and Melton 2011), engineering stands alone in awarding high entry-level salaries and long-term prospects of career vitality.

In those environments that support out-of-class services for underserved students and reward faculty "champions of diversity" (who do not trade off research for teaching, but rather integrate these functions as mentors and coaches), minority degree recipients increase. And the engineers among them, as this collection attests, are those likely to traverse the various sectors of the economy and types of academic institutions. This is how, albeit slowly, practices and policies alter institutional cultures.

## A Final Thought

The stories recounted in the preceding chapters are shared by many who have lived pre– and post–Great Society and have engaged in higher education teaching and administration, business leadership, and public policy service. Anyone who has participated in these roles has come to recognize that culture and habit die hard. Thus, the law becomes a creative intervention and not just a punitive force (Coleman et al. 2012a).

As we celebrate the 60th anniversary of *Brown v. Board of Education*, we must also acknowledge that despite the Supreme Court's ruling in *Fisher v. University of Texas*, diversity remains a "compelling interest" (thanks to the 1978 *Bakke* decision) and an American value. Race still matters. Class does not trump race. And "post-racial society" remains an empty slogan.

To paraphrase Randall Kennedy (2013), if there were some way to guarantee that all American citizens, regardless of race, ZIP code, income, or status, could receive a world-class K–12 education, the racial playing field might become level on its own. If so, we would no longer need programs (rigorously evaluated) and policies that offer broad solutions for racial inequality. In the clear absence of such a guarantee, however, we must continue to look to the law as an assurance that racial justice is still possible. Still, legal activist Ted Shaw cautions that "litigation can change things in the moment, but unless people on the ground embrace those changes, the impact of litigation is going to be ephemeral" (in Monaghan 2014, A17).

The glacial pace of cultural change frames the history of minority participation in engineering education. Those who have steadfastly stayed the course are owed a debt of gratitude for their insights, their vision, and, above all, their deeds.

## References

Bandura, A. 1977. "Self-Efficacy: Toward a Unifying Theory of Behavioral Change." *Psychological Review* 84 (2): 191–215.

BEST (Building Engineering and Science Talent). 2004. "A Bridge for All: Higher Education Design Principles to Broaden Participation in Science, Technology, Engineering and Mathematics." www.bestworkforce.org/PDFdocs/BEST_High _Ed_Rep_48pg_02_25.pdf.

Broad, W. J. 2014. "Billionaires with Big Ideas Are Privatizing American Science." *New York Times*, March 15. www.nytimes.com/2014/03/16/science/billionaires -with-big-ideas-are-privatizing-american-science.html?_r=0.

Carnevale, A. P., N. Smith, and M. Melton. 2011. *STEM*. Georgetown University Center on Education and the Workforce. https://georgetown.app.box.com/s /tlfsn8vah390yb42tpyi.

Chemers, M., E. Zurbriggen, M. Syed, B. Goza, and S. Bearman. 2011. "The Role of Efficacy and Identity in Science Career Commitment among Underrepresented

Minority Students." *Journal of Social Issues* 67 (3): 469–491. http://dx.doi.org /10.1111/j.1540-4560.2011.01710.x.

Chubin, D. E. 2013. "Where Are the Other Leaders of STEM Workforce Diversity?" *AAAS MemberCentral*, December 16. http://membercentral.aaas.org/blogs /driving-force/where-are-other-leaders-stem-workforce-diversity.

Chubin, D. E. 2014. "Undermatching vs. Mismatch: Undermining College Access." *AAAS MemberCentral*, February 10. http://membercentral.aaas.org /blogs/driving-force/undermatching-vs-mismatch-undermining-college -access.

Chubin, D. E., and A. L. DePass, eds. 2014. *Understanding Interventions That Broaden Participation in Research Careers*. Vol. 5, *Intervening to Critical Mass*. www.understandinginterventions.org.

Chubin, D. E., G. S. May, and E. L. Babco. 2005. "Diversifying the Engineering Workforce." *Journal of Engineering Education* 94 (1): 73–86.

Ciment, J. 2007. "The GI Bill and African Americans." In *Atlas of African American History*. Rev. ed. New York: Facts on File Inc.

Coates, T.-N. 2014. "The Case for Reparations." *Atlantic*, June.

Coleman, A. L., J. L. Keith, and D. E. Chubin. 2012a. *Summary and Highlights of the Handbook on Diversity and the Law: Navigating a Complex Landscape to Foster Greater Faculty and Student Diversity in Higher Education*. 2nd ed. Washington: DC: American Association for the Advancement of Science. www .aaas.org/report/summary-and-highlights-handbook-diversity-and-law -navigating-complex-landscape-foster-greater.

Coleman, A. L., K. E. Lipper, J. L. Keith, D. E. Chubin, and T. E. Taylor. 2012b. *The Smart Grid for Institutions of Higher Education and the Students They Serve*. Washington, DC: American Association for the Advancement of Science. www.aaas.org/sites/default/files/SmartGrid.pdf.

Czekalinski, S. 2014. "Does Your STEM Degree Have a Shelf Life?" *National Journal*, March 12. www.nationaljournal.com/next-economy/big-questions/does -your-stem-degree-have-a-shelf-life-20140312.

Devarics, C. 2014. "30 Years In: The Evolving Federal Role in Higher Education." *Diverse: Issues in Higher Education*, March 18. http://diverseeducation.com /article/61264/.

Esters, L. I., and D. Mosby. 2007. "Disappearing Acts: The Vanishing Black Male on Community College Campuses." *Diverse: Issues in Higher Education*, August 23. http://diverseeducation.com/article/9184/.

Field, K. 2013. "In Victory for HBCUs, Department to Reconsider a Policy Change on Parent PLUS Loans." *Chronicle of Higher Education*, August 15. http:// chronicle.com/article/In-Victory-for-HBCUs/141133/.

Frierson, H. T., W. Pearson, Jr., and J. H. Wyche, eds. 2009. *Black American Males in Higher Education: Diminishing Proportions. Diversity in Higher Education*, vol. 6. Bingley, UK: Emerald Books.

Golland, D. H. 2011. "Arthur Allen Fletcher: 'The Father of Affirmative Action.'" www.blackpast.org/perspectives/arthur-allen-fletcher-father-affirmative -action.

Kahlenberg, R. 2012. "Online Fisher Symposium: Race-Neutral Alternatives Work." September 4. www.scotusblog.com/2012/09/online-fisher-symposium -race-neutral-alternatives-work/.

Kennedy, R. 2013. *For Discrimination: Race, Affirmative Action, and the Law.* New York: Pantheon Books.

Kurashige, S. 2014. "In Diversity Gap at Michigan Flagship, Signs of a Lost Public Mission." *Chronicle of Higher Education*, March 3. https://chronicle.com /article/In-Diversity-Gap-at-Michigan/145057/.

Lederman, D., and S. Jaschik. 2014. "Federal Accountability and Financial Pressure: A Survey of Presidents." *Inside HigherEd*, March 7, www.insidehighered .com/news/survey/federal-accountability-and-financial-pressure-survey -presidents.

Malcom, S. M., D. E. Chubin, and J. K. Jesse. 2004. *Standing Our Ground: A Guidebook for STEM Educators in the Post-Michigan Age.* Washington, DC: American Association for the Advancement of Science. www.aaas.org/sites /default/files/Capacity_Center/Standing_Our_Ground.pdf.

Maton, K. I., S. A. Pollard, T. V. M. Weise, and F. A. Hrabowski III. 2012. "Meyerhoff Scholars Program: A Strengths-Based, Institution-Wide Approach to Increasing Diversity in Science, Technology, Engineering and Mathematics." *Mount Sinai Journal of Medicine* 79:610–623.

Minor, J. T. 2008. "A Contemporary Perspective on the Role of Public HBCUs: Perspicacity from Mississippi." *Journal of Negro Education* 77:323–335.

Monaghan, P. 2014. "Law Professor Follows His Late Friend in Directing Civil-Rights Center." *Chronicle of Higher Education*, June 20, A17.

Moskos, C. C., and J. S. Butler. 1996. *All That We Can Be: Black Leadership and Racial Integration the Army Way.* New York: Basic Books.

National Research Council. 2011. *Expanding Underrepresented Minority Participation: America's Science and Technology Talent at the Crossroads.* Washington, DC: National Academies Press.

National Science Board. 2014. *Science and Engineering Indicators 2014.* NSB 14-01. Arlington, VA: National Science Foundation.

Office for Civil Rights. 1991. "Historically Black Colleges and Universities and Higher Education Desegregation." US Department of Education. www2.ed .gov/about/offices/list/ocr/docs/hq9511.html.

Olivas, M. A. 2013. "Not Quite an End to Affirmative Action." *Insider Higher Ed*, June 25. www.insidehighered.com/views/2013/06/25/essay-supreme-court -ruling-affirmative-action.

Patel, V. 2014. "Graduate-School Debt Is Raising Questions about Degrees' Worth." *Chronicle of Higher Education*, March 17. http://chronicle.com.article /Graduate-School-Debt-Is/145347/?cid=at.

Sedlacek, W. E. 1989. "Noncognitive Indicators of Student Success." *Journal of College Admissions* 1:2–9.

Steele, C. M., and J. Aronson. 1995. "Stereotype Threat and the Intellectual Test Performance of African Americans." *Journal of Personality and Social Psychology* 69:797–811.

Stewart, P. 2014. "Many Institutions Continue to Struggle with Gender Imbalance." *Diverse: Issues in Higher Education*, March 12. http://diverseeducation.com/article/61276/.

Tatum, B. D. 2003. *Why Are All the Black Kids Sitting Together in the Cafeteria: And Other Conversations about Race*. New York: Basic Books.

# CONTRIBUTORS

*Editors*

JOHN BROOKS SLAUGHTER is a former director of the National Science Foundation; chancellor of the University of Maryland, College Park; and president of Occidental College. Dr. Slaughter has served for many years as a leader in the education, engineering, and scientific communities. He is well known for his commitment to increasing diversity in higher education with a special focus on the science, technology, engineering, and mathematics disciplines.

A member of the National Academy of Engineering—where he has served on the Committee on Minorities in Engineering, cochaired its Action Forum on Engineering Workforce Diversity, and served two terms on the NAE Council—he is also the recipient of the academy's Arthur M. Bueche Award in 2004. A Fellow of the American Association for the Advancement of Science, the Institute of Electrical and Electronic Engineers, and the American Academy of Arts and Sciences, he was elected to the Tau Beta Pi honorary engineering society and was named Eminent Member of the Eta Kappa Nu honorary electrical engineering association. He is a member of Phi Beta Kappa, and in 1993 Dr. Slaughter was named to the American Society for Engineering Education Hall of Fame and was the recipient of the society's Centennial Medal. He received the University of California, Los Angeles (UCLA), Medal of Excellence in 1989; was elected to the Kansas State University Engineering Hall of Fame in 1990; received the Roger Revelle Award from the University of California, San Diego (UCSD), in 1991; and was named that institution's Alumnus of the Year in 1982.

Dr. Slaughter, a licensed professional engineer, began his career as an electronics engineer at General Dynamics and later served for 15 years at the US Navy Electronics Laboratory in San Diego, where he became head of the Information Systems Technology Department. He has also been director of the Applied Physics Laboratory and professor of electrical engineering at the University of Washington, academic vice president and provost at Washington State University, the Irving R. Melbo Professor of Leadership in Education at the University of Southern California, and, until his return to USC in January 2010, president and CEO of the National Action Council for Minorities in Engineering Inc. from August 2000 to September 2009.

Among the boards of directors on which he has served are IBM, Northrop Grumman, Monsanto, Baltimore Gas and Electric Co., Sovran Bank, Union Bank, Avery Dennison, Atlantic Richfield, and Solutia Inc. He was appointed by President Jimmy Carter as assistant director and, later, as director of the National Science Foundation. President George W. Bush appointed him a member of the President's Council of Advisors on Science and Technology (PCAST) and, in 2015, he received a Presidential Award for Excellence in Science, Mathematics and Engineering Mentoring (PAESMEM) from President Barack Obama.

Dr. Slaughter earned a PhD in engineering science from UCSD, an MS in engineering from UCLA, and a BS in electrical engineering from Kansas State University. He holds honorary degrees from 30 colleges and universities. Recipient of the first US Black Engineer of the Year Award in 1987, Dr. Slaughter was awarded the Martin Luther King Jr. National Award in 1997.

He and his wife, Dr. Ida Bernice Slaughter, herself an educational leader, have two children: a son, John Brooks Slaughter II, DVM, and a daughter, Jacqueline Michelle Randall, a university administrator and teacher.

YU TAO is an assistant professor of sociology and director of gender and cultural studies at Stevens Institute of Technology in Hoboken, New Jersey. Her research interests include science, technology, engineering, and mathematics education and workforce. More specifically, Dr. Tao investigates how gender and race/ethnicity affect educational outcomes and career choices, performance, and advancement in STEM. She also examines STEM education and workforce from an international perspective. Her most recent research investigates gender differences in career outcomes (e.g., publication productivity) among scientists and engineering in the United States and China.

Dr. Tao's research has been published in books and journals such as *International Journal of Gender, Science and Technology*, and *Journal of Women and Minorities in Science and Engineering*. She also serves as a reviewer for journals, including *American Journal of Sociology*,

*Journal of Women and Minorities in Science and Engineering*, and *Science, Technology and Human Values*. Dr. Tao's work has had both university and government support, including grants from the National Science Foundation, the Walter B. Jones Foundation Fellowship from Georgia Tech, and travel grants from the National Academies.

Dr. Tao received her BA in English from East China Normal University in Shanghai, China, her EdM in educational media and technology from Boston University, and her MS and PhD in sociology of science and technology from the Georgia Institute of Technology.

WILLIE PEARSON, JR., is a professor of sociology at the Georgia Institute of Technology in Atlanta, Georgia. Prior to joining the faculty at Georgia Tech as chair, School of History, Technology and Society in July 2001, he held a distinguished appointment as professor at Wake Forest University and adjunct in medical education at Wake Forest University School of Medicine. Dr. Pearson received his PhD in sociology (with a specialty in the sociology of science) from Southern Illinois University at Carbondale in 1981. He has held postdoctoral fellowships at the Educational Testing Services and the Office of Technology Assessment, Congress of the United States. Since 1994, Dr. Pearson has served on the editorial board of *Science and Engineering Ethics*. In 2001, he was elected a National Associate of the National Academy of Sciences. Dr. Pearson was elected as an American Association for the Advancement of Sciences Fellow (Social, Economic and Political Sciences) in 2005. In 2013, he was selected to represent the Class of 1988–1989, 40th Anniversary of the Science and Technology Fellows Program, American Association for the Advancement of Science. In 2010, he was appointed by President Barack Obama to the Presidential Board of Advisors, White House Initiative on Historically Black Colleges and Universities. In 2013, he was selected as a Distinguished Member, 125th anniversary, Sigma Xi. Pearson's research has centered on the career patterns of PhD scientists—particularly African Americans—and human resource issues in science and engineering. His publications include numerous articles and chapters in refereed journals and books. He is the author and coauthor of 11 books and monographs, including *Blacks, Education and American Science* (1989), *Who Will Do Science? Educating the Next Generation* (1994), *Beyond Small Numbers: Voices of African American Ph.D. Chemists* (2004), and *Advancing Women in Science: An International Perspective* (2014).

## Authors

RODNEY ADKINS is president of 3RAM Group LLC, a privately held company specializing in capital investments, business consulting services, and property management. Formerly, Mr. Adkins was senior vice president of IBM, having served in that position from 2007 until 2014. Mr. Adkins is the company's first African American senior vice president and corporate officer. In his more than 33-year career, he has held a number of operational and management roles spanning across strategy, technology, systems, and supply chain. In 2005, Mr. Adkins was inducted into the National Academy of Engineering. He currently serves on the board of directors for United Parcel Service, Grainger, PPL Corporation, and the national board of the Smithsonian Institution. He is a trustee of the Georgia Institute of Technology and Rollins College. Mr. Adkins also previously served on the board of Pitney Bowes, PeopleClick Inc., and the National Action Council for Minorities in Engineering. Mr. Adkins holds a bachelor of arts degree with an emphasis in physics from Rollins College, as well as bachelor of science and master of science degrees in electrical engineering from Georgia Tech. He has been awarded honorary doctoral degrees from Georgia Tech and the University of Maryland, Baltimore County.

LENELL ALLEN has over 25 years of experience in increasing the transition of students from secondary to postsecondary education. Dr. Allen earned her bachelor of science degree in mechanical engineering from the University of Kansas. She received a master of arts degree in higher and adult education and her doctor of philosophy in educational leadership and policy analysis with a concentration in engineering education from the University of Missouri–Columbia in 1994 and 2001, respectively. Dr. Allen currently serves as the director for the Aerospace Research and Career Development Program, at the National Aeronautics Space Administration Office of Education.

SYBRINA Y. ATWATERS received her PhD in sociology of technology and science and bachelor's degree in electrical engineering from Georgia Institute of Technology. Dr. Atwaters's research focuses on the sociology of technology (examining religious practices and knowledge production in 3D virtual worlds) and social inequality (examining patterns of inequality within science, engineering, and higher education).

GILDA A. BARABINO is dean of the Grove School of Engineering of the City College of New York. She previously served as associate chair for graduate studies and professor in the Wallace H. Coulter Department of Biomedical Engineering at Georgia Tech and Emory University. She joined the department in 2007 after an 18-year career at Northeastern University, where she rose to the rank of professor of chemical engineering and served as the vice provost for undergraduate education.

TIFFANY BEASON is a doctoral student at the University of Maryland, Baltimore County, in the Human Services Psychology program, with a concentration in Clinical and Community Applied Social Psychology. In 2012, Tiffany was awarded the UMBC Meyerhoff Graduate Fellowship. She is currently conducting master's thesis research on the topic of academic and career outcomes for immigrant college scholars majoring in science, technology, engineering, and mathematics fields.

RETINA Q. BURTON is a native of Atlanta, Georgia, who holds a BS in electrical engineering from Morgan State University (C/O 1998) and an MS in technical management from Embry Riddle Aeronautical University (C/O 2005). In September of 2005, Ms. Burton began her career in the field of academe in her current position as the director of the Dual-Degree Engineering Program / Office of Science, Engineering, and Technical Careers.

DARYL E. CHUBIN is an independent consultant living in Savannah, Georgia. Formerly senior advisor to the American Association for the Advancement of Science, in 2004 he became founding director of the AAAS Center for Advancing Science & Engineering Capacity. Prior to that, he was senior vice president for research, policy, and programs at the National Action Council for Minorities in Engineering, after nearly 15 years in federal service.

LANGO DEEN has been technology editor at Career Communications Group Inc. since 2005. She joined the media services company as a staff writer in 2002, working with contributing editors such as Michael A. Fletcher, Gale Horton Gay, Marvin V. Greene, Imani Carter, and Rayondon Kennedy to develop ideas for columns and features and produce content on leaders in science, technology, engineering, and mathematics.

SANDRA L. HANSON is professor of sociology and research associate at the Institute for Policy Research and Catholic Studies at Catholic University.

She received her PhD from Pennsylvania State University. Her research examines the gender structure of educational and occupational systems in a comparative context. Sandra's research on gender, race/ethnicity, and science has been funded by multiple grants from the National Science Foundation.

FREEMAN A. HRABOWSKI III has served as president of the University of Maryland, Baltimore County, since 1992. His research and publications focus on science and math education, with special emphasis on minority participation and performance. He chairs the Obama Commission on Educational Excellence for African Americans. He is a fellow of the American Academy of Arts and Sciences and the American Association for the Advancement of Science and serves on the board of the Alfred P. Sloan Foundation.

JENNIFER S. JOHNSON is the science, technology, engineering, and mathematics education outreach manager at Spelman College, focusing on increasing the number of students enrolling, graduating, and pursuing careers in STEM disciplines. In this role, Ms. Johnson is heavily engaged in outreach to local K–12 schools and Spelman undergraduates through various student development programs.

CHERYL B. LEGGON is an associate professor in the School of Public Policy at the Georgia Institute of Technology. Dr. Leggon's research underscores the criticality of disaggregating data by race/ethnicity *and* gender to develop policy, programs, and practices that enhance the quality of the United States' science and engineering labor forces. She was elected a Fellow of the American Association for the Advancement of Science for her work on women of color—African American, Mexican American, Puerto Rican, and Native American—in science and engineering in the United States. In 2006, she was elected to membership in Sigma Xi.

JOHN D. LEONARD II serves as associate dean for finance and administration for the College of Engineering at the Georgia Institute of Technology. Dr. Leonard is also a member of the Institute for Transportation Engineers, the Association for Institutional Research, and the American Society of Engineering Education.

SHIRLEY M. MALCOM is head of the Directorate for Education and Human Resources Programs of the American Association for the Advancement of Science. The directorate includes AAAS programs in education, activities for underrepresented groups, and public understanding of science and technology. Dr. Malcom serves on several boards—including the Heinz Endowments—and is an honorary trustee of the American Museum of Natural History.

LINDSEY E. MALCOM-PIQUEUX is an assistant professor of higher education administration in the Department of Educational Leadership in the Graduate School of Education and Human Development at George Washington University. She has authored numerous publications appearing in the *Review of Higher Education*, the *Harvard Educational Review*, *Educational Researcher*, the National Academy of Engineering journal, the *Bridge*, and the *Journal of African American History*.

KENNETH I. MATON is professor of psychology and affiliate professor of public policy at University of Maryland, Baltimore County. Dr. Maton received his PhD in community-clinical psychology from the University of Illinois at Urbana-Champaign in 1985. He is a past president of the Society for Community Research and Action and winner of its Distinguished Theory and Research award.

IRVING PRESSLEY MCPHAIL was named the sixth president and CEO of the National Action Council for Minorities in Engineering Inc. on September 1, 2009 (www.nacme.org). He joined NACME in 2007 as executive vice president and chief operating officer. He served 15 years as a college president or chancellor at the Community College of Baltimore County, St. Louis Community College at Florissant Valley, and LeMoyne-Owen College.

CARL S. PERSON retired from the National Aeronautics and Space Administration (program manager) in 2012. In his 14-year career at NASA, Person's final role was as the director of aerospace research and career development in NASA's Office of Education. He was responsible for working with a national network of colleges and universities to expand opportunities for Americans to understand and participate in NASA's aeronautics and space projects by supporting and enhancing science and engineering education, research, and public outreach efforts.

PERCY A. PIERRE is vice president emeritus and professor of electrical and computer engineering at Michigan State University. He also directs programs to recruit and mentor domestic graduate students in the College of Engineering, with an emphasis on underrepresented groups, and collaborates on research programs with other faculty members in the college. His specific research interests are in the area of applications of stochastic models in engineering systems.

TAFAYA RANSOM is special assistant to the provost at Morehouse College, where she leads a range of college-wide strategic initiatives around the governance and utilization of data as an asset and vehicle for improving student success and institutional performance. Prior to Morehouse, Tafaya was a US Department of Education Institute for Education Sciences predoctoral fellow.

CARMEN K. SIDBURY is currently associate provost for research at Spelman College. She has a pivotal leadership role in the college's science, technology, engineering, and mathematics education outreach initiatives, including overseeing the STEM Education Outreach Program, which assists Spelman students with creating and facilitating special projects geared toward exposing K–12 students to STEM. Dr. Sidbury also provides leadership in the cultivation of faculty research capabilities, the coordination of activities associated with undergraduate student research, and research training programs.

TYRONE D. TABORN is publisher, chairman, and CEO of Career Communications Group Inc. CCG is a minority-owned media services company that promotes significant achievements of minorities in science, technology, engineering, and mathematics through its magazines and conferences. Taborn is also founder of the Foundation for Educational Development, which aims to raise awareness of technology literacy in minority communities across America.

GARLAND L. THOMPSON is an author and newspaper and broadcast journalist, long known as a writer and editor for *US Black Engineer and Information Technology*. He began his career as a Navy submarine and amphibious sailor during the Vietnam War years. After working 10 years on advanced technologies in the Navy and the telephone system, Thompson shifted to journalism to help explain the communications revolution to lay audiences.

ANGELICQUE TUCKER-BLACKMON is the president and CEO of Innovative Learning Concepts, a full-service premier science and mathematics educational coaching, consulting, and tutorial firm. She has been CEO and director of research and evaluation for 11 years, responsible for growth and expansion in the southeastern region and for providing research and evaluation services for federally funded programs.

KAREN M. WATKINS-LEWIS is a lecturer and research associate at the University of Maryland, Baltimore County, where she teaches courses in statistics, research methods, and sociocultural psychology. She earned both her BS in mechanical engineering and her PhD in developmental psychology (with concentrations in socio-culture and education) from Howard University in Washington, DC.

DARRYL N. WILLIAMS is the associate dean for recruitment, retention, and community engagement and the director of the Center for STEM Diversity in Tufts University's School of Engineering. He has served as a program director for the National Science Foundation in the area of engineering education research and led the Innovative Technology Experiences for Students and Teachers program, while working across NSF to support K–16 engineering curricular frameworks and broadening participation of underrepresented groups in STEM.

# INDEX

Page numbers in italics signify figures and tables.

Change the Equation, 281
chemical engineering, 68, 95, 102, 161, 295; and field choice, 171–175, 244
Chen, X., 2
Chevron, 222
Chrysler, 274
Chubin, Daryl E., 154, 389–407, 413
Cinque, 208
City College of New York (CCNY), 356
City University of New York, 219
civil engineering, 102, 161, 177; and field choice, 171–175
Civil Rights Act of 1964, 391
civil rights movement, 90, 390–391
Clark, Garvey, 30
Clark College, 19
Clay, Roy, 237–238
Claytor, W. W. Schiefflin, 193
Cobb, Charles E., 319
Cochran, Donnie, 261–262
Code2040, 236
College of Staten Island, 219
Committee on Diversity in the Engineering Workforce, 309
Committee on Engineering Education, 309
Committee on Institutional Cooperation (CIC), 28–29
Committee on Science, Engineering, and Public Policy, 57, 81
Common Core State Standards (CCSS), 298
communalism, 253
community building, 253
Community College Research Center (CCRC), 319–320
community colleges, 305–331, 397, 398; and African American students' knowledge base, 312–313; attendance at, 305, 306–307, 308, 329; and developmental mathematics, 319–322; innovations in teaching and learning at, 317–319; NACME strategy on, 324–327; partnerships with four-year universities by, 309, 314–315, 324, 329; as pathway to engineering, 95–98, 327–329, 330; pre-engineering at, 313–315; and STEM education,

315–316, 330; student support services and retention at, 315–316; transfer and articulation at, 322–324
community service, 274, 358
Compaq Computer Corp., 238
computer-aided design (CAD), 279
Computer Attitude Questionnaire (CAQ), 289
computer engineering, 93, 102, 161, 173, 244
Concannon, James, 153
*Confronting the "New" American Dilemma*, 309
Congressional Black Caucus, 399–400
Conliffe, Calvin, 21
ConnectEd: The California Center for College and Career, 320
*Contamination Control in Trace Analysis* (Zief and Mitchell), 279
contextualized instructional models, 320
contextualized learning programs, 320
Cooley, Keith, 18
Cornell University, 18
Council for Opportunity in Education (COE), 298
Council on Jobs and Competitiveness, 230
counseling, 316, 339; career, 152, 324; MSP and, 357, 362, 380
course instructors, 373, 376. *See also* faculty
court cases, 15, 30, 390–391, 400, 404
Craft, Elaine L., 315
Crawford, Amanda J., 271
Crenshaw, Kimberle, 61–62
critical mass, 111, 269, 339, 380, 393; as concept, 401
critical race theory (CRT), 323–324
Cross-Disciplinary Initiative for Minority Women Engineering Faculty (XD): conceptualization of, 244–246; as interdisciplinary initiative, 241–242; lessons learned from, 252–254; primary objectives of, 243; reframing issues by, 242; year 1 of, 246–250; year 2 of, 250–251; year 3 of, 251–252

Integrated Postsecondary Education Data System (IPEDS) Completion Survey, 104, 124, 125, 132, 134, 157, 232; about, 63, 83–84
international collaboration, 403
*International Journal of Computers and Electrical Engineering,* 263
international students, 22, 377, 397
International Technology and Engineering Educators Association (ITEEA), 299
Internet, 229
intersectionality, 61–63, 394
Iowa State University at Ames, 277–278

Jackson, Shirley Ann, xi–xiii, 3, 260, 268–270
Jacoby, Monte, 336
Jain, Dimpal, 323–324
Jamaica Train, 227
Jemison, Mae, 196–197
Jenkins, Harriet G., 193–194
Jennings, Thomas, 256
Jessie Smith Noyes Foundation, 338
*Jet,* 194
Jet Propulsion Laboratory (JPL), 195–196, 395
Johnson, Christyl, 212–214
Johnson, James, 16
Johnson, Jennifer S., 335–353, 414
Johnson, Katherine, 193
Johnson, L. R., 226
Johnson, Lyndon B., 192
Johnson, Norman, 18
Johnson Space Center, 198, 205, 206
Jones, Frederick M., 238
Jones, Kevin L., 219
Jones, Reginald H., 23, 26, 29
Junior Engineering Technical Society (JETS), 20
Just, Ernest E., 258

Kane, E. R., 29
Kane, Michael A., 315
Kaufman, Howard C., 33
Kennedy, Randall, 404
King, Martin Luther, Jr., 208
Kreidler, Robert, 27, 28
Kuh, Ernest S., 30

Lam, David, 272
Landis, Raymond, 18–19
Langley Research Center, 215–216, 217–218
Latimer, Lewis, 238, 256
Latinos. *See* Hispanics
Laudise, Robert A., 278–279
"leaky pipeline" concept, 354–355
Lear, W. Edward, 19
learning theory, 318–319
Leggon, Cheryl B., 6, 150, 151, 247n, 414; chapters by, 39–56, 241–255
LeMoyne–Owen College, 20
Leonard, John D., III, 149–188, 414
Lincoln University, 20, 212, 270
Lockheed Martin, 235, 263, 264–265, 275–276
Lord, Susan M., 152, 153, 177
Louisiana State University, 15, 155
low-income students, 106, 142, 222, 296, 319, 324, 397
Lumina Foundation for Education, 320
Lyles, Lester L., 260

Mack, Lynn G., 315
MacLachlan, Anne, 247n
*MacLaurin v. Oklahoma State Regents,* 390
Malcom, Shirley M., 6, 90–119, 152, 269, 414–415
Malcom-Piqueux, Lindsey E., 90–119, 152, 415
Management Advisory and Review Committee (MARC), 33
managerial skills, 253
Manley, Albert E., 336, 338
Marshall, Robert, 18, 24
Martin, Danny, 153
Martin, Thomas, 18
Maryland Business Roundtable for Education, 264
Mason, Kelvin R., 260
Massachusetts Institute of Technology (MIT), 19, 216, 268, 269, 272, 357
massive online open courses (MOOCs), 114
master's colleges and universities (MCUs), 165–167

Office of Management and Budget (OMB), 32
Office of Postsecondary Education (OPE), 84
Ohland, Matthew, 151
Olin Corporation, 336
out-of-school time (OST) programs, 287, 288–290, 393–394
OWN, 221

Packard, Becky W., 323
Padulo, Louis, 19–20, 25, 28, 29, 336
Pajares, Frank, 345
Parent PLUS loans, 399
Parker, Alice H., 270
partnerships, 106, 111, 112, 114, 403; with community colleges, 309, 314–315, 324, 329; inter-university, 142, 169, 170, 296; with NASA, 339–340; and precollege programs, 328, 337; with private industry, 154, 180, 276, 295–296, 300, 324–325, 346; to promote women in engineering, 339, 340
patents, 2, 227, 228–229, 256, 270
Payload Operations Control Center (POCC), 205
Payne v. LSU, 15
Pearson, Willie, Jr., 2, 6, 149–188, 411
Peeples, Fred N., 19
peer-led team learning (PLTL), 45, 46
peers: as barrier, 377–378; as mentors, 358; as support, 374
Pell Grants, 399
Pennsylvania State University, 212, 382
Perkins, Courtland, 26
Perna, Laura W., 343
Person, Carl S., 191–224, 415
Phelps, Lucius, 228
Pierre, Percy A., 13–35, 415–416
Plessy v. Ferguson, 390
Powell, Colin, 234
Prairie View A&M University, 16, 132, 397
precollege programs: emergence of, 298–299; empirical evidence from, 288–290; evaluations of, 298–300; funding for, 31–32; importance of, 393–394; models of successful,

293–295; pinpointing success for, 295–297; summer bridge programs, 357, 380
predominantly White institutions (PWIs), 115; engineering degrees awarded by, 100–101, 102, 104; HBCUs and, 108, 396; minority engineering efforts at, 18–19; as pathway to engineering, 99–101, 103, 105; shift toward African American enrollment in, 99, 394
President's Council of Advisors on Science and Technology (PCAST), xiii
Prince George's Community College Foundation, 264
problem-based learning (PBL), 320–322
Proctor, Martin, 260
professional associations, 52–53, 54
professional socialization, 241
Project Lead the Way (PLTW), 299, 300, 328
Proudfoot, Steven L., 106
Purdue University, 18
Purnamasari, Agustina V., 314
purposive sampling, 41

QEM (Quality Education for Minorities) Network, 400
Qualcomm, 324–325
"quiet crisis," xii, 3, 59, 83

racism, 60, 61, 153; US legacy of, 58–59, 309, 390–392
Ransom, Tafaya, 120–148, 416
Rechtin, Eberhardt, 273
recruitment, 18, 22, 45, 47, 328; from community colleges, 313–315; literature on, 151–152; MSP and, 357; by NASA, 192, 194, 222
Redmond, William, 260–261
Rendon, Laura I., 323
Rensselaer Polytechnic Institute, 268
research activities, 82, 358, 400–402
research universities, 65, 71, 77, 106–107, 400; undergraduate degrees awarded at, 164–167
retention, 45, 46, 47, 328, 356; community colleges and, 315–317, 323; failure in, 108; Georgia Tech case

University of California, Los Angeles (UCLA), 123, 196, 263
University of California, San Diego (UCSD), 263, 324–325
University of Dayton, 219
University of Denver, 229
University of Detroit, 31
University of Florida, 104
University of Houston, 356
University of Maryland, Baltimore County (UMBC), 276, 356, 357, 398
University of Maryland, College Park, 262, 356
University of Massachusetts, Amherst, 260
University of Michigan, 18, 142, 219, 357, 400
University of New Mexico, 28
University of North Carolina, Chapel Hill, 382
University of Pittsburgh, 273
University of Richmond, 214
University of Southern California, 273
University of Tennessee, 19, 205, 271
University of Virginia, 214
University of Washington, 262
University of Wisconsin, 18
*US Black Engineer & Information Technology,* 257, 259, 261, 262, 263, 264, 267, 275
*US News and World Report,* 230

Van Depoele, Charles, 228
venture capital, 236, 237–238
Virginia Tech, 215
VisiCalc, 233
von Braun, Wernher, 192

Walker, Charles, 259
Walker, Tristan, 236
*Wall Street Journal,* 234
Washington, Booker T., 258
Washington Mutual Investors Fund, 272
Watford, Bevlee A., 325
Watkins, Charles, 16
Watkins-Lewis, Karen M., 354–386, 417
Wayne State University, 31
Webb, James, 192

WebCASPAR, 63, 84
Weinberger, Catherine, 125
Welch, John F., Jr., 22–23
Wenner, George, 60
West Virginia University, 193
Whitlow, Woodrow, 216–218
Whyms, Robert, 260
Wilberforce University, 20
Wilburn, Adolph Y., 125
will.i.am, 221
Williams, Darryl N., 287–304, 417
Williams, Pharrell, 221
Wilson, Prince, 336
Wingard, Raymond, 33
women, African American, 241–254, 336, 337; and baccalaureate degrees, *102,* 128, 130, 164–165; and engineering doctorates, 46–47, 67–74; and "female advantage," 91, 93, 394; and female undergraduate majority, 92–95; HBCUs and, 99–102, 107, 127, 128, 130, 136–140; in STEM fields, 113–114, 338–340; underrepresentation of, 62, 93, 136, 152, 342–43. *See also* gender gap
Women in Science and Engineering (WISE), 339, 345–346
Women of Color STEM Award, 260, 274, 280
Woods, Granville T., 226, 228–229, 238, 256
Woods, Lyates, 228
Woodson, Carter G., xi
*The World Is Flat* (Friedman), 4
WPL Holdings, 266–267
Wulf, W. A., 5

Xavier University, 20
XD. *See* Cross-Disciplinary Initiative for Minority Women Engineering Faculty
Xerox Corporation, 346

Yale University, 259
Young, Whitney M., Jr., 194
Youngstown State University, 264

Zhang, Guili, 151
Zinser, Richard W., 322–323, 324
Zuckerman, Mortimer B., 230